ALIGNMENT MATTERS
THE FIRST FIVE YEARS OF *KATY SAYS*

KATY BOWMAN, M.S.

ALIGNMENT MATTERS
THE FIRST FIVE YEARS OF *KATY SAYS*

KATY BOWMAN, M.S.

PROPRIOMETRICS
PRESS

Printed in the United States of America

Fourth Printing, 2015

ISBN 978-0-9896539-0-9

Propriometrics Press
2419 E. Harbor Blvd. #130
Ventura, CA 93001

wwww.propriometricspress.com

Cover and Interior Design: Zsofi Koller, www.zsofikoller.com
Author photo: Nancy DeLucrezia

The information in this book should not be used for diagnosis or treatment, or as a substitute for professional medical care. Please consult with your health care provider prior to attempting any treatment on yourself or another individual.

Publisher's Cataloging-In-Publication Data
(Prepared by The Donohue Group, Inc.)

Bowman, Katy.
  Alignment matters : the first five years of Katy Says / Katy Bowman.

  p. : ill. ; cm.

  Includes index.
  ISBN: 978-0-9896539-0-9

  1. Posture. 2. Human mechanics. 3. Human locomotion. 4. Physical fitness. 5. Chronic pain--Treatment. 6. Essays. I. Title.

RA781.5 .B69 2013
613.7/8

To the Alpha RES Warriors:

Nancy Burns
Brian Campbell
Ann Gallimore
Breena Maggio
Patricia Pasternak
Diane Rennell
Karen Sullivan
Roxanne VanPelt

They took a chance on a course and affected the course of many.

ABOUT THIS BOOK

Alignment Matters is a collection of blog posts from 2007 to 2011. It is not just the good ones. We've put the bad posts in as well. We've organized them to make the content easy to search when you have a question.

My blog has no rhyme or reason when it comes to organization. I post on whatever inspires me. Some days it is the pelvis, and some days I write about the hair follicles inside the nose. And, because there are years of posts, to read through it strictly chronologically is a challenge. Through the miracle of super-duper editing, you can now sit and read through years of posts organized into chapters by topic (and chronologically within chapters). And, if you're looking for something in particular, you can search the chapter names or the super-awesome index. By using these tools, you can easily find the topics that interest you.

TABLE OF CONTENTS

chapter 1

INTRODUCTIONS

INTRODUCTION

In 2007 I started a blog. I didn't know what a blog was, but like many things I don't understand completely, I began enthusiastically. I posted twice, and then nothing, until 2008. At that point in my career as a biomechanist, I was on my third Restorative Exercise™ facility and was teaching fifty new individuals a week. Our alignment specialist certification program (Restorative Exercise Specialist™) had been going on for two years. There were a lot of people who had been exposed to a very simple but profound idea: Alignment Matters.

Posture and alignment have been components of various treatments, like physiotherapy, for a long time, but significant issues persist. The biggest is that there is no standard set of measures for evaluating body position. Well, there is, actually, but the language is so unspecific that even those trained in the science of anatomical position have become very casual in their assessment and as a result have lost the precision that the body's geometry truly requires.

There came a point in my academic career when, after designing and testing a (very successful) program for reducing risk factors for pelvic floor disorder, my professors sat me down. "Would you like to do research full-time? Because people are going to start calling you to come to their university once your information is out."

Now, you have to realize that I have a very—insanely, actually—high level of respect for science. Science, logic, and mathematical proof are beacons for me. But because I had participated in the academic scene for a few years, I was aware that most research—and I *really* mean most of it—is never dispensed for actual use. You may see a pop-culture magazine pull some of it out to misreport on, but that's it. There is no standard practice of health professionals going to research journals to figure out how to better be a practitioner. There is no requirement that you stay up-to-date on research. To participate in academia in that way for the prestige of being published was not what I wanted then, and it's not what I want today.

So I declined the traditional academic route and decided to take science directly to the people who need it. And because I found a huge lack of science education in the general population (even in health professionals), I decided to create a program that would teach biomechanics and how to apply it, today, to anyone who wanted to learn.

Teaching people directly was a success. It turns out that there are millions of people who want to be better, but don't know how. And their ability to know the difference between something that makes sense and something that just doesn't feel right is keen.

Once you're exposed to a very specific, mathematical approach to the functions of the body, everything begins to make sense—the pain. The injury. The cramping. The birthing outcome.

My methods of objectively quantifying body position are so simple that anyone can learn them. Which, in this age of health care, is extremely important. People should know how to evaluate and repair themselves—and, better yet, know how to use their bodies in a way that prevents the affluent ailments. It is through my career path—my time at university, my interest in all things scientific, the many people who had questions, and my passion for offering simple-yet-specific lessons—that this blog was born.

I was never, and still am not, a reader of much online. I dabble here and there, of course, usually when searching for answers to questions like: Why is there blood in my stool? (Answered, always, about six minutes into the search: Because you ate beets last night.) So I can understand how so many people stumble onto my blog. If you're looking into the body, the information is there…somewhere.

The KatySays blog was not started as an income generator and still isn't a few years later. I don't do advertising. I'll refer you to the right exercise DVD or program that I've already created—but this is because the information you seek is much more complex than my blog-time can handle. The blog is still my favorite hobby. The posts have absolutely no form whatsoever. Some are two hundred words and some are three thousand. Every once in a while I will post heavily referenced scientific papers of mine and sometimes there are videos of me teaching a hip movement in a dinosaur costume. Each is a genuine representation of me, and how I am in regular life.

Science education should be fun. It should be "shoot milk out your nose" fun. And someday, I'm truly hoping that I write a post so funny, it makes a lactating mother shoot milk out her nose. Without drinking it, I mean. So funny that the milk reverses direction and comes out the…whatever. You know what I mean.

Health is getting worse by the year. I don't mean to be a downer, but it is. The good news is, the reasons are easy to identify scientifically. The not-so-good news is, it is taking the people responsible for bringing you the information (medical media) a long time to catch up. Mostly because, as I said earlier, actually reading and using research is not a requirement.

So, until it is, I will continue to write out what I deem to be helpful to those interested in doing the work and learning the stuff wellness is made of.

MY MANIFESTO

Okay, so I read this great article on the twenty-six blogs every blogger should write because blogging, while I love it, is basically a hobby I do when I'm not working. And I'm never not working, so I take all the help I can get when it comes to blog suggestion. Number twenty-six on the list is Write Your Manifesto.

Sounds important, right?

Wikipedia says that "a manifesto is a public declaration of principles and intentions, often political in nature."

A political statement also "feels right" the day after Independence Day. This is a new twist on a post from a while back (I'm a better writer now!) and is also the first page out of the new Whole-Body certification manual. I hope it inspires you. It inspires me to work every day, on every level.

Bio: Life; living organism

Mechanics: Classical mechanics is a branch of physics concerned with mathematically describing the forces that create motion of bodies and the effect of these motions on their environment.

Biomechanics began at the beginning—not the beginning of the century, not the start of the Christian Era, nor during the Ancient Olympics. The movement forces began with life, as the very definition of life includes mechanical concepts of work (movement):

Life: The property or quality that distinguishes living organisms from dead organisms and inanimate matter, manifested in functions such as metabolism, growth, reproduction, and response to stimuli or adaptation to the environment originating from within the organism.

The written history of biomechanics is generally recognized to begin with a series of Aristotle's essays *De Motu Animalium* (*Movement of Animals*), in which he describes the not-yet-termed mechanical concept of ground action force as a starting point to deliberate where movement comes from. The body? The spirit?

These are good questions.

Biomechanics as a United States university program began just after the First World War. The first-time widespread use of explosives in battle resulted in unprecedented numbers of amputees. These weren't old men who could live out the rest of their lives with a peg for a leg. These men needed to start and support families, take over family businesses, and fully participate in the country they had just defended. The use of heavy explosives created a situation in which military technology had exceeded medical technology. Medicine was obliged to deal with the situation, and had to adapt.

Designing a prosthetic wasn't a medical emergency nor a chemistry problem. Doctors had to turn to mechanical engineers in the university system for help. Engineers took a look at the arm or leg from a mechanical perspective and attempted to create a replacement with as much of the functionality of the original limb as possible. At this point, biomechanics became a university study option for mechanical engineers. In modern science, this was the first marriage between the biological and physical sciences, and required a mastery of anatomy, physiology, physics, and mathematics.

Exercise

As the effects of the Industrial Revolution set in, Americans became more and more sedentary, which gave rise to the concept of "exercise," where to compensate for the decrease in regular, daily movement, we would intensely move the whole body in short bursts. Fifty years later, biological movements like walking, squatting to birth or defecate, flexing, extending, and rotating all the joints on a regular basis were decreased to the point that new generations had no handed-down knowledge of how people moved before the population had stopped moving (if you watch Disney's *WALL-E*, you can get a nice visual of this phenomenon).

Around the sixties and seventies, programs of kinesiology (technically, the science of human movement) began appearing in university curriculum, but these programs were incorrectly named. The departments of kinesiology should have been called sport science, as the programs offered contained only those movements used in modern athletic and sport-related activities—very different from the science of movements humans have been doing for the last 200,000 years. Biomechanical programs, following the trend of kinesiology, have since focused the bulk of their curriculum and funding to study "exercise" instead of movement, and athletics (golf, golf, golf!) instead of health.

Both science and research have shown that fitness does not equal health, not in the heart and not in the joints. This is confusing to many. If the cause of most ailments affecting affluent populations—ailments like osteoarthritis, diabetes, osteoporosis—is lack of movement, shouldn't exercise be the solution? The answer is no.

Exercise is not the flip side of the sedentary coin—movement is. While the difference may seem like an argument in semantics, these two habits are quite different.

Movement, specific to the requirements of biological survival, not only keeps us fit in the terms set by pop culture, but also matches the mechanical requirements of human tissues. These are movements like walking long distances, squatting to bathroom and birth, hauling your weight up and over, or lowering full body weight. Natural motion respects tissue's threshold for loads and requirements for vibration, and provides other mechanical necessities like gravity-assisted functions of the internal physiology. Doing natural, reflex-driven movements from birth utilizes not only the large muscle groupings we commonly think of at the gym, but the other five hundred muscles as well, including the muscles between the ribs that open and close to inflate the lungs, the intrinsic muscles of the feet that create the arch shape of the foot, or the constant force-generating sheets that make the pelvic floor.

Exercise, while having much of what makes movement good, carries with it elements that make it far less superior, such as small quantities, high intensities, and large unnatural and repetitive forces in the joints. It is these characteristics that leave exercise a poor substitute for movement, and explain why professional-level athletes, weekend warriors, and gym rats do not have better health (in such terms as surgeries, medications, and death from cardiovascular disease) than those who sit and do nothing at all.

As for any natural organism functioning in a natural world, balance is the key to human survival. Survival of the population, survival of the individual, and survival of the cellular structures that make up said individual all depend on the delicate balance of forces within the body. Movement, when mimicking the habits of those humans long passed, provides us with the required mechanical stimulation without which we die. While we tend to think of death only in terms of the whole person, we allow small deaths like that of cartilage, bone, or parts of organs to fall under other names and categories. It does not occur to us that a lack of movement is the cause of these deaths, but blame other, uncontrollable things like age or genetics.

It is very easy to understand how whole-body movement keeps our body in a state of re-generation (and not de-generation), through the application of simple geometry and physics to known processes of physiology and anatomy. Our physiology functions much like a self-winding clock. Actually, we are comprised of more than six hundred self-winding clocks, for each muscle has its own responsibility to feed itself and the tissues of that area. The "original blueprints" for the human, no matter the source, surely couldn't have accounted for a time when body-owners would spend so much time ignoring biological signals of stress, hunger, and fatigue; when the owner of a human-machine would render itself inert by choice first, and then eventually by habit.

Since Aristotle and Newton, we as a population have become less concerned with our understanding of the laws of nature and natural movement and more familiar with cardio machines, hand weights, and high-tech footwear. Movement has been a void in our lives for so long that to become moving creatures (as opposed to exercising ones) seems impossible. Rearranging our lives to accommodate less—less work, less stress, less furniture, less driving, less sitting, less convenience—seems too much to ask...at first.

This platform that you are about to study is the ever-elusive human manual. A guide to the nuts and bolts of the biological machinery under the influence of mechanics, this book shall first be a service to one's self. Before taking this course as a practitioner, you must first take this course for yourself. Get to know yourself and then spend the rest of your life mastering yourself. If there is time, you can work on yourself with others, allowing them to learn via your example.

In both Aristotle's writings and Isaac Newton's equations, full stability (a fixed point at each joint) gives a human being the ability to live using every single one of their muscles...all six hundred of them. But the lack of moving for years (or, as a population, for a few hundred years) has left us overusing a few joints, while the rest of the body sits dormant and inert. Leonardo da Vinci would likely say that the result would be decreased output and function of the machinery. Cells would die. Disease would ensue. Aristotle would say that, under sub-optimal conditions, one's spirit and life force would be unable to express.

Both would probably be correct.

There is a solution. Freedom from disease is attainable, by using the whole body in a biological, reflex-driven way. First, we must be careful not to force the concept of movement into an inaccurate paradigm. To isolate parts of the body when strengthening or to think of strength as something any less than a whole-body event is to miss the point. You were designed to be a strong-yet-supple dynamic creature of endurance. It's time to start acting like one.

Can I get an Amen?

1.   rightmixmarketing.com/right-mix-blog/blog-ideas-2-types-of-posts-to-write-be-fore-your-competitors-do/

July 26 2010 ALIGNMENT MATTERS

I get a lot of emails asking "So, what does good alignment LOOK like?" Well, here you go. Have a look-see at the cover of our certification manual:

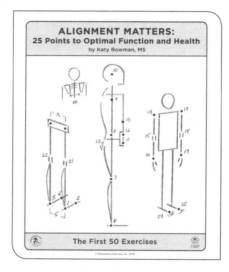

Each one of the points refers to a bony marker (something you can clearly see with your eyes, no matter your body composition). Each point needs to be in the correct spot, relative to other points, and relative to the gravitational force, in order to have optimal health. If your tissue tension does not allow the points to be in this orientation, then there is a resulting issue with the neurological, lymphatic, and blood flow. These points are not approximations (like "this is kinda a good way to be") but the actual mechanical position your joints should be in for optimal flow and minimal degeneration. When you are walking (which you need to be doing, multiple miles per day, to drain your lymph), the only markers that should be changing are the swinging of the arms and legs (and there are specific places those need to be too!)

The physics of the body is a fascinating science, especially when you realize that things we believe to be major diseases are simple things, like points 6, 7, and 8 not being vertical. Or 11 and 12 not being vertical. Given the choice, would you rather have surgery on your spine, or just get point 12 over point 11? Chronic headaches? I'd check point 20. Thyroid slowing down? Check 20, 10, 14, and 17. High blood pressure? You've got more than a few points to work on, but there are points still the same! Knee osteoarthritis? ACL damage? We always start with 1–5. Adjusting 1–5 is free, takes less than five seconds, and is waaaay more convenient than a knee replacement, I think. But, what do YOU think?

Getting this information out is essential for the longevity of the naturally free (free to move!) and mobile human race. Without this information, there is only one direction we can go: bed- and chair-ridden. Or, like the police officer I saw the other day (who was less than forty years old), Segway-ridden. Hey criminals, want to be successful in the next twenty years? Stretch and use your legs and steal things from places cars or Segways can't go. (I'm not condoning criminal activity, of course, but "the survival of the fittest" will always be true, so we'd better change our mindset, fast!)

We are looking for paradigm-shifting health professionals, those looking to transform their community starting with their family, teachers wanting to undo poor alignment practices in the classroom, birthing professionals wanting to reduce baby- or mama-damage due to delivery, and you, sitting at home right now, knowing that you want this information, to learn the science, and pass it on.

We are looking for people who demand their health care makes sense.

October 8, 2010 — SCIENCE OF WELLNESS

I'm going to explain a little what is missing from the field of human science, but first, a few definitions.

Biological Science: A natural science concerned with the study of life and living organisms, including their structure, function, growth, origin, evolution, distribution, and taxonomy.

Physical Science: An encompassing term for the branches of natural science and science that study non-living systems, in contrast to the biological sciences. However, the term "physical" creates an unintended, somewhat arbitrary distinction, since many branches of physical science also study biological phenomena.

Physiology: The science of the function of living systems. It is a subcategory of biology.

Physics: A natural science that involves the study of matter and its motion through spacetime, as well as all applicable concepts, including energy and force. More broadly, it is the general analysis of nature conducted in order to understand how the universe behaves.

Anatomy: A branch of biology and medicine that is the consideration of the structure of living things.

Geometry: "Earth-measuring" is a branch of mathematics concerned with questions of shape, size, relative position of figures, and the properties of space.

Kinematics: In biomechanics and kinesiology, *kinematics* is a term for the branch of classical mechanics that describes the motion of bodies without considering the forces (how the motion was created).

Kinetics: In physics and engineering, *kinetics* is a term for the branch of classical mechanics that is concerned with the relationship between the motion of bodies and the motion's cause, namely forces and torques.

Engineering: The discipline, art, and profession of acquiring and applying scientific, mathematical, economic, social, and practical knowledge to design and build structures, machines, devices, systems, materials, and processes that safely realize solutions to the needs of society.

Medicine: The science and art of healing humans. It includes a variety of health care practices evolved to maintain and restore health by the prevention and treatment of illness.

Don't worry, there won't be a quiz.

So here's my point. The current academics in charge of researching disease, setting parameters for health, outlining function of human tissues, and engineering products for all these categories, are almost entirely students of the biological group of sciences. They are missing a fluency in mathematics, physics, engineering, human movement science, kinetics, and kinematics. Because of the very unique way we have compartmentalized our education system, we have segmented "life science" from "things in the universe" science, meaning the average health-care scientist (keep in mind this is a very, VERY educated person) has minimal comprehension of how flow, pressure, force, electricity, heat, and energy (just to name a few) affect cellular processes. All forces in the universe impact cellular processes, just like pharmaceuticals do. Not knowing these other, non-biological sciences places a limitation on our ability to figure out possible causes and solutions to ailments. We dynamic creatures are entire universes of physical matter, subject to physical laws, suffering the biological consequences when, frankly, we don't know how to operate.

Enter biomechanics (do you hear angels singing? I know I do!), an academic option that requires all of the sciences above. That explores how the human body works under the physical laws. When you know all of the information, the optimal level of health and beyond (what most of us like to call WELLNESS) is attainable. I'm happy to see that at least one major university is beginning to offer a Medical Biomechanics option, where the graduates are required to know ALL of the sciences that govern the human body, instead of half of them. I want the person in charge of my body to know a hundred percent of the science.

Oh wait. I'm the person in charge of my body. And YOU are the person in charge of YOUR body. The time has come to no longer hand off your health to someone you see five times a year. It is time to become an expert in human science, because, well, you're all you really have.

I'm finally (finally!) offering our Whole-Body Alignment and Propriometry course online, to make it accessible to everyone instead of only those local to California and our institute, as it has been up until now. Take it for yourself, or take it to round out your health-care education. There is more information that you don't know than you could ever imagine. And you can use it to help yourself, your family, or others. I like to say that:

WELLNESS = Physics + Physiology

You MUST know both sciences to be a fully functioning human (get your whole body involved in the living experience!) And, P.S., you already know a lot more physics than your high school experience led you to believe. You are a Master of the Universe. You've got physics down, really, it just needs to be brought to the front of your mind.

 ALIGNMENT MATTERS, REALLY

I'd agree that starting any sort of document with a definition is not only extremely third grade, but also terribly passé. Which is why I do it a lot. I also like plaid, short pants, and mismatched socks. So here goes.

From Dictionary.com:

A·lign·ment: The proper adjustment of the components of an electronic circuit, machine, etc., for coordinated functioning; as in: The front wheels of the car are out of alignment.

Mat·ters: To be of importance, signify; as in: It matters a lot.

I felt compelled to start writing a monthly educational piece because of all of the fad-based health information showing up in magazine articles and poorly understood science making its way into the headlines.

We currently have an epidemic of poor health. But the interesting thing is, this poor health is not only in the populations we expect (couch potatoes, Kit-Kat lovers, and cigarette buyers), but in those that are eating well and exercising. The reasons, we are told, are beyond our control. Genetics and age are unavoidable, right? When you end up with a diagnosis that doesn't make sense to you, then, it is easy to agree with the statement: This ailment (this pelvic floor disorder, or high blood pressure, this osteoarthritis in my knees and hips, and the degeneration of my spinal disks) is not a result of anything I have done and there is nothing I can do about it.

Only problem is, this is not correct. You did do something to cause it and you can do something to fix it.

While every event that occurs is multi-factorial in causation, the way we use our body is what dictates wear patterns and contributes to the very ailment you are dealing with now. Even the resulting injuries from acute traumas (like car accidents and falls) are affected by the stress risers (patterns of wear) that have been developing since your first step.

When you are experiencing poor health, of any kind, keep this in mind: Your alignment matters. It matters more than anything else, because everything else you are doing (exercising, eating well, medicating, hydrating) depends on basic principles of physics to work correctly, and these principles are affected by your geometry.

I will be walking you (and anyone else you want to bring along) down the path to restoration. Let's put your body back not where it was (you can do better than that!), but where it CAN be.

This ReAlignment ReEvolution requires a quote:

Watch your habits, for they become your posture.
Watch your posture, for it creates your boundaries.
Watch your boundaries, for they restrict your growth.
Watch your restrictions, for they create immobility.

Watch your immobility, for it becomes your illness.
And yes, I just wrote that, but feel free to paste it on Facebook. :)

I hope you continue to read along with me. Read through my newsletter to find links to blog posts, recorded lectures, contests, and a community! I am producing educational materials just about every day in various formats. The closer you follow, the more you will learn. :)

October 6 2011 ABOUT THIS BLOG

This blog was originally set up for students and graduates of our whole-body alignment and biomechanics course at the Restorative Exercise Institute. Back then, everyone reading it already had a certain amount of information—at least seven months of listening to me yap on and on. Fun, right? Well, over the last couple of years, we have grown to a blog being read by over 100,000 people a month. This is great, but what is happening is, no one knows what is going on. And by no one, I really mean myself.

I thought I'd do a clarification piece, for the both of us.

This is a blog about natural movement.

What is natural movement? All the movement humans used to do to get around, live life, etc. before there were any cars, any bikes, any clothes. Before there were any pilates classes, any yoga practice, any shoes. Any belly dancing, any kegels, any fitness. Any Monsanto, any Whole Foods, any farms. Any houses, countries, heating systems. Even before there were any blogs written about health. Way back.

This is a blog about reflex-driven movement.

You probably get what I mean when I say natural movement. Reflex-driven movement might be a bit more complicated. You come with all of the hardware to move in alignment. You come with all the software to move in alignment. Alignment is a reflex. It is part of the package you came in. These reflexes are continuously expressing themselves like background computer programs, but are being stifled by newer patterns we learned via our exposure to modern habits. These newer movement patterns are the ones that you have to undo, so that your natural, brain-controlled, hardwired, human alignment can happen.

This is a blog about alignment.

Why is alignment so stinking important? Well first, you need to understand that alignment is not posture. This comes up a lot, so I'll type it again: Alignment is not posture. Alignment is the position something needs to be in, in order to work correctly. Natural body alignment would be the required position for your body to work correctly as a body.

When you ask me the correct alignment for riding your bike, I can answer how to best ride a bike to not jack up your body. But being on a bike is not natural movement. Habitually riding a bike is interfering with the physiological homeostasis. So I can tell you how to best ride your bike, but really what I want to say is, you need to stop riding your bike. Riding your bike is jacking up your body. Actually I do say that sometimes, but I know you love your bike, so I don't say it very often. So then I'll say, here's a better way to ride your bike than how you were before, and P.S. riding your bike didn't create the quantity of natural movement you needed today, so make sure when you're done riding, you still walk a few miles, squat a few times, and swing from a monkey bar or two.

Then, check if your standing and walking body hasn't adapted to your cycling habit. If your hips don't open and your back doesn't straighten, make sure you add an hour of working towards alignment, putting pieces back to where they should be. Your body will adapt in shape to the activities you are doing. And that shape is associated with disease.

And P.P.S., if you email me about belly dancing, or the T-Tapp program, or your triathlon practice, I'll say the same thing. Do what you love. But know that your fitness exercise does not count toward your daily requirement of natural movement. If you only do your joy, you'll be jacking up your body. Your fun fitness or sport is dessert. You have to eat your veggies, too. The part of your veggies is being played by a whole-body walking pattern, using each joint to its full extent, and maximizing your strength-to-weight ratio.

This blog is about biomechanics.

Biomechanics is a science. When you have thorough knowledge of the physical forces (like pressure, torque, friction, etc.) and the physiological processes of regeneration, it becomes very clear, very quickly what the body should and shouldn't be doing if mechanical longevity is your goal (healthy knees, hips, back, etc. into old age). I like to give little lessons on it via the blog, but to really understand, you need a much more robust education. I try to simplify stuff without making it incorrect (which happens a lot in the exercise science field) and I feel that you are learning. You let me know.

On a semantic note: Biomechanics is the study of the structure and function of biological systems by means of the methods of mechanics. The term has become a more and more popular (let's hear it for the nerds!) term in exercise circles. Like, *Improve Your Biomechanics!* or *Adjust Your Bike Seat for Safer Biomechanics.* I just got an email from someone saying they have good biomechanics, so they can't figure out why their ___ hurts. (That blank is there because I can't remember, not because it was dirty.)

Biomechanics is a science, you can't really "have it" or "improve yours." It's an excellent system and tool for analysis.

This blog is about responsibility.

As a society, our health is getting worse by the day. Our children, who have learned from us (body posture and gait pattern included), will have even worse health than we do. It's time to learn some academic content behind how your body works, so you can fix yourself. Ninety percent of YOUR ailments, right now, are self-induced through the way you move— too much or too little, at friction-generating angles, with too-high loads. There is no one who can fix your ailment (nor your alignment, for that matter) but you. There are many who can relieve one set of symptoms temporarily, while underlying mechanical issues worsen with time. This isn't the best-case scenario, and I'll bet you'll remember reading this twenty years from now.

This blog is about fun.

Learning science as it pertains to your body is fun. It is awesome, actually. Farting can be embarrassing or it can be cool. Pick cool. Learning that you have a habit of thrusting your pelvis or lifting your ribs is AMAZING. How is it that you've tuned out of your mind something you physically do every minute of every day? And there are a thousand more patterns your mind is ignoring right now. This is the definition of mindless. Mindless has become the acceptable human condition. Mindlessness is not natural.

This blog is about time.

The time is now. There is no later. There are Now and Laters, the candy, but that makes it confusing. So let's pretend that there is only now.

Wellness = physics + physiology.

chapter 2

FEET AND SHOES

(and what gets stacked onto them)

AH, THOSE GOLDEN ARCHES...

This is a response I gave to an MD colleague who asked why she was seeing an increase in flat feet in her patients and what she, as a doctor, could do about it. I thought I'd share this information with you, as there is very little literature on the *cause* of flat feet—only the pain and injury that results after walking so many years on a foot without muscle activity.

From a biophysics point of view, the arch of the foot is created by the simultaneous innervation of the intrinsic foot musculature (as opposed to extrinsic) and the external rotators of the femurs (pelvic musculature—the obturators being the most significant).

1. The above-mentioned musculature can only contract with the necessary quantity of force required to maintain the foot arch if the sarcomeres (segments of muscle fiber) are at the correct length. The quantity of force produced is dependent on the correct distance between muscle attachment sites. Since the attachment site distances are dependent on skeletal alignment, parents/teachers/pediatricians should be taught about the objective markers that they in turn must teach children about. However, children learn gait via observation—so new parents need to be demonstrating correct gait as well, in order for the child to pattern it. More simple: There are three to four basic assessment postures one can use to determine the effectiveness of the musculature and corresponding exercises every child/adult can do to restore function.

2. The ideal footwear is "none" for any human. A shoe will weaken the function of the musculature within the foot (intrinsic) by limiting the motion to the ankle. This all being said, the foot needs to be protected in our bacterial and potentially puncturing environments. Select shoes that move most like the foot (again, this is an entirely different structure than the ankle!). Think lighter, less structured, and flexible at multiple segments along the foot bed. Absolutely NO HEELS—which includes athletic shoes with excessive padding underneath the heel—as a heel elevated above the toe box will increase posterior tilt (further weakening the pelvic floor and increasing plantarflexion, which increases tension in the plantar fascia and also weakens the intrinsic foot muscles). I like negative-heeled footwear due to the fact that most folks are also dealing with an anterior shift in total mass.

Arch support cannot strengthen musculature of the foot any more than a sling can increase the strength of the biceps group. Quite the opposite, right?

Regular barefoot walking is good (in a safe place) and NO FLIP-FLOPS or mis-sized footwear where the patient has to grip the toes to keep it on. Toe flexion is NOT a desired action for the foot—it is extremely overused due to the toe gripping most do in response to lack of balance during a typical gait pattern. Developing children should not be in footwear unless required for school. For kids under eight who are bed-wetters or have back pain, scoliosis, or flat feet, I recommend ballet slippers and socks as much as possible. There are also more and more foot-sleeve companies popping up (as an alternative to shoes).

3. Again from an engineering perspective, an orthotic should be used only to supplement a gait- and muscle strength–restoration program. Just as a sling weakens shoulder and arm musculature, an inert device that gives shape to the arch will only weaken the muscles' ability to contract. A "flat foot" is simply a foot with non-innervated foot and hip musculature. An arch support should be fifty percent of the treatment protocol as a developing child or athlete will be at risk for further injury due to the fact that the ligaments of the knee are also out of place to resist normal lateral and torsional forces (especially ACL) and any condition arising out of low tone in the pelvic floor (bed-wetting, hernia, severe menstrual cramps, sacral pain, to name a few…).

July 28 2009 SODAS AND FOOTWEAR AND BONES, OH MY!

I recently wrote a quick tip on Twitter, about habits and bone generation.

"Here is an Osteoporosis Quick Tip: Ditch the heels and the diet soda if you want to generate bone."

(Is it rude, or at least redundant, to quote yourself while blogging?)

A bit more information on the soda: In all the soda studies, only cola seems to have an impact on bone mineral density. What's the big deal? The phosphoric acid (PA) content in diet (and less so in regular) soda quickly increases the body's level of PA, affecting the balance of calcium and PA in the blood. The body will take calcium from your bones in order to balance the levels. If you've got extra calcium lying around, then by all means, use it to metabolize your soda.

That's a joke, by the way.

Now, a quick note about high heels. Most of us understand that the skeleton needs to be weight-bearing in order to generate bone. What does weight-bearing mean? It means the bones below are feeling the total weight of the body above. On the cellular level, the fluid inside the bone cells need to be compressed or squeezed in order to receive the signal "there's weight here we need to support! Quick, make the bone dense… put some calcium right here. No, right HERE! And there. And maybe over there." It's like rearranging your living room furniture. So, let us go back to the heels.

Let's fetch a book. Better yet, find a book with a thickness equal to the height of your favorite heels. Now, slide that book underneath one end of a bookshelf that is about your height. What you should see is a leaning bookshelf. The more narrow the bookshelf, the greater a "lean" you will see. If you had a bookshelf as narrow as your foot is long, the one-, one-and-a-half-, or (gasp!) two-inch heel would project the top of your bookshelf sideways over two feet in distance. Your high-heeled bookshelf is now no longer "weight-bearing." Now, if *you* are wearing the heels, you have all sorts of things you can do to offset that pitch. You can bend your knees, you can flex the ankle, you can tuck your pelvis, and you can arch your back. These things will all bring your body upright so that it looks straight. But your bony skeleton is no longer stacked vertical to the floor. It is now a series of zig-zagging lines, which reduces the quantity of compression on the bone cells as well as the direction you are loading the bone. Osteoblasts are very sensitive to the angle of load and the rate at which they are being stimulated. If you want bone to generate, you have to load your skeleton very particularly and very "perpendicularly" to the floor. It may seem like just a nice way to finish an outfit, but the effect a shoe can have on the mechanical way your structure aligns is instantaneous and extensive.

Until I have a sponsor to grant funds for the bone-generating research I'd like to do, you'll have to peek at these studies…

American Journal of Clinical Nutrition 2006; 84: 936–942. "Colas, but not other carbonated beverages, are associated with low bone mineral density in older women: The Framingham Osteoporosis Study." Tucker, KL, et al.

Journal of Biomechanics. 2009 Jul 20. "Calcium response in single osteocytes to locally applied mechanical stimulus: Differences in cell process and cell body." Adachi T, Aonuma Y, Tanaka M, Hojo M, Takano-Yamamoto T, Kamioka H.

Biomechics. 2007; 40(6): 1246–55. Epub 2006 Aug 2. "Measurement of local strain on cell membrane at initiation point of calcium signaling response to applied mechanical stimulus in osteoblastic cells." Sato K, Adachi T, Ueda D, Hojo M, Tomita Y.

WEDNESDAY IS "WIGGLE YOUR TOES" DAY

"The human foot is a masterpiece of engineering and a work of art."
—Leonardo da Vinci

Well, while no one is breaking down the door to paint a picture of *my* feet, I truly marvel every day at the intricate musculature and fine motor skill we were gifted with from the ankles down.

In the last two years, the notion that we may be completely neglecting our feet has been well-journaled in the media. The *LA Times* featured the entire January 1 health section on the foot two years ago, and in March stated that "over the last 15 years, the U.S. rate of foot amputations from complications of diabetes has soared, approaching 100,000 annually, according to studies and government statistics." Last year, *New York Magazine* ran an amazing article about the impact footwear has on reducing the function of the overall body.

Glad everyone is finally coming around!

Your feet are extremely complex. About twenty-five percent of the number of bones and muscles in your body are located below the ankles. And because we slap shoes on at such a young age, most of us have never used the bulk of muscles located there. It's a shame, as these muscles provide us with balance, keep the nerves healthy, and are part of a complex lever system that is required for the hip to work correctly.

If you are interested in increasing the health of the feet, knees, hips, and pelvis, try this exercise:

Lift ONLY the big toe

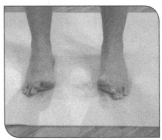

Once you have that, try to lift the rest up one at a time

You should be able to do this, no sweat…but when was the last time you asked your feet to move this way? Imagine what kind of movement your fingers would have if you had worn mittens your entire life. The muscles are all there, they've just never been asked to contract. Another skill to practice is spreading your toes away from each other. It's completely normal to feel cramping in these muscles. And if it's been a REALLY long time since you've used these muscles, you might have to clear away the cobwebs from between your toes. ☺

For excellent foot health, remember to spend as much time as you can without footwear. This allows the muscles to be active instead of making the ankle do all of the work. When you do wear shoes (and of course you are going to have to) I recommend the most minimal footwear you can get. I know there is a lot of "healthy footwear" coming onto the market—which is a great thing! People are starting to choose clothing based on how it impacts their well-being—I'll never argue with that! But from an alignment perspective, the footwear that is the least damaging to the joints in the body is the one without a heel.

The more you get your weight back over your heels and off the front of the foot, the healthier your feet (and your overall skeletal system) will be!

August 23 2009 | SHHHHHHHH, MY FEET ARE KILLING ME!

Two things happened yesterday. Well, a lot of things happened yesterday, but two that I am deeming important enough to influence this blog. In the morning I was walking and chatting with my friend Sandy K. She is a massage therapist and was describing a session with one of her clients. While working on Client X, she found a huge knot or mass in X's foot.

"What's this?" she exclaimed, as loud as you need to exclaim over quiet-and-relaxing music/waves/gong bells/chanting/trickling water, etc.

"That's from the five-inch heels I wore last night." (I'm hoping that Sandy misheard, due to the face that X's face was buried in a fluffy face pillow.)

When Sandy inquired as to how X could actually walk in five-inch heels and cope throughout the night, X said that after a while her foot would go numb and she currently was unable to feel the ball of her foot.

I am using "Client X" to protect the client's identity. So far there is no plan to protect her from herself. I remembered my grandmother's mantra was *sacrifice for beauty*. I assumed she meant sleep, or money, or maybe even comfort. But the nerves in the body? Wow!

That same evening I took the rare opportunity to go and listen to music with some friends and co-workers at a new-and-fabulous hangout spot in Ventura. I danced. I even had three olives out of a martini. As I looked around at the smorgasbord of shoes and feet, obvious were the shuffles, small strides, and downright LIMPS of people dressing up. And it made me wonder…

WHO HAS A SHOE HANGOVER this morning? Forget the wine-induced haze and tequila headache! What do your FEET feel like today?? Is it all thick and comfy socks, Epsom-salt soaks, and ankle braces? Elevated limbs, deep tissue massage, and Tylenol?

Inquiring minds want to know!

 ## WHY YES, I **HAVE** DONE A MARATHON!

Nov 24 2009

The National Institutes of Health, in monitoring obesity and overall public health, has announced the impact of "holiday weight gain" on the long-term issue of obesity. Are the five to seven extra pounds between Thanksgiving and Christmas really an issue? No, not really. Most people will take the initiative after the new year and get most of it off. But it's the *most* that's the problem. There seems to be about one extra pound that lingers each year, and that yearly pound is beginning to look like a possible cause of the slow, age-related (upward) movement of the scale.

So, what can you do?

Keep on walking. Not walking? Start today. In fact, beginning an exercise program during the holidays can have greater impact than waiting until this festive time is over! Weight gain isn't caused by calorie intake alone. Stress can also play a huge role. Shopping lines, shrinking wallets, and…er, *excessive* family time can take even the most Zen Betty to the brink! Take a chillout lap around the block to keep stress hormones to a minimum and keep your metabolism up.

Believe it or not, there is no significant calorie difference between walking and running. Keeping the intensity down will reduce sweating, which means you can leave the "I don't have time to exercise because then

I have to take a shower" excuse right next to your high heels.

High heels? Yep. Not only do they wreak havoc on your feet, hips, and spine, these puppies can decrease your metabolism by reducing joint range of motion while walking. The number of calories burned throughout the day (your basal metabolic rate) is determined by the electrical activity of the muscles. The amount of flowing electricity depends on how flexible your muscles are. Heeled shoes cause a shorter stride (or is it the tight skirt...hmmm?), which means less action of the calves and hamstrings.

According to an economist at the University of Washington, YOU are going to go shopping 5.4 times between the day after Thanksgiving and Christmas Eve, for a minimum of two hours each time! (Evidently, on one of your shopping excursions, you will become lost on the way there and only make it 0.4 of the way to the mall). By my calculations, you are going to spend the equivalent to ten hours or so—essentially the time it takes to walk a marathon—on your feet this holiday season! A perfect time to match your athletic feat with athletic feet. Get it? (Author apologizes for terrible sense of humor. It's genetic, we've been tested as a family, and apparently we shouldn't be allowed to participate in public speaking forums or blog on the internet.)

What I mean is, you have an easy solution to what could be a national problem. Pick better footwear (I recommend negative-heeled footwear to everyone at the institute) and change what could be a potentially stressful time into a calorie-burning, life-saving experience.

(Author's note: Research has yet to have been found showing shopping as a direct cause of health and wellbeing, but the 450,000 people waiting in line at 4:00 a.m. on Black Friday just seem to be happy people, don't they?)

Nov 25 2009 IS FOOT PAIN BIGGER IN TEXAS?

Volunteer to make Thanksgiving meal for friends and family? Check.

Find out you need to fly out Thanksgiving afternoon to make TV appearance on Friday? Check.

Find out that TV appearance is live from a mall at 5:20 in the morning? Check.

Realize you have to get up at 4:00 a.m. to make said appearance? Check.

No, there's not going to be a professional to do my makeup at 4:00 in the morning? Ouch.

Oh, you want me to fly to Texas the Monday *before* Thanksgiving to do a Tuesday a.m. segment too? Okay.

Wait, do I have to get up at 4:00 a.m. Tuesday morning? Yup.

Car trouble on the way to the airport? Check.

Finding out about death in the family just after touchdown? Check.

Wait, am I still cooking TG meal on Thursday? You betcha.

The thing is, life never stops, so you just better hang on and enjoy every moment. And it works out every step of the way, and I've been given the best of the best to spend my days with. Thank you, every one of you.

I just got back from doing a segment about feet and healthy footwear on Good Morning Texas! In case you missed it (and why wouldn't you have missed it?), click here[1] if you want to brag to your friends that you've seen it.

In memory of Jo Jo, because you had quite the experience, and through you, the adventure goes on.

---

1. youtube.com/watch?v=GazKfBoT4cg

 ## Dec 1 2009 — THE CAPITAL OF FOOT PAIN

I wonder if Governor Arnold S. got a chance to watch my *Good Day Sacramento* segments on skeletal health and footwear.

And if you happen to wonder why I am posing as both the expert and the interviewer in these clips, it is due to the fact that some people didn't want to go to work at 4:00 a.m. on the day after Thanksgiving. Can you believe that?

The best part about the whole experience (excluding the fact that the manager of Cinnabon came over to bring me a free, fully loaded bun in between segments) was how enthusiastic and friendly people were at this ungodly hour! Thanks to everyone at Arden Faire Mall!

Healthy Foot/Footwear Science[1]

Holiday Shoe Makeover[2]

Short blog today, I know. But don't worry, I'll Be Back.

1. youtube.com/watch?v=Gtvf4kAOaXM
2. youtube.com/watch?v=1j_Wj3lhp1Y

January 5 2010 | "HEART" MY PELVIC FLOOR

Stiletto: A short knife or dagger with a long slender blade of various designs primarily used as a stabbing weapon. Its narrow shape, ending in a rigid pointed end, allows it to penetrate deeply.

This morning you picked shoes that best matched your outfit. You know...the pair that makes you feel taller, thinner, and just a little more fantastic than you did before you got dressed. But what if I told you that the footwear you selected not only affected your inner fashionista, but also impacted the nerves in your feet, the pain in your arch, the ability to balance as you get older, your bladder control, and get this...your bone density. You may be struggling with health issues that you didn't even realize were related to what you wear on your feet. How you feel may be a result of the accumulated, daily affect of footwear!

I'm writing this for the twenty-five percent of women currently suffering from debilitating foot pain right this minute, the eighty (!) percent of women suffering from Pelvic Floor Disorder, and the millions of women realizing they have lost a substantial amount of bone mineral density. This is also for the woman interested in preventing or treating the failure of the structures most often injured in women: the pelvic floor, the bony skeleton, and the foot. The staggering facts show that hundreds of millions of women have not one, but two or more of these issues—with the number of consumer dollars being spent in search of a solution adding up to BILLIONS.

I'll let you in on a secret. Research[1] on footwear shows that for every one degree of heel you wear, there is one degree of spine vertebrae, hip joint, or knee joint reaction. A one-inch-high shoe (that's not even a high heel, more like a man's shoe or a good walking shoe) is twenty degrees! That little inch can be increasing the wear on your joint cartilage, be contributing to disk degeneration, and absolutely be decreasing the weight-bearing status of your bones.

Women are making health hard for each other, and it needs to stop... now!

1. "The Effect of Heel Height on Gait and Posture: A Review of the Literature." Cowley et al. J Am Podiatr Med Assoc. 2009; 99: 512-518. Emma E. Cowley, MSc, Thierry L. Chevalier, BSc, and Nachiappan Chockalingam, PhD

April 8 2010 DO **YOU** HAVE HAPPY FEET?

One in four Americans now has chronic foot pain. These are insane numbers! Because my Aligned and Well program offers such simple solutions, I am often asked to write or speak on foot health. In doing so, I came across an excellent product that helps with alignment passively. Being passive (which means you only have to exert enough energy to put your socks on) is good sometimes, because it is hard to be mindful all the time, and correcting posture takes diligence!

My-Happy Feet foot-alignment socks help stretch the muscles in between the toes (the adductors) that squish toes together. If you've been wearing heels, or shoes with narrow toe boxes, these tight muscles can create or exacerbate bunions, hammer toes, and plantar fasciitis. I have committed to writing a twelve-month foot-health e-news publication for this company, to help them educate the masses. I have included the first article below.

Happy Feet Science: Alleviate Foot Pain with 4 Simple Ways to Soothe Your Feet

Since your very first step, you have been moving around your life on your own two feet. Remarkable as that is, it's even more amazing how little most of us know about how feet work, and how they affect the rest of the body. When your feet are not working to their full potential, the effects can be felt throughout the body. While poor foot mechanics can lead to many issues that manifest throughout the skeleton, you don't have to head north of the ankles to start the list. Bunions, plantar fasciitis, and hammertoes can all be created or affected by weak muscles inside the foot called the intrinsic muscles. Shoe choices, gait habits and posture all contribute to weakening of these muscles, which in turn creates tension on ligaments, tendons and fascia—all common sources of foot pain.

Here are four simple things you can do improve your foot mechanics immediately:

1. Fix Your Foot Position

Because every step you take is a result of how you push off of your feet, foot position is a critical factor in how your knees and hips deal with force. Most of us can classify ourselves as "pigeon-toed" or "duck-footed," common slang for the positioning of the toes pointed toward each other, or away from each other. Very few of us walk with the bones of our feet pointed directly in front of us, rendering the "cables and pulleys" of our muscle and ligament systems largely unable to do the work they are designed to do.

The simplest way to align your feet objectively is to find a straight edge, such as a tile floor in your home or a yoga mat. Line up the outside edge of the foot so that it matches the straight line. Now your foot is straight. It is common for this to feel very abnormal, even pigeon-toed! Now adjust your foot into this position every time you can remember.

2. Barefoot Walking

Your human machine was not designed to wear shoes. Did you know that one quarter of the bones in your body are from your ankle down? The same is true for one quarter of your muscles. What does all this mean? It means that your feet are designed to be very dexterous. They have the potential to be just as dexterous as your hands, as a matter of fact, but very rarely do we challenge ourselves to utilize this potential. Instead, we keep them bound up in shoes, and put the ankle in charge of all the refined hinging and balancing movements that the foot is designed to do. Imagine how ineffectual you would become by wearing your shoes on your hands when doing your daily chores! We habitually handicap our feet in exactly this same way, every day.

When you take your shoes off and begin reactivating the muscles of your feet, you also reactivate large segments of your brain. As with your muscles, your brain and nervous system need to be used or the unused cells gradually die off.

Incorporate barefoot time into your schedule. Consider making your home a shoe-free zone, or at least doing some yoga poses that press the soles of your feet to the ground. Local school sports fields are often pet- and debris-free, and can provide good places to do some barefoot walking.

3. Shoe Choice

If you're not yet convinced to donate all your shoes to your dog as chew toys, then you should know that all shoes are not created equal. Your body has to compensate for different shoes in different ways.

One truth is simple: every time you wear high heels, you are damaging your body. Keeping in mind that being barefoot is your natural state, any shoes that elevate the heels provide undue strain on the body—the same effect as walking down a hill every time you go anywhere. This is true even if the heel is only slightly higher than the ball of the foot, which is the case even in most shoes perceived as "flat."

If you have to wear shoes, you would do well to shop from a company that considers the health effects of its line. Choosing a negative-heeled shoe, where the heel is slightly lower than the ball of the foot, puts the body's mass back over the heels where it belongs and away from the front of the foot.

And if you're still not ready to give up your high heels, then it becomes all the more important to reverse some of the damage by using foot alignment socks.

4. Foot Alignment Socks

Cramped foot muscles due to tight shoes, excessive impact from running, and flip-flop use can get relief by simply sliding on My-Happy Feet. The gentle stretching motion of toe abduction (the fancy word for toe spreading) actually stretches and strengthens the small muscles inside the foot, increasing circulation!

5. (A bonus tip!) Exercise Therapy

Absolutely! Most chronic foot issues can be corrected with specific foot stretching and strengthening.[1]

---

1. This seems like a good time to remind you that there are a whole lot of resources available at restorativeexercise.com, including my book & DVDs about foot health.

April 8 2010

12 BILLION DOLLAR HIGH HEELS

This just in the *LA Times*:

High-heeled shoes your feet won't hate you for[1]

From corsets to sky-high stilettos, women have been suffering in the name of fashion for years. Thanks to the Beverly Hills-based footwear line Dana Davis, named for its founder and chief executive, high-heeled-shoe lovers can breathe a sigh of relief. In its third season and selling well locally at Nordstrom South Coast Plaza and at danadavis.com, the label features chic styles with invisible comfort technology such as customized arch supports and strategic cushioning. Imagine heels that may actually be good for you.

I agree, *LA Times*, it *would* be nice to imagine a heel may actually be good for you. It would also be nice to imagine that an eclair is good for you too. Or that a daily cigarette is good because it causes you to exercise your lips. When you start associating wearing a stiletto (even one with a squishy foot bed) with taking a daily vitamin, I hate to tell you, you're reaching. Reaching up high with your four-inch heels.

Although the new shoe line features wider toe boxes to minimize foot squish (YAY!) and a softer sole, the most body-tissue-deforming characteristic of a shoe is the positive heel. *Any* positive heel. Now I know that you ladies, especially the ladies of Hollywood, are reluctant to give up their heels, so I'm not stressing over that. What I AM freaking about is the fact that Dana Davis is a Type 1 diabetic, whose habit of wearing high heels (even when she was teaching in a classroom all day long!) has resulted in "8 corrective foot surgeries." PICTURE RED FLAG WAVING HERE.

Lower-leg amputations are responsible for $12 billion national health-care dollars per year (according to the *Journal of the American Podiatric Medical Association*). Because diabetics are having the bulk of these surgeries, shame on the podiatrist who doesn't clearly communicate the relationship between compromised muscle, neurological, and circulatory function and the height of the heel. If you have Type 1 or Type 2 Diabetes, please know that a high-heeled shoe is having a direct, instant, and negative impact on the health of your feet. Can you wear them down the red carpet? Of course. Can you wear them to weddings? Go for it. Should you wear them while you teach Kindergarten? Duh.

I know high heels and long legs are sexy. But trust me, no one looks smokin' hot with a self-induced limp. And, P.S., the whole *"I feel sexy when I _____"* argument has already been used for smoking. We think it's hilarious when we see old movies and people are smoking (I love it when the doctor is smoking while talking to his patients). Just wait twenty to thirty years, we'll be saying the same thing about heels.

Dana's shoes are $260–$450 a pair. They are lovely. But the heels are stilettos, the wedges are the height of regular heels, and the "flat" even has a small heel. After eight surgeries, I can tell you that the price of wearing heeled-shoes for Dana Davis (or her insurance company) is over a hundred thousand dollars, easy.

Dana, email me and I'll send you our Fix Your Feet DVD and a pair of My-Happy Feet. And, like all the people I pick out (not pick on!) for my blog, I am happy to give you a private session to teach you the foot alignment essentials. Bonus!

My favorite part of the article is the mention of the "invisible comfort technology." I think someone should invent some "invisible nutritional technology" and put it in a donut. I'd buy it.

1 latimes.com/features/image/la-ig-danadavis-20100321,0,5795903.story

 April 27 2010 THE "BUNION BLOG"

I decided to launch my New Katy Says with The (rhymes with thee) Bunion Blog. The epic explanation of Where Bunions Come From. The counter-argument to the statement "My Bunions are Genetic." The Blog Heard Around the World. My Big-Toe Odyssey. I'm a bit excited, you can tell.

Let's start with some basic anatomy.

Hallux is the Latin word for the big toe. It was derived from the Greek verb "I spring or leap." The big toe is very much like the pole plant of a pole vaulter. The tip of the big toe is the last part of your body to leave the ground after your pelvis (where your center of mass is) vaults over into your next step. Your big toe may also be called the Pollex Pedis, or the Thumb of the Foot, but until I have to hitchhike without my hands, I'm sticking with Hallux.

Hallux Valgus is the term for a big toe that, instead of pointing forward, begins to reach laterally (toward the other toes). After a while, the big toe can be pulled so heavily toward the pinkie toe that it becomes perpendicular to the other toes. Not only does Hallux Valgus reduce the function of the foot, it is not very comfortable to walk on, especially if the tissues on the side of the joint swell into what we call a bunion.

The term *bunion* also comes from the Greek language, meaning small hill, mound, or heap. The growth of the bunion is a response to unnatural loading occurring at the joint of the Hallux. *In every body*, the bunion is caused by inappropriate loading (using the joint in the wrong position). The only thing that can make it "genetic" is the collagen content of the tissues that stabilize the joint. Some people have conditions in which their collagen content is genetically proportioned differently than others. But the percentages of the population that have this condition are very low (about three to ten percent, depending on which literature I looked at). If you were in this low percentage, then you would have a collagen problem in all your tissues and joints. If you don't have a whole-body collagen problem, your bunions are most likely caused by habits you picked up along the way.

What makes bunions *seem* genetic is the fact that your parents (usually mom) had them, as did her mom, and her mom's mom, which seems like a lot of moms. Before there is a bunion, there can be the valgus (outward displacement of the big toe) OR there can be an incorrect loading of the joint first, causing bony growth that pushes the big toe out of the way. Let's look at both situations.

Shoes (of the fashiony, high-heeled kind) not only push the big toe toward the other toes, they place the weight of the whole body right smack over the big-toe joint. Now you've got your joint in the incorrect position (muscles tightening in between the toes) and you've got your weight right over it (sending signals to beef up the bone in the area, also known as "the bone spur signal"). Wear shoes like this for a night and you can feel soreness in your feet. Do this over a lifetime and you dislocate your big toe and can no longer walk without debilitating pain!

Let us take a look at the space you should have between your toes when you are standing, and how footwear with a narrow toe box (see it push my big toe towards my other toes?) can affect the foot while just standing around (never mind the damage wearing these puppies to work every day!).

Do you see the natural space between my toes BEFORE I put on the shoe?　*Now my big toe is being forced toward my other toes. A little bunion-to-be!*

If you have been helping your bunions form, there are two simple things you can do right now to fix your bunion (yes, you can absolutely minimize the hump of your big toe, but it requires diligence!). First, stop wearing shoes with toe boxes that are too small for your foot. Even having low collagen doesn't mean that you WILL get bunions. It means that IF you push your toes together, the joint doesn't have the ability to resist your shoe habits for long. If your foot is wider than your shoe (see pic one) you are damaging the first toe joint. Stop it. (This works on most animals, so I'm trying it out here...)

Bunion Exercises:

When your feet are all squished together, the adductors (muscles between the toes) get very tight. Begin with a daily stretch of all toes, giving extra attention to the big guy. Placing your fingers between all of the toes will help restore adductor muscle length and joint range of motion. If you have a pretty significant angle on the hallux, spend some time pulling it away from the toes by itself.

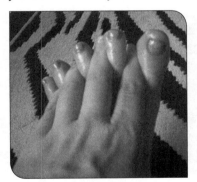

Place a finger in between each toe, gently!

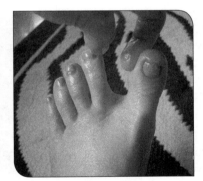

A little extra stretch for the hitchhiking thumb.

Now let us assume that you have never worn shoes that pinched your feet. Let's say you've never worn shoes at all. One of the greatest arguments in support of the "genetic bunion" is the fact that hunter-gathering populations who have never worn shoes still have a few people that get bunions. Genetics, perhaps. But bunions are made in other ways too.

Foot Position

The lever of your foot should point straight ahead, but most of us have developed a slight to exaggerated turnout of the feet over the last two hundred years. It's called "turnout" if you were a ballerina. It's called Duck Foot if you weren't. It's the same thing, though. Your feet not pointing straight ahead. How can you tell? Find a straight edge on a rug, or in this case, I used the spine of a book. Line up the outer edges of both feet along a straight edge to make sure the feet are straight (the toes may point in a bit when you first start).

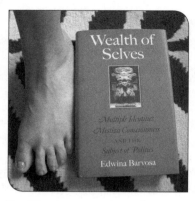

Feel pigeon-toed? That's normal in the beginning. Most people have been walking in turnout from early on, because their parents were turned out. Their parents were turned out because they learned to walk by watching *their* parents move during a time period when military and dance styles reinforced the military "heels together, toes out" culture. Gait patterns are handed down from many, many generations without conscious effort. It just happens!

(Note: The book used in the photograph, *Wealth of Selves: Multiple Identities, Mestiza Consciousness, and the Subject of Politics*, by Edwina Barvosa, was selected to send subliminal messages to you readers out there interested in where notions like "I have to wear heels to feel worthy" come from. Thank you to scholars like Barvosa. Read it!)

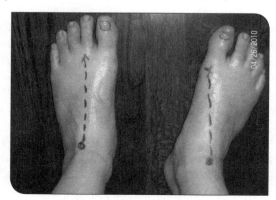

Left foot is straight. Right foot is in slight turnout.

When you walk around in turnout, the side of the big toe is loaded incorrectly. The dotted lines in the picture below show where the "pole vaulter's pole" hits the ground before the weight in the pelvis moves over the foot. When the foot is straight (meaning lined up as shown in the photo above) the weight is distributed well over the entire forefoot (front of the foot). When then foot is turned out (as it is on the right foot), you can see how the outer edge (bunion area) of the joint presses into the ground with each step!

The side of the joint is the side. That sounds profound, I know, but joints are designed with the same engineering laws used in building your house. The walls don't have the same weight-bearing quality as your foundation. When you walk in turnout, you are walking on the "walls" of the joint instead of the foundation. Your body responds by "building up" the walls, which is the bony protrusion you see as the bunion develops.

You now have a few habit modifications to begin immediately. And I, evidently, need to give myself a pedicure. ☺

 ARE HIGH HEELS THE CIGARETTES OF THE FUTURE?

May 12 2010

Yesterday I ran across a blog asking if fast food was becoming the cigarettes of the future. I disagreed and suggested that it is positive-heeled footwear that was, in fact, creating the expensive (and painful!) conditions suffered by many.

My post was:

I don't think fast food has become the "new smoking" as much as high heels have. Once upon a time, the rationale for smoking was the sex appeal and the positive effect on weight management. Because it is hard to rationalize the choice of donuts as a positive one, poor food choices don't fit as smoothly into the analogy.

Bone density–decreasing, nerve-damaging, and arthritis-causing high heels are probably being worn by your favorite OBGYN during your annual exam, eerily reminiscent of a 1950s doctor's visit, where Mr. Doctor chain-smoked throughout your entire exam (anyone out there watch *Mad Men*?).

Blind to the detrimental effects footwear has on health, or the correlation of footwear to the prolapsing bladder waiting in her office, Ms. High-

Heel-Wearing Women's Health Specialist doesn't know, in the exact same way Mr. Smoking MD didn't know, that when you look at the heeled shoe, you are looking at the cigarettes of the future.

I submitted a book proposal for curing the high-heel hangover—a book about alignment, gait patterns, and footwear—to a publisher that told me "Women will *never* give up their heels, no matter how much they wreck their health."

Is this true? Even though a positive heel compresses and degenerates the disks in the spine, weakens the pelvic floor, decreases the mineral density in the bones, and kills the nerves in the feet, would a woman *still* choose to wear that shoe? And teach her daughter, her students, or, heaven forbid, her patients to do the same?

This (female!) executive is saying that women are okay suppressing their (and their offspring's) health.

Why would a woman choose to pass on women's diseases to the next generation for the illusion of height, longer legs, or a leaner appearance? Do we feel that poorly about ourselves, or, is it because no one has clearly stated that Your Shoes, Ma'am, Are Making You Sick?

I'd like to hear from you on this one. What's your honest answer, and why?!

May 24 2010 CAR ALIGNMENT, PART ONE

I don't usually drive more than five to ten minutes per day, but sitting in traffic today kept me behind the wheel longer than normal (over an hour). I was sitting there bored out of my mind until I noticed that driving was making the big toe joint on my right foot ache. Yay! Something to do!

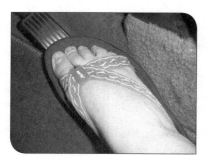

I was shocked, and I mean SHOCKED when I noticed that I was tightly gripping my toes when pushing down the accelerator. Why am I doing that? Scrunching my toes doesn't make the traffic any better. I even popped my shoe off to check if the shoe type was influencing the toe clenching. It wasn't. Why was I shocked? Because this means that my driving time (up until now!) has been strengthening my right foot and toe muscles in a way that buckles the toe bones (hammertoes, anyone?) and pulls my big toe toward my pinkie toe (bunions!). What happened to my lovely toe abduction (spreading)?

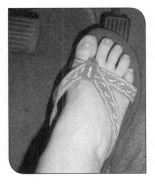

Car Exercise: Can you press the accelerator without gripping your toes? Now how about toe spreading?

And here's a second observation for all you out there with sciatic pain and low back and SI joint issues. Driving usually causes one leg to be actively reaching forward from the foot all the way up to the hip. This reaching motion places the pelvis under constant torsional forces (rotation) that tax and fatigue the lumbar spine. But guess what? The car manufacturers know that your hips should be ergonomically balanced at the wheel, so they put a "balancing pedal" on the left side of the floorboard. Check out my lousy, lazy driving posture before...

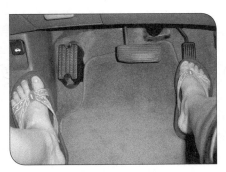

and after! Better! Now my hips are square to the front of the car.

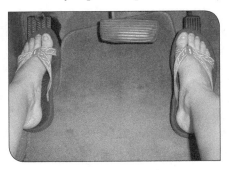

Keeping your hips even requires a lot of work. You'll quickly drop right back into your old habits if you stop paying attention. And let me tell you, it was rather challenging to align my hips, take a picture of the aforementioned foot position, *and* talk on the phone. What?!

June 1 2010 NO SHOES, NO PROBLEM!

Today is National Barefoot Day. I say we make this a day off too, whaddaya think?

If you are living under a rock, you may have missed the fact that a barefoot movement is happening across the U.S. You may have missed the author of the bestseller *Born to Run* on every major media talk show. You may have missed ten-page articles on footwear minimalism in *New York Magazine* and the *LA Times*. You also might have missed some of the healthy foot posts on this very site. Or maybe, you are sitting there reading this in your Five Finger shoes, My-Happy Feet, or in your barenaked dogs. Whether you are just hearing about it now, or had the date circled on your calendar, there are some things you should consider before baring it all.

Barefoot walking is a natural occurrence and to get maximal health from your body, barefoot walking from birth is probably required. However, the terrain that is optimal for walking upon is not manmade turf. Not only is friction and traction greater in the urban jungle, the interaction between asphalt or cement and the tissues of the feet are higher than, say, firmly packed dirt. If you are a walker (1.5 G-forces) or runner (2–3 G-forces) and are on manmade terrain, opt for a manmade solution like an extremely thin and flexible sole layer that creates the force-dampening effect dirt would, while still leaving your foot free to fire muscles needed to support its own structure.

If you want to join the barefoot movement, you have to prepare properly. If you've been wearing footwear the bulk of your lifetime, the muscles in the feet, the circulation to the skin, and the communication of the the nerves to this area are going to be compromised. Pretend your feet are like your eighty-year-old grandmother. You're not going to take her to the gym, throw her onto the chest press machine, and load her up. You'd start with the basics. Like getting her to stand up without holding on to anything. And then progressing to a walk around the block. When we try to go back to "natural ways," we have to remember that we are not, in fact, living in a natural world. Your foot should have had a lifetime to prepare naturally, with your muscles naturally strengthening as your weight and height generally increased. To make the transition, here are some guidelines:

Start by being shoeless in the home as much as possible (you can even keep your socks on at this stage).

If you've been wearing positive-heeled shoes for a lifetime, begin a daily practice of foot stretching and strengthening.

Runners, learn to barefoot walk first. Removing the shoe is like taking a cast off after breaking an arm. Before you do hand stands, you first need remedial therapy. To "wake up" the nerves and tiny intrinsic foot muscles, start by walking around the block, then progress to a quarter-mile. Increase your distances by a quarter of a mile per week, decreasing your distance if you experience any heel, arch, or joint pain. Only progress when you are foot-strong.

Calluses are good things. An interesting thing about the biomechanics of skin is the high level of health, strength, and circulation callused skin has when compared to uncallused skin. Calluses are indications of healthy, robust cellular turnover. A callus becomes a bummer when only one small area thickens because of friction with a shoe or when toes rub together. This tiny hardened area creates more pressure (like a pea for you and your princess feet) and the sensations from the callus area stick out like a sore thumb; which is ironic on the foot, no?

The more time your entire foot skin spends interacting with the ground, the faster the foot cells beef themselves up uniformly, to become a protective layer to reduce skin puncturing and tearing. Now here's more irony. Wearing shoes has, in fact, made us too sensitive for our environment. In the meantime, we have allowed the environment to become too aggressive for our biology.

The solution: slowly and safely restore the health of your feet in whatever manner makes you feel most comfortable. If you're not ready to bare all, just kick off your shoes when you get home. It's all a step in the right direction... ☺

IF THE SHOE FITS

Here's another article I wrote on foot/pelvis/balance (are you bored yet?). This one is from July's *LA Yoga Magazine*[1]. Perhaps the most important piece of the article is the 60/40 "rule" of foot weight bearing (which is not quite correct) explained. Enjoy!

If The Shoe Fits...

by Katy Bowman

After you've twisted, lunged, lengthened and aligned, focused, prepared, repaired and reset, you must get off the mat. The notion of bringing a yoga practice to the rest of your daily life (the part that takes place between classes) can improve personal relationships, success at work, *and* your shoe rack by improving your relationship with your feet.

Postures that require grounding action in the feet, increased range of motion of the toes, and full lengthening of the arch are a delicious part of any practice. These foot movements are both challenging and especially rich in sensation, because of the sheer quantity of time we spend not using our feet. "The human foot," according to Leonardo da Vinci, "is a masterpiece of engineering and a work of art." Made up of twenty-five percent of the body's bones and muscles, and articulating around thirty-three joints, our feet have the potential to subtly change shape at each joint, thus sending valuable information to the body's center of mass (located in the pelvis). The tiny stretches in between every one of each foot's twenty-six bones are a gold mine of proprioception that allow the pelvis to make three-dimensional positional adjustments.

Optimal foot health, however, has been compromised in a very large way, due to the heavy use of footwear over a lifetime. Different characteristics of various types of footwear have unique effects on physiology and biomechanics; the limitations footwear places on motion of the foot (along with motion of the ankle, knee, hip, and sacrum) are not equal across all types of shoes. The healthiest footwear is one that interferes little with your natural body movements.

Barefoot is Best.

It would be unfair to discuss footwear without clearly stating that optimal foot health is ultimately reached through full, shoeless interaction between nature and foot. Foot health can become compromised, however, when you walk on unyielding, manmade surfaces that may or may not be speckled with broken glass and other dangerous items. Small doses of being barefoot (wearing socks is okay, too) can be done in your home; just make sure the jacks are picked up before you do. Minimalist footwear brands like Vibram Five Fingers are also great options for using your feet.

Before you go baring it all, keep in mind the supporting structures of the feet have been, for the most part, inert for the bulk of your life. Loading fresh arches on long walks after removing a lifetime of support can stress and strain tissues. It is important to think about building strength in the musculature of the feet just as we would with any other part of our body. Start with smaller doses of barefoot walking and make sure you do lots of foot stretching in between walking sessions. Pamper your feet, which will help them be happier as they cart you around: A coconut oil foot massage and nontoxic pedicure can be a mini-vacation as Southern California heats up and dries out this summer.

Eliminate the Heel.

Not quite ready to go au naturel? Shoes that closely mimic barefoot walking, such as Earth Footwear's negative heel technology, can really drive home the "weight should be in your heel" feeling. Podiatric journals have recently become riddled with articles illustrating that for every positive degree of heel (for a point of reference, the one inch found on a man's dress shoe creates an average angle of twelve degrees) there is a resulting angle of deformation in the lumbar spine, pelvis, knees, and/or ankle. There is no footwear characteristic that jars one out of whole-body alignment faster than the positive heel. If a dress shoe creates twelve positive degrees, just think about what a stiletto can do to deform the rest of the body. Pause and consider choosing well before selecting footwear that undoes all your hard work and increases mechanical stress on a cellular level.

Spread the toes.

Toe abduction, or a movement of the toes away from each other, is a normal part of a healthy gait pattern. Years of carrying weight too far forward on the feet, wearing too-small shoes as a kid (anyone else out there live in hand-me-downs?), and narrow toe-boxes on certain shoe styles have really limited the toe-spreading motion. Selecting footwear that provides ample room to splay your toes when walking is the healthiest choice; ideally, our toes should spread just as the fingers do.

If tight toes have become a habit, foot alignment socks (my-happyfeet. com) can work on spreading them for you. The perfect product for the hard-core alignment freak (me!), you can load muscle and fascial tissues while you sleep. Brilliant.

Be attached to your shoes.

It is surprising how quickly the flip-flop has moved beyond the favorite pool accessory to the ultimate fashion staple. In California, I totally get it, although I think the New York professional scene may still be in shock. Flip-flops are usually enjoyed for their lack of restriction—lots of fresh air and minimal friction. They're also time-friendly. Don't have time to find matching socks and bend down to tie your shoes? Flip-flop may be your guy.

The only negative to the bikini of footwear is the fact that it doesn't stay on your foot without some major muscle clenching and bony alteration. Research on gait patterns and poorly attached shoes demonstrates increased risk for hammertoes, plantar fasciitis, and knee pain. I say keep your favorite flops around for water and beach activities and invest in the newer Roman-style sandals that offer the same open-air feel but with better binding.

Weight in Your Feet.

Where is the best place to carry the weight in your feet for optimal foot health? The oft-given instruction for correct weight-bearing in the foot usually cites the 60/40 rule, which is widely misinterpreted as: Sixty percent of the body's weight back in the heel and forty percent towards the front of the foot. This weight distribution, however, is not actually correct, and the reason comes down to the actual scientific definitions of commonly used terminology.

The term "weight" means "result of the vertical force (gravity) acting on the body's mass." In order to achieve the 60/40 weight distribution, your center of mass would have to shift forward, removing the plumb alignment of the hips, knees, and ankles. This forward motion increases the torque not only on the ankle but also on the lower spine and SI joints.

The optimal place to carry one's body weight is actually toward the center of the heel bone. Keeping the weight of the pelvis over the ankle joint is the only way to ensure a straight leg and a healthy lumbar curve. However, having one hundred percent of your "weight" over the heel does not mean the front of the foot is inactive. With the pelvis centered on a plumbline relative to the ankles, the forefoot (but not the toes!) can now actively press into the ground. The action of backing up the hips to ground the heels while simultaneously pressing of the forefoot into the ground creates an active, force-generating interaction with the earth. This is so much better for the body than passively thrusting the hips forward.

An effective yoga practice is one that improves one's mindfulness, not only on the mat, but for the countless choices one has to make every day. When you understand the impact your shoes can have not only on your feet but on your entire body, then choosing mindfully means selecting the footwear (or lack of footwear!) that is most appropriate to your highest goals for yourself.

When I envision my highest goals, healthy feet, knees, hips, and spine are always in the picture, for as long as I'm going to be using this body. That doesn't mean you won't catch me in flip-flops when I'm down by the beach in Maui. It just means that every step I take will honor the works of art that I am stepping on.

---

1 layogamagazine.com/content/index.php?option=com_content&task=view&id=6 63&Itemid=1

August 3 2010 FEET: THE SIXTH SENSE

In addition to hanging out with my sister's thirty-seven kids while on vacation, I also hiked my butt off. In California, people don't tend to be hikers in the way people in Oregon or Colorado are hikers. In fact, one of my favorite authors, Pam Houston, sums the reason up quite nicely, saying that "when hiking alone in Colorado, about one out of a hundred encounters makes me nervous, and when hiking alone in California, that number is one in three." Amen. Especially when most of those encounters have four wheels and 300 horsepower.

Being a pseudo hunter-gatherer, I attempted to wear my Vibram Five Fingers (shoeless shoes) throughout my entire vacation, something that was going quite smoothly until my Most Fit Friend suggested a twelve-mile hike up behind the Columbia River. I was totally in. What a great chance to take my VFF into the rough.

This wasn't my first time wearing my non-shoes, in fact, I had already logged hundreds of "shoeless" urban miles on the asphalt and concrete of city sidewalks. That's why I was completely unprepared for the fact I had to tap out around mile four. Starting off on the trail, my flexible feet felt awesome. I felt so healthy. So biomechanical. So...smug. My feet were mobile while walking, just as they were designed to be, and I could feel every lump of dirt, pebble, and ow...SLATE...jabbing into my foot. I toughed it out until I realized that while I had slightly increased the flexibility of my foot over the last year by minimizing shoe use, my intrinsic muscles (the muscles between the bones of the feet) hadn't really worked due to the artificially flat and debris-free surfaces I frequent.

The mobility of the foot is extremely complex. The foot's thirty-three joints allow the foot 8.6×10^{36} unique positions. Just in case math isn't your thing, this means that there are more than a zillion (really!) motor programs you foot *could* have, each one needing additional brain/body communication. Our current biomechanical, medical, anatomical, and podiatric texts identify *three* motions. That's how stiff our feet have become. Just to be clear, *you have the potential for billions and billions of unique foot motions, but we typically talk about three.* Wow. That's a low bar.

I teach the physics and biomechanics of the foot because of its enormous responsibility to whole-body wellbeing. Think of it in this way—the foot's ability to deform is a method of data collection for the body. The way the foot bends in tiny amounts to form to a surface creates a picture of what is underneath the foot. A healthy foot creates an "image" in the brain (very similar to sonar) that helps the body's center of mass position itself perfectly over the surface's contours for optimal balance. When the feet are stiff and tight (and constantly in shoes), the center of mass (in the pelvis) also becomes stiff and immobile. The center of mass is never being told where to go by the feet. Eventually you get lumbering, lurching, and unbalanced movements reminiscent of an old or injured body—which essentially the body has become, due to minimal sensory input.

I know all of these things in an academic kind of way, but what I hadn't really experienced was taking *my* new foot muscles (new because they had never been over so many rocks for such a long distance) on a three-mile off-road hike that challenged the sensory input of my feet. My feet were baby feet. These muscles had never been used and were experiencing fatigue. You wouldn't take a newborn on a three-mile hike, but that's what I did. I almost wanted to cry as each step pushed into tired and sore muscles. My solution? Putting shoes (still super-flexible, light, and no heel) back on to finish the hike. I brought my trusty GOLD Earth sandals (really, they are metallic gold with little faux diamonds on them) and finished nine more miles, no sweat. My ankles and regularly used foot muscles were fine, it was just the little guys in between my foot bones that were tired!

Through the rest of my ten-day trip, I hiked miles in the Olympic mountains and walked tons all around town. There was a huge difference in the muscles used on urban terrain (flat and hard) as compared to natural terrain with rocks and uneven ground. If you can't tell if you're "in nature," just use comedian Demetri Martin's definition, *"Hiking is just walking...where you can pee."* Oh, and P.S. I peed A LOT in the woods, just because I could. Love, love, LOVE those squats! Other notable mentions were my post-walk super-open hips and hamstrings (I didn't do any stretching, but let the natural movements be my program and unwind my fascia under the clear-blue sky), and I had ZERO menstrual cramps. Hello, open pelvis!

There is a very large barefoot movement happening now, which is a wonderful thing. However, we Westerners have a habit of picking a "natural" habit and jamming it into our unnatural lives, i.e. walking long distances on cement and asphalt. This takes a good thing (natural foot movement) and creates an unnatural vibrational load that could potentially lead to fractures in overloaded foot bones. Train smart. Be logical. If you want natural foot movements for optimal health, walk in natural environments. Shoes have been protecting us from our over-rigid environment for too long, and it takes time (years, even) to restore function. I also strongly suggest a plan to move toward increasing foot health via increasing range of motion and function of the intrinsic foot musculature.

1. Start by daily stretching and massage of the heels, mid-foot, forefoot, and toes.

2. Do your foot exercises.

3. Understand that the position of the foot is maintained by the muscles of the hips and make sure you optimize lateral hip (IT Band), hamstring, gluteal, and adductor (inner thigh) strength with full range of motion. Tight hips limit foot function.

4. Get a super-flexible shoe with minimal (or better yet, no) heel.

5. Before jumping into non-shoe shoes, deal with your whole-body alignment and gait mechanics. Podiatrists are seeing a huge increase in forefoot fractures from people (even highly-experienced runners!) who land with excessive force on the front of the foot. Walking should be heel to toe, not landing on the front of the foot. Running should be done on natural surfaces with body weight stacked correctly (without the torso leaning forward).

6. If you get non-shoe shoes, be a walker. But if you do choose to run, build up your walking mileage for a year before even considering running in them.

7. Log your miles on a natural surface, with elevation changes and rocky obstacles. The urban jungle is not a natural walking surface, and the friction and traction of this surface coupled with the lack of its yield can be quite damaging to the human body.

Hiking in minimal shoes can yield amazing results. For all of you body nerds out there, this is the way to tap into a greater portion of your brain/body connection. Do the work!

Sept 14 2010 ALIGNMENT...AT THE EMMYS!

As you may know, I was so fortunate enough to attend a pre-Emmy celebrity gift lounge event two weeks ago. I didn't know what to wear. I didn't know what to expect. I didn't know whom to expect, as I don't actually have a television. I do have Netflix, though, so whoever came through from 1) *The Office*, 2) *Glee*, or 3) *ThirtySomething* (FYI, I rented the first season and I have to say it makes a LOT more sense watching it when you're actually *in* your thirties as compared to when you're a pre-teen), I would be sure to recognize them.

First off, I have to say that I stayed in Beverly Hills—at a super-swanky hotel, even. Only on closer inspection, there was neither a bathtub nor a teakettle. Evidently if I wanted to soak, I should go up to the spa and get a service. Or, if I wanted tea I should order it from room service. For eleven dollars. Out of my league. I saw a bunch of supermodel statues posed in the lobby and out front to give it that Hollywood vibe. More close inspection. Nope, those were actual supermodels, hanging out in the lobby and around the outdoor fireplace surrounded by a library. Upon closer inspection, the library was fake. Real people. Fake books. I was starting to feel a little out of my "zone." I decided to make myself feel better by having the valet (which was mandatory—he chased me down) park my limping Honda Civic. With dirt all over it. And a squeak in the wheel well. Does anyone out there know what I'm talking about right now? But, I got over feeling out of place and got prepped for the Tic Tac™ Celebrity Cocktail Party.

I am bummed for not taking a picture to share of the Tic Tac Celebrity Wall. It was covered, and I mean COVERED, entirely with Tic Tacs. The words and images were all created by different colored Tic Tacs. This was super-cool the first night. By the next morning, however, the wall had gotten slightly wet and attracted a giant swarm of bees. Bees with Minty Fresh Breath. They had to call in an emergency beekeeper from Hollywood to round 'em up while breathmint artists got to work repairing the wall. Nature does Beverly Hills...pandemonium!

The cocktail party was fun—we all got to roam around checking out all of the stuff celebs would get over the next two days. There were clothes, jewels, cosmetics, trips to the Caribbean (!), cookbooks, hair extensions, and my booth...Curing the High Heel Hangover. In an effort to save the stars from themselves, we were fully stocked with My-Happy Feet alignment socks and some Fix Your Feet DVDs, and a half dome, where celebs could hop on to sample the calf stretch if they were so inclined. Out of all the ladies at the party I only saw TWO pairs of flats. One pair was on my feet and the other was on a teenager. After about an hour, this (tall!) lady sits down next to me on a bench and says "I don't want anybody to see this, but I have to change my shoes." No joke! She then pulled out an old beat-up pair of slippers from her bag. At the cocktail party. About two hours into the party, all these women were walking around with flip-flops and their lovely dresses. I don't get it. Why not buy a nice pair of lovely flats that actually go with the outfit, and look smashing all night? Everyone loved my new Earth SWANK suede boots (flat) and kept saying "I wish I had worn better shoes!" Here's the High Heel Hangover booth...

The next day celebs came—a lot of them—from many shows that I'd never seen and a few I had! Luckily, I didn't have to see their show to know if they needed the alignment socks...I could see their feet! These poor people wear super-high heels EVERY DAY and had the feet to prove it. I met so many super-great folks.

 DETECTIVE

Sept 28 2010

This seventy-year-old gentleman came into the institute about a year ago to get some post-rehab alignment work after a very complicated foot, ankle, and knee surgery. We started as we always start, with the feet. I pointed out the position of his feet—the left foot careening off into the lateral atmosphere and the right somewhere in the northeasterly quadrant. It went the same way it always does:

Me: You see your feet? Would you drive in a car that had tires that pointed like that?

Him: My feet are like this because of the surgery. They got all stiff like this and that's why I can barely walk.

Me: Doubt it. Surgeons don't move your joints around into opposite hemispheres. They're kind of perfectionists that way.

Him: What are you saying, that my feet were like this before?

Me: Yes, I'd bet money on it. Walking on feet like that would lead to the kind of surgery you needed.

Him: No way, man. The Doc told me I had bone spurs.

Me: Yes, walking on your feet that way would stimulate bone spurs to develop. They don't grow spontaneously.

Him: No way, man. My dad had bone spurs and Doc says they're genetic.

Me: No way, man. Bone spurs aren't genetic. They're a result of overloading a tissue and that is all a bone spur is—your body moving joints in incorrect planes of motion. And you mimic your parents' walking patterns, so I'd bet Dad walked like a disaster too. No offense.

Him: I'm paying you for this?

Me: You bet you are.

Him, one week later: You're never going to guess what I did this weekend. I was looking through my old pictures and in every one, me at ten years old to pictures of me and my kids, my feet are exactly like they are now!

Me: Yeah, that's what I told you. Now, stop walking like that, stretch your calves, and you are going to feel about twenty years younger in a week. (And it took about two weeks before he ditched his cane!)

So, the moral of the story is, the state of your physical body is the sum total of how you have used it. The moral to the moral is, you don't have to take my word for it. If you've got photos, you can be your own alignment detective. To illustrate my point, let's take a walk in my shoes for a quick second.

Here I am at about age one. Notice the straight feet found in almost all kids once they get upright. Also notice that while I am currently thirty-four years old, I was, apparently, born in the fifties. Also note the petti-coat. It will make a comeback, I promise.

Already reading at age one, I manage to still have straight feet despite them being bound into what appears like too-small shoes and too-frilly socks. Hey Mom, don't my ankles look a little puffy?? Parents: Please don't put your kids in shoes. Especially when you have gold shag carpet. Check out my dad rocking the orange beanbag chair. He's still cool like that.

Here we start to notice a preference for the left leg and the turnout of the right. Something I've picked up between year one and year three. I like that haircut. And so do all three of the musketeers. (Image below left.)

Awesome jumper, and yes, my preference for the weight-on-the-left turn out of the right seems to be consistent. Hair is improving. (Image above right.)

How many kids do you know that can smoke and eat a sandwich at the same time? I learned that from my dad. The stance, as the pictures will show, I've picked up from my mom. (Image below left.)

Almost non-sequitur, but then again I'm all about my left leg. Wanted to add this to demonstrate BEST Christmas present over. No, not the doll with the freaky eyes, but the socks. The REVERSIBLE socks. You can still wear them, but they don't have to match. Also note the seventeen layered nightgowns ensemble. It was cold where I grew up…in Santa Cruz, Ca. It was a record low of 59 that year. And what's with the shoes, Imelda? You're in your pajamas! (Image above right.)

It's baaa-aack. Here's my mom and me at the junction. Petticoat Junction. Today is my mom's birthday. HAPPY BIRTHDAY, MOM! She thought it would be fun to have matching clothes. And hair. And glasses (which she wisely removed for the photo). Also check out how she stands on her left foot and turns out the right. Where have we seen that before…

hmmm? This pic is from the five years our family square danced. I don't want to talk about it. I did have about twenty bitchin' p-coats though. Maybe this is where I gathered my fondness for unruffled clothing. If it has any frill, I'm out of there. Out of there on my left leg, obviously.

Please, check out the feet and not the outfit. This is a picture I like to call:

Happy Birthday Mom, I Love You Anyways. (Image bottom right on previous page.)

Kids, when your mom gives you something that she promises is So Cute, Just Put it On, and you know better, make sure there's no one around with a camera. Dang Christmas paparazzi.

So now you've seen just a little of my postural evolution, and I assure you, it's something I have to work on every day. If I don't, years of that habit would lead to osteoarthritis in the knee and hip, increased stress on the instep of my foot, sacral and pelvic floor issues, and neuropathy, for sure.

Spend a little time looking at your pictures beyond the clothes and haircuts. Beyond the memories and nostalgia. Use photos as a tool to reprogram your way of thinking about your own personal (learned!) postural habits and see if you can pick out two or three things you've always done and, just, well, *stop it.*

Oh, and just in case you want to check your alignment from every angle, make sure you have sneaky relatives following you around with a camera for posteriority. Posterity, I mean.

One of the great things about living by the beach in Southern California is that while it's foggy and cold June, July, and this year, August, you can depend on dry heat October and November. And probably December too, if people don't stop drinking out of plastic water bottles.

I just got back from an hour-long barefoot walk on the beach.

This is what I observed:

1. Good song on my iPod = Duck Footprint. Not paying attention.

2. Post-observation of duck-footed position = Straight-foot foot-prints. Bad news—I have a lot of realllly good songs on my iPod. Why else would one have an iPod? I've just written the word "iPod" four times. I wonder if Apple will send me a free one now.

3. I spend a lot of time looking at other people's footprints instead of concentrating on my own. Some were good. Some were okay.

4. There was one family of kids in bathing suits, but in shoes. Why?

5. There was a woman jogging on the path pushing her four-year-old in a stroller while the kid ate Cheetos.

6. The winner of Best Footprint goes to the birds. Not ducks, obviously, but straight, perfect little bird-prints.

New topic: I watched *Splash* last night. If you've never seen it, you may be missing the best movie ever.

I used to want to be a mermaid so badly, when in water I would bind my feet together with a plastic ring so it forced me to swim dolphin-style. Oh, and I used to crimp my hair so it would look more mermaid like.

This movie is really good.

I also had vegetable sushi last night (no fish!) to keep with the theme.

When I say vegetable su-shi, I mean jalapeno-stuffed-with-cream-cheese-and-then-tempura-battered sushi.

This was me at the beach today:

I was hoping that if I put my Pisces legs back into the brine, they would know what to do, just like in the movie. I imagine my tail is blue, green, grey, orange, and gold. Not like yellow, but like the metal. My legs stayed the same, but the rest of me got really wet.

After looking at how much a human has to pay attention to maintain alignment versus the animals, I can't help but think—this whole thing would have been a lot easier if I had been born a mermaid.

The end.

7 (FOOT) SIGNS SUMMER HAS ARRIVED

1. You just bought a new (thirty-seven-dollar) bottle of O.P.I. nail polish called: "I just may quit my job."

2. You can tell from your tan lines what shoes you were wearing yesterday.

3. The sand that was between your toes is now on your yoga mat. And on the floor mats of your car. And in your house.

4. Your relaxed feet are not that interested in putting on your work shoes.

5. On your big toenail, there is a butterfly, or a rosebud, or something adorned by faux-diamonds.

6. Your bunion is looking just a little less buniony with a pool behind it.

7. While "hanging out" by the pool (or BBQ or ocean), you impress your friends with your amazing intrinsic foot muscle skills by doing your individual toe lifts. Like this:

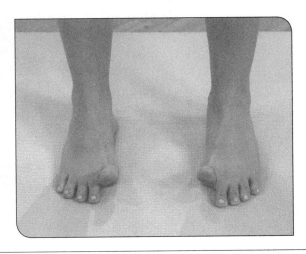

Human feet, like tires on a car, are designed to point forward. It is in this position that the hinge of the ankle works with the least friction, the (hopefully) arch shape created by the foot bones can support the most weight, and the toes articulate properly.

So, in classes, seminars, and to people I randomly meet on the street, I have them adjust their stance until it looks like this:

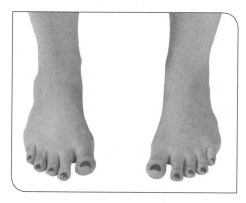

And then the person usually says "but now look at my toes! They are pointed inward! Now I have pigeon toes!"

And then I say "have you ever seen a pigeon's toes?" Because they look like this:

And then I try to explain that, although closely positioned, your toes are separate structures from your feet and the muscles of the toes can pull the toes wherever the heck you want. And the cool thing is, if you are a duck footed (feet turned out):

(Wait, is this a picture of a pigeon-toed duck? I'm so confused! Who came up with these names?)

As I was saying, if you feet are turned out when you walk, the crooked action of the ankle during gait is similar to the wrist action of cracking a whip, which accelerates the toes forward—even though the foot is pointed outward.

So when you make your feet straight, you may see that your toes have been pulling medially (towards the midline) for years.

"Don't worry," I say. It's just muscle and you can stretch that out.

And then I give them exercises. And I tell them to get alignment socks because I love them.

And then about one out of every four people says, "But I was told in my yoga class that the correct position of the foot was to line up the second toe."

So then I say: Yes. If your foot had been positioned correctly the entire time you've been up and walking around, the second toe would have been parallel to the outside of the foot, see?:

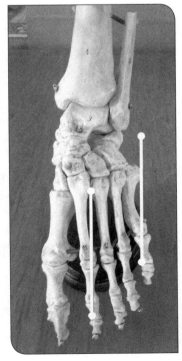

But, because you have the habit of walking around in turnout, that second-toe guideline doesn't apply to you. And most of us are walking in turnout, so you need a better, objective way of seeing where your feet are.

If I could talk to all of the yoga teachers out there about biomechanics (and someday, maybe I will!) I would say that the much more accurate and objective straight-foot marker is the outside of the foot, NOT the second toe. Toes are not attached to the foot in a fixed way, where they could be used to determine foot position.

Toes are like teenagers. They can do what they want. ☺

July 28 2011 PLANTAR FASCIITIS & TIGHT HAMS

Super (super) excited about this month's issue of *Foot & Ankle Specialist*! (Don't you get this magazine too?)

Well, I'm not really a subscriber, but I have a lot of awesome friends who are on the lookout for articles that I may want to blog about.

Incidentally, everyone should get a podiatrist friend. Mine is awesome, because not only is she an excellent doctor, she is also really good at cribbage. Almost as good as me. And if I ever get a cribbage injury, she'll probably treat me for free. Although she won't because she's not allowed to treat anything but the feet. So if I get a cribbage injury in the lower leg—say I accidentally step on a cribbage peg—I'll surely be saving some cash money. I love my friends.

Anyway.

The CLINICAL RESEARCH section in this month's *Foot & Ankle* is about plantar fasciitis.

And in case you were wondering:

"Plantar fasciitis is an overuse syndrome characterized by localized inflammation or degeneration of the plantar fascia at its anatomical insertion on the calcaneus."

And in case you were wondering:

Localized means that the whole structure is not suffering from inflammation, just at a particular point, *anatomical insertion* means where the tissue attaches to bone, and *the calcaneus* is the name of the heel bone.

The structure looks like this (image on top of right page)

Well, it *kind of* looks like this. I just made this using the "shapes" feature on my computer. I don't think the big toe is typically twice the size of the heel, but beggars can't be choosers. Or is it bloggers can't be artists?

Anyway.

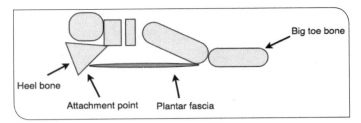

Big toe bone

Heel bone

Attachment point   Plantar fascia

What got me so excited is that this is the first study that has looked at tension in the hamstrings (the large muscles down the back of the thighs) and the relief patients are getting from plantar fasciitis with focused hamstring stretching.

I am excited, but I'm also like, duh. (While the titles of many studies differ, they often tend to read "Common Sense Proved, Yet Again.")

The say a picture is worth a thousand words. Let's see if they are right.

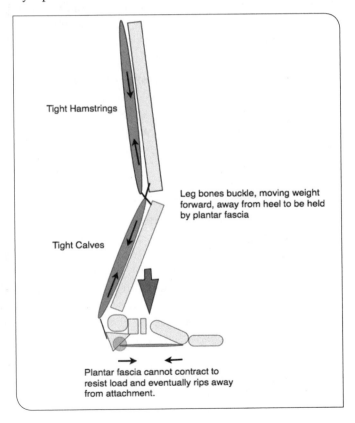

Tight Hamstrings

Leg bones buckle, moving weight forward, away from heel to be held by plantar fascia

Tight Calves

Plantar fascia cannot contract to resist load and eventually rips away from attachment.

Make sense?

Moral of the story: Tight muscles eventually lead to damage of non-force-generating tissues like bone, ligaments, and fascia. Next week: HOW to stretch the hamstrings correctly.

Labovitz, J.M.; Yu, J., and Kim, C. "The Role of Hamstring Tightness in Plantar Fasciitis," *Foot & Ankle Specialist*, June, 2011, vol. 4, no. 3, pages 141–144.

Nov 7 2011 THE DIABETIC FOOT: TO SHOE OR NOT TO SHOE

I feel especially passionate about foot care in the diabetic population. Of the sixteen million Americans with diabetes, twenty-five percent will develop foot problems related to the disease. While these statistics blow my mind, even more mind-blowing is the number of dollars accompanying these foot problems. Lower-leg amputations, the bulk stemming from "diabetic feet," are responsible for twelve billion national health-care dollars per year.

Repeat that slowly: Twelve. Billion. Dollars. Per. Year.

Shameless Promotion

As you know, I regularly create movement programs for various populations. For those with diabetes, it is essential to address the health of the foot muscles, not only for the strength of the feet, but for the sake of the nerves. Weak muscles promote nerve deterioration. Simply put, healthy nerves require healthy muscle, and vice versa.

There is a quandary in exercise prescription for those with neuropathy or diabetes. On one hand, when neuropathy has progressed, barefoot training—a required part of a foot-strengthening program—creates risk for the desensitized foot. There is a huge risk of infection when feet with a poor blood supply step on tissue-damaging items that you can't feel. Infection in feet with poor innervation can be the first step in the twelve-billion-dollar problem mentioned above. On the other hand, the always-shod foot is heading towards tissue death if the numerous foot muscles are left unused.

To resolve this issue, barefoot training and exercise are needed, and extra (extra) caution must be taken with the environment. To create the best of both worlds, use these guidelines from my new book on foot pain when creating a safe in-home or professional foot training space:

1. Run a vacuum over your exercise space. A good vacuum will pick up smaller, hard-to-see items like sewing needles, pins, and tacks.

2. Clear enough space to roll out a yoga mat or a large bath towel. This will give you an extra protective layer.

3. For extra-sensitive feet, place your exercise mat on carpet, or layer a blanket or multiple towels to cushion your exercise space.

4. Do a final, visual check for any potentially missed items.

Safe and gentle exercises can be found in the Fix Your Feet DVD, the Easy Rx...ercise for Diabetes DVD, and *Every Woman's Guide to Foot Pain Relief: The New Science of Healthy Feet*, out this month!

Diabetes aside, the cost of unhealthy feet is higher than you could possibly imagine. Make sure you are being part of the solution!

 HAPPY FOOT

Nov 21 2011

I'm getting ready to be gone for two weeks. I'm super excited to be on a lecture tour for the book and to visit my family, but on the other hand, I'm reluctant to leave our new state as we've just settled in to our new place. Although, as I type this, I don't even really buy it because I LOVE traveling to new places, meeting new people, and opening their minds and bodies to mind-blowing information.

Here's the book cover!

To get ready this tour, I decided to paint my toes. Actually, I decided to paint my toes after finding my toe-alignment socks in the back of my drawer. I stopped wearing them all the time once I started wearing my Vibrams all of the time. A mistake, I realized, once I put the socks on. Or, sock, actually, because I can only find one. Them toes were still tighter than they should be. And it's a good thing my left foot is tighter than my right, since that's the one I could find.

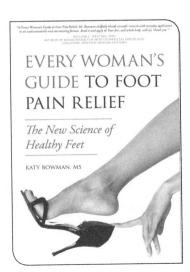

EVERY WOMAN'S GUIDE TO FOOT PAIN RELIEF

The New Science of Healthy Feet

KATY BOWMAN, MS

Alignment socks help combat the narrow toe boxes of most shoes (who wears these pointy shoes???). (Image at right.)

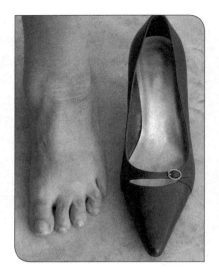

They do that by stretching the adductors (muscles that pull the toes together). Stretching the adductors allows the toe abductors (muscles that move the toes apart) to work once again. You can also do this with your hands, but I often have other things to do with my fingers besides putting them between my toes for five hours a day, like type on the computer. And practice speaking Italian.

If you're working on toe-spreading (a toe-abductor strengthening exercise) you have to undo years of between-the-toe tension. A little toe exercise here and there is a drop in the bucket compared to how many thousands of hours you have smashed your toes together. That's why I love these socks. They allow me to work on toe stretching for hours. I'm all about time management these days. Now if only someone would invent some sort of sock that went around the neck, pushing the shoulders down from the ears. An alignment turtleneck.

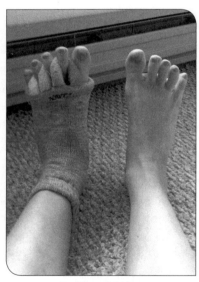

Toe bliss, right here.

Dear My-Happy Feet Alignment Sock People,

Could you please start working on My-Happy Neck Alignment Turtlenecks please? I'd like a multi-colored one. And I'd like it to be made from organic cotton or from the hair of humanely raised (and shorn), free-range unicorns, that are only fed rainbows and non-GMO plants.

Sincerely,

Your best customer

That is something I'd buy for sure.

And, if you are wondering what the name of this polish color, it is OPI's Barefoot in Barcelona, which sounds a lot better than Popsicle Toes or Thirty-Degree Weather Make Your Feet Bluish White. Am I Right?

I'd also like to take this post as an opportunity to thank those of you who, a year ago, supported the book effort with your foot-ailment testimonials. I told you that, if I used your story, I would send you an autographed copy of the book! Here are the stories that I used! If you recognize your comment and your name below, please send me an email with your mailing address and I'll get your book out in the mail!

My first job in college was as a salesclerk at a department store. Women were required to wear skirts (this was in '95), so of course I bought my first pair of heels to wear on my first day—a "sensible" brown pair with a two inch heel. We were not allowed to sit down except during breaks, so my first eight-hour shift was excruciating! When I got home after work, I pried off my shoes and discovered that *both* of my pinky toenails were completely gone. I did not learn my lesson, though, and wore heels throughout college—and my pinky toenails didn't grow back until I got a desk job and found a nice pair of dress flats. —Linda

I got my first pair of high heels at age thirteen and I thought I was just too cute, until an older guy laughed at me for being so wobbly. Devastated but determined, I practiced wearing them and finally succeeded, I even was able to run in them. I couldn't figure out why I kept getting backaches and finally herniated L5-S1 disc (duh). Went to chiropractic college to learn about the spine and learned about those wicked, but seductive, high heels. —Gayle D.C.

I have been walking on four toes ever since I was little as my next-to-pinky toe on each foot never grew out past a certain age in my childhood, and that toe has made it hard to find shoes that fit. They either had to have the top front of the shoe go low below that toe because it sticks up (the pinky toe has curled underneath that toe to support it) or go high enough that it doesn't rub that toe and make it sore because it is too tight. Can you imagine cramming that toe into heels and walking without it hurting? Wasn't a good idea but I have done it anyway since high school. Your exercises and Earth shoes have helped me so much to be able to spread my toes now and the shoes don't rub the top of my toe next to my pinky toe and make it sore. No more sore and aching feet! I am forever grateful! —Lanene W.

My first pair of high heels were the plastic ones they used to, and maybe still do, sell in the "girl" kits you'd find in the drugstore toy section. They often came with gaudy, clip-on earrings, lipgloss, and such. I was about seven years old and wore them to bed the first couple of nights. That phase died out in me pretty quickly, thankfully, since I was one of the taller girls in school, not to mention a creature of comfort. Those darn things hurt! These days I prefer to be barefoot, which makes me truly feel like a little girl again. —Amber N.

I am here to say that not only high heels but even toescrunching flats are painful for me. My mother put us in sensible brown oxfords all through grade school and we hated her for it. When we reached the ripe old age of eleven, we were allowed to pick out and wear whatever shoe types we wanted. I was in love with a pair of light blue flatties, which I bought in a size $7^{1/2}$—only my foot measured a size $8^{1/2}$!!!!! I paid a dear price for trying to have my feet appear smaller! Bunions, hammertoes, fallen arches, and very bad arthritic knees! Thank you, Katy, for introducing me to REXI and Earth shoes so that at sixty-five years of age, I'm finally able to spread my toes, lessen my bunions and hammertoes, and walk for at least an hour with no knee pain!!!! —Chris D.

I have a bone spur on the top of my foot at the base of the big toe. It was very painful, red and swollen and limited my walking because it hurt all the time. The medical term for it is Hallux rigidus. I started making changes with my stance and walking, specifically feet straight ahead, weight on my heels, and changed my shoes to no or negative heels. Very soon after this my pain and the redness went away and with no swelling I was able to enjoy walking again pain free. I went to the podiatrist for his input and he told me the pain would "Continue to intensify until a year from now I would be begging him to do surgery to correct it." Well, I have continued to follow the REXI guidelines and still no pain after nine months! It even seems to be a bit smaller! —Diane L, PT

chapter 3

KNEES, HIPS, AND BACK

| January 23 2009 | NOW I **REALLY** NEED A MASSAGE! |

Q: After I receive a massage, inevitably my low back is very sore when I rise from the massage table. Why??

A: Before I answer this, I'd like to let everyone know that extensive research is really required, so, if anyone's offering…oh, so to answer your question! One possibility could be that your psoas (hip flexor) is tight and your pelvis may be somewhat fixed in a post-tilted position. When you lie down on your stomach, the pelvis isn't moving relative to your leg, so you have to extend your low back to lie flat (*ouch* on the lumbar vertebrae). I know a lot of people are sore in the low back when they sleep on their bellies. Ideally your hip flexion should cease when sleeping, but it is such an engrained pattern—especially because most people spend more hours sitting than they do lying down. For your next massage, try adding a small, thinly rolled blanket under the ASIS (front of the pelvis) to encourage your pubic symphysis to drop toward the table, decompressing your spine while your face down.

Or, you can quit your desk job, try to get hired on with a standing job (UPS or comedienne), and book a twice-weekly quad massage. Sounds like you could use it!

| Sept 8 2009 | PLANES, TRAINS, AND…A SORE BACK? |

Rose is off to Washington DC!

This is exciting! My client Rose is, this minute, flying to Washington D.C., where she will tour the White House vegetable garden and meet with the movers and shakers at the frontline of national food policy. Go, Rose! I am super excited for her as well as super excited that I am *not* the one traveling via car, plane, shuttle, plane, taxi, limo, repeat. Not that I don't adore traveling. I would gladly take the next free trip to rural Africa, thank you very much! It is the sitting for hours upon hours I detest.

I have worked hard to make the "movement" component of my life reflect that of yesteryear, minimizing car, gas, and gear-powered travel, opting to use my fabulous hips instead. Hips are, by definition, the greatest joint you have. Greatest joint in size, greatest joint in terms of engineering (check out all those muscles that attach there!), and the GREATEST joint, by far, in your pants. Really…if you don't have those muscles in your hips innervating, you probably have a fantastic belt collection!

So for Rose (or whomever sits at a desk or spent this weekend driving) I recommend this great stretch for your piriformis muscle. This hip muscle often aggravates sciatic pain in the sacral (low back) area and can limit the health of the hip joint. I call it the Number 4 stretch, because when you get into it, you kind of resemble the number 4. The most simple version is: while perched on the front of a chair, cross your ankle over your opposite knee. Try to reach forward to the floor… yikes, that's tight! P.S. If you have a hip or knee replacement, please avoid this stretch. If you want to avoid a knee or hip replacement, do the Number 4 stretch often!

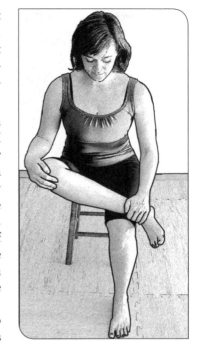

Thank you, Shelah W., for making me a cartoon! (Image top right.)

I do this one a lot when flying—both in the airport and in the back of the plane. You can do it in a chair, against a wall, or if you have very good balance, you can do it without any support. While filming our DVD line, the camera operator was having sciatic pain, so I showed him this stretch and the entire crew did it in their chairs,

standing, and using the set walls. (Image bottom right.)

Take 5—Five minutes to stretch, that is!

I'm also fairly positive you can do it while reading a blog…hint hint.

SAVE THE KNEES!

At this year's Radiological Society of North America's annual meeting (did any of you make it?), a study on exercise and knee health was presented.

The question: Does activity level (low, medium, or high) impact the health of the knee joint?

The activity level: *Low* was deemed as sedentary to exercising less that once per week. *Medium* was exercising regularly, but only low to non-impact activities. *High* activity levels featured both high-impact activities as well as large quantities of movement (multiple hours per day of high-impact or repetitive-motion type exercises).

The results: Looking at MRIs of 236 adults, ages 45–55, showed that ninety-three percent of the high-activity level group had knee cartilage damage as opposed to sixty percent in the low-activity group. The high-activity group's knee damage was three times more severe than that of the couch potatoes (note: not the group's official name).

The conclusion: Excessive mileage and impact forces are contributing to the increasing levels of osteoarthritis, knee surgeries, and knee replacements in the United States.

But before you go logically extrapolating the fact that your La-Z-Boy is a healthy choice, allow me to comment.

Firstly, we need to keep in mind that while "movement" is a physical requirement to physiological function, "exercise" is a modern creation designed to deal with the waning health of our immobile population. Cramming a whole day's worth of movement into a one- or two-hour-long intense bout has the same physiological impact of starving yourself all day then following it up with a 2500-calorie meal. Your body doesn't work that way!

Secondly, please understand that *alignment matters*. The position of your toes, feet, ankles, knees, hips, chest, and head all dictate the loading in the knee (and all the joints, if we're going to be correct). If you are not balanced, the incorrect and non-symmetrical wearing patterns will result in an injury. If you are more active (timewise) and choose activities with high loads (running vs. walking, or riding hills vs. flats), the damage will accumulate faster.

The moral of today's story? It is much better for the health of your joints (and, surprise surprise, your metabolism) if you move throughout the day as opposed to all at once. The weight-loss effect of exercise is the same (in fact, it's actually easier to lose weight by getting your movement in smaller doses but more often), and your joints have time to adapt to the loading. Imagine your knees, completely unloaded while you sit in your desk chair/car/recliner and then all of a sudden you get up and start jamming the bones into the cartilage at 3-G forces. Not very considerate. Also, get yourself aligned. If you have one foot that points slightly out to the side, a knee that doesn't bend the same as the other, or a torso that tends to lean, don't make the problem worse by jumping up and down on it...FIX IT! We align the tires on our cars because we want our car to last. Extend the same courtesy to the human body.

And...she's off the soapbox.

June 28 2010 KNEE BONES CONNECTED TO THE PF BONE

Last week I was teaching a squat class at the institute when a physical therapist asked, "How do we protect the knees in such a deep squat?" I found it such a telling question, because the primary reason our country's knee health is so poor (yes, we've got the highest incidence of osteoarthritis in the world) is because we *don't* squat. The most true answer to the "how to we protect the knees" is *squat more often*. The bigger issue is, once you have "bad knees" (I really dislike this term. What did your knees do, rob a bank?) it seems like you can no longer squat without pain, right? Well, that, we can do something about.

One of the first things you should know is, OSTEOARTHRITIS IS NOT A DISEASE. I can't type this big enough (I need bigger font!). One of the largest disservices done by the medical community is not clearly explaining this fact. Osteoarthritis is not an autoimmune disorder like regular arthritis, which means a joint inflames for no reason. Osteoarthritis is the damage caused by user-induced friction of the joint. What does this mean? Let's make a little math formula:

Damage: Ripping, tearing, burning up, or acidic damage to joint tissues like cartilage, meniscus, bony surfaces, etc.

User-Induced: That's YOU, your gait pattern (quality of movement), and the amount you move (quantity of movement).

Friction: Put your palms together and, pressing hard, rub them together for one minute. The interaction of the palm surfaces is friction. The heat in your hands after rubbing them together for one minute is the result of friction.

User+Gait Pattern & Tight Muscles=Friction=Heat=Joint Damage

Or, more simply,

User+Gait Pattern & Tight Muscles=Joint Damage

If you vigorously rubbed your palms together every time you took a step, you would eventually tear into the flesh of your hands. This doesn't happen because you know how to pull your hands apart. What you don't know how to do is pull the joint surfaces in the knees apart, which is why it's so easy to think of OA as a disease. 'Cuz you can't do anything about the space in your joints...it's just genetic, right? Wrong! Joint space is determined by the tension in the muscles that cross over the knee joint. How tight are your calves and hamstrings? If they are really tight, that's awesome! It means that with just a little stretching you can increase space, decrease friction, and decrease heat. It means you can get better whenever you choose to.

When you have osteoarthritis of the knee, what you really have is the constant rubbing of the bones in the knees. This is not a disease, but a mechanical situation that can be changed at any time. Elevated heels on shoes, extreme tension down the backs of the legs, and the habit of tucking the tailbone under have arranged muscle fibers into short, tight muscles that increase friction in the knees.

Did you notice the mention of the tailbone being tucked? Chances are, if you have pain in the knees you may also be having issues of the pelvic floor (as mentioned in other squatting and pelvic floor blogs on this site). One of the reasons pelvic floor exercises like the kegel have been invented is because we aren't walking with the correct "pushing-off" (hip-extension) motion. This pushing-back motion tones the glutes (naturally), which (naturally) keeps the pelvic floor long and contracting eccentrically (instead of shortening during contraction), and minimizes the

falling-forward action most people do when they think they are walking (using the knee's cartilage for each crash landing is a good way to burn through the knees quickly). Can you see how it's all connected? The good news is, the very same movements that get the pelvic bones aligned and the PF strong are also the same stretches needed to increase joint space in the knee (and in the hips, too).

If your muscles have been super-tight for a super-long time, begin to open your joint space with these exercises. They can be used for knee and pelvic floor issues and are a gentle way to prepare your body for a future squat.

Exercise 1: Double Calf Stretch

Do you sit all day? Wear heels? Tuck your pelvis under because it is the "proper" way to stand? However your pelvis became tucked, the double calf stretch is a great way to see if your pelvis can even move! Place your hands on the seat of a kitchen or desk chair and step your feet up onto a thick, rolled towel. (Image bottom of previous page.) Line up the outside edges of your feet and straighten your knees all the way. Your weight should be back in your heels and all your toes liftable. See if you can lift your tailbone up to the ceiling without bending your knees. No? Do this exercise a few times a day, holding up to a minute, until you can.

Exercise 2: Seated "Number 4" Stretch

This is a great way to stretch your piriformis, and you can do it any-where you sit! Cross your right ankle over your left knee. Try to lower the right knee to the same height as the right ankle. Keep both cheeks on the seat and try to move your pelvis from a tucked position (below left) to an untucked pelvis (below right). If you have a hip or knee replacement,

only stretch the leg that has not been replaced.

Exercise 3: Prone Inner Thigh Stretch

If you have always been a pelvis "tucker," the groin muscles can become very tight. Start with this groin stretch before moving on to the more advanced Legs on the Wall shown later. Lie down with your belly flat on the floor. Reach one leg out to the side without bending the knees! Relax your head and neck on your hands. Try to bring your leg up until it is at 90°.

If you have a hip replacement, only stretch the leg that has not been replaced.

Exercise 4: Inner-Thigh Opening

Do not do this exercise with an artificial hip.

This is another great exercise for stretching the adductors while the hip joint is externally rotated a bit. Lying on the floor, place the soles of the feet together and let the knees drop open to the sides. The height of the legs above the floor indicates the tension in the groin and hips.

How high are your knees from the ground?

If the stretch invokes too much sensation, rest, and then resume the stretch when you're ready!

Exercise 5: Legs on the Wall

Do not do this exercise if you have an artificial hip.

Place your legs up a wall, keeping your entire pelvis on the ground, but a bit of space under your waistband.

Keeping your knees straight, allow your legs to open away from each other to stretch the adductors (the muscles between your knees and hips). If you can't keep your legs straight due to the tension in the back of the legs, scoot away from the wall a bit. This is an intense exercise. Rest when you need to and resume when you are able.

Other suggestions: Years of fat-free dieting, chronic, whole-body inflammation, and chronic stress can all reduce the healing cycle of joints within the body. Find an anti-inflammatory guide to eating you can follow (your diet should be a minimum of thirty percent good-quality fats!) and check out my Knees and Hips video. The most important exercise for the knees is learning how to release the knee caps, but that's for another time!

 MINIMALLY INVASIVE

August 17 2010

I am always getting excellent news stories and media tidbits from you, Awesome Readers. Today was no exception. Elizabeth sent me (thanks Elizabeth!) a link to a story on ABC News tonight, "Minimally invasive surgery could relieve patients of back pain."[1]

The story focuses on the eighty percent of Americans that will suffer back pain in their lifetime, and a particular type of back pain—SI Joint (sacroiliac) Syndrome. Now I wouldn't consider SI joint pain a "back" or spinal issue, because the muscles that act upon the sacrum belong to the pelvic and hip category, but I'm not even going to be picky on this point. I am, however, going to nitpick the term "minimally invasive" when talking about the new surgery, SI joint fusion.

Yep. With just a couple of rods and three incisions, your sacrum can be *permanently connected to your pelvis* for reduction in pain.

"The procedure works because it stops the joint from moving, and it's only movement which brings on the pain," says Dr. Graham Smith.

Correct me if I'm wrong, but this sounds a lot like the *Hey Doc, It Hurts When I Do This, Then Don't Do That* philosophy of medicine. Anyone else get that?

Let's back up a second to an experience you may have had with a broken bone and cast or sprain and sling. For those who have had rigid structures limiting their joint movement while healing, do you remember the day the cast came off and how the muscle atrophy (shrinkage) was so easy to see when comparing the "fused" and mobile appendages? Here's the thing with muscles—they need movement at the joints in order to maintain tone, tissue health, and fluid content. No joint movement, no muscle tone.

Now let's talk about the sacroiliac joint. This joint is supposed to have free, non-friction-generating movement. Just like movement of the knee keeps the quadriceps, hamstrings, and calves healthy, the movement of the sacrum keeps the glutes and pelvic floor healthy. In fact, sacroiliac pain comes from unbalanced muscular force between the glutes and pelvic floor, and the resulting pelvic floor hypertension on the sacrum. The solution is restoring the function to minimize friction, not fusion. Fusing this area will not only create core-musculature atrophy, it will cease natural motion of a body part. Natural motion that is probably important, if the human body developed a joint there. Important movements like childbirth, reproductive organ support, pelvic and abdominal support, digestion, and walking. You know, those types of things. Do you have SI joint pain? Check your butt. Don't see one? Get one. Your SI joint will thank you.

Sooo, back to "minimally invasive." This term is a misnomer. An incision may be small, but the long-term effects of a fusion result in a progressive and accelerated degeneration of all tissues around the location of the fusion. Most people do not clearly understand the role of the pelvic floor, sacrum, and gluteal muscles in supporting practically ALL human body function. If they did, they would not be so quick to fuse this area, creating a "limp" in the core muscles that lasts FOREVER. "Minimally invasive," my butt. My *strong* butt, at that.

Dr. Graham is not incorrect when he says that movement brings on the pain, but not moving is not an option for those of us who know that our body is a self-winding clock. Fuse a joint, limit your time. Pain with movement is a signal to be heeded. The signal is saying, *The way you are moving is doing you harm. The muscles on your frame are not supporting you.* You can fix the way you move, or you can just lie down now. Which makes more sense?

1. wndu.com/mmm/headlines/99969099.html

October 4 2010   5 SIMPLE WAYS TO AVOID BACK PAIN

I'm stuck in the airport because my flight is delayed. To kill a little time, I thought I'd share five back-pain tips that I donated to *Natural Awakenings Magazine* this month. When I say simple, I mean SIMPLE, yet most people who email, call, or come into the institute have a hard time making these basic changes. Let me share something with you: Simple doesn't mean easy. Not eating sugar, simple. Not eating sugar, reallllly hard to do. Solutions to most things are simple, yet they take great rearrangements of your life, i.e., how you spend your time and money, what you prioritize over your health (everyone says health is their first priority, yet the facts show something different!), and how deeply your habits are ingrained. Take a stab at these and see how things go. Which one is the most difficult for you? I can tell you that for me, it is getting down onto the floor and doing an hour of spinal twisting. I actually LOVE to do this, but I have to be in a class to do it, it seems. That's crazy!

I've already done Legs on the Wall here in the airport. Also, I usually travel with a pelvis in my carry-on. I've also been "randomly screened" every time I fly, except for the time I accidentally had a Swiss Army knife in my purse, which I didn't even know I had and only realized it once I got to my hotel. They didn't stop me for that, which is weird, right?

Thanks for keeping me entertained—here you go!

1. Lose the high heels. The scientific consensus is that high heels compress and damage the lumbar spine, increasing osteoarthritis and degenerative disk disease in the low back.

2. Let the feet point the way. Just like the wheels on a car, your feet should point straight ahead when you're walking. Military or dance training, or an ankle or back injury, can sometimes result in a sort of duck walk. Line up the outsides of the feet along the straight edge of a carpet or tile floor and walk along it to practice.

3. Stretch the calves. Tight calves are a major contributor to back pain. The tighter the lower leg, the more one's gait pattern whips the upper back forward and contributes to curling of the upper spine. Adding a daily calf stretch to any exercise routine helps to better align the spine.

4. Do the twist. Each vertebra in the spine not only bends forward and backward and from side-to-side, it also rotates. Of all these natural motions, the twisting of the torso is the least used in our culture. Incorporating a yoga spinal twist into an exercise routine will gently reintroduce rotation back into our movement repertoire.

5. Get a better butt. The main culprit of low back pain is weak butt muscles. Gluteal muscles not only stabilize the tailbone, they help support the function of the low back muscles. If the glutes are weak, the low back muscles have to work harder than normal, which makes them fatigued and sore. Squats work well to strengthen the butt.

 Nov 12 2010 RUA RIB THRUSTER?

I think it's the perfect time, in this political climate to ask: *Where do you stand?*

Try this. Stand with your heels three to four inches from a wall. Press your thigh bones toward the wall, letting your tailbone relax. Next, bring your shoulders, arms, and back of the head against the wall. You should have a small space underneath your waist where your low back naturally curves in, but *your middle back (the ribs/bra strap/heart rate monitor area) should also be touching the wall.*

Ahhhh! A rib thruster! (Image below left.)

Ribs are all fixed. (Image above right.) Notice the pelvis DIDN'T tuck when fixing them? That's because they are different body parts!

If your middle back is not touching the wall, drop your ribs (allow your head and shoulders to move forward if you need to) until your mid-back touches. If your head has to come off the wall in order to get your ribs back, this is how much excessive curve you have in your thoracic spine (called kyphosis.) Many people "hide" their kyphosis by thrusting their ribs forward, but it is better to keep your ribs down and work on stretching these tight muscles (instead of manipulating the skeleton to hide them!)

If your waist is *entirely* on the wall, you are a tucker and are overly tucking your pelvis under, pulling the lumbar spine and the pelvic floor out of alignment. Adjust this by tilting your tailbone out toward the wall until the space reappears, and follow the above directions to notice the curvature in your upper spine.

I'd like to point out that a very large amount of people are told in physical therapy, the orthopaedist's office, and by their chiropractor that they are "sway backed" (see picture on the left) and to fix it by tucking the pelvis forward, i.e. bringing it forward to match the ribs. Using the wall as an alignment marker, it becomes clear (hopefully) that what most people are doing is THRUSTING the ribs, not overly sticking out their bottom. While the curvature of the spine looks excessive (because it is!) and the pressure on the low back feels painful (because it is!), this is an UPPER BODY ISSUE. Tucking your pelvis under to decrease the pressure is a short-term solution that will give some relief but end up creating foot, knee, hip, and pelvic floor issues. Mother tucker!

To decrease lumbar disk degeneration, keep your perfect pelvis neutral (not tucked) and, using an objective alignment marker (like a wall), introduce yourself to the real alignment instigator, your shoulder girdle. And then do this:

And work on dropping your ribs as you stretch your chest muscles, okay?

 ## ALL ABOUT YOUR KNEES

Nov 16 2010

How do I answer fifty emails asking the same question? With a video blog, baby, that's how.[1]

Here are the top sellers of the Aligned and Well DVD line: "Down There" for Women, Biomechanics for Bad Backs, and Knees and Hips. Really. Eighty percent of the people you know are dealing with at least one, if not all three, of these issues.

The common theme of all of these ailments is this:

Not only are people walking around with non-straight (and non-innervated) feet, they are also walking around with non-active deep hip musculature. These deep hip muscles not only affect the tone of the piriformis (that muscle that often tenses up and causes radiating sciatic pain), but also the pelvic floor. Wellness requires full-body participation. The muscles in the pelvic girdle not only literally support your organs, they also support organ function, sacral stabilization, and...KNEE HEALTH. If your knees rub, bang, burn, scrape, or slough against each other while you're walking, if you're experiencing pelvic floor issues, or if you are having lower back or hip pain, check out my second-ever Quick Tip Video Blog and see if you not only end up smarter, but better...all around. All around the thighs, that is!

If you were one who emailed me about their (or their husband's or kid's or dog's) knees, would you be so kind to leave a comment saying "Thanks, I get it now!" or "You look great four months pregnant," or something like that? You'd be surprised at how many emails I answer, fully, and never hear anything back. I have a life, you know.

Okay, I don't really have that much of a life, but I do have technical difficulties with my email sometimes. ☺

Test, Test. Is this thing on?????

---

1. youtu.be/qcGPY4BMdIU. A roughly four-minute video describing how to bring both your feet and your knees into proper alignment when you're standing. It contains some pretty cute sound effects, too.

August 3 2011 WHEN IN DOUBT, STICK YOUR BUTT **OUT**

No, this is not a blog about your awesome dance moves, but about how to move things (like heavy boxes) without causing a spasm in your back.

You've heard "lift with your legs, not with your back," but of course, guidelines for lifting correctly need to be more specific than that. Using the backs of the legs (hamstrings) allows heavy loads to be lifted with the massive muscles in the backs of the legs. If you are using the fronts of the thighs (quadriceps), then sorry, you are still going to strain your low back.

What causes this low-back overload is the tucked position of the pelvis. Anytime your low back is in flexion (see movie for a visual), the disks are primed for damage. Couple flexion with twisting and you've placed the spine in an extremely vulnerable position. Add a heavy load (piece of furniture, bag of potting soil, or box of books) and BAM! the psoas goes into a spasm to protect this area.

I made this quickie video so you can see the difference.[1] Mostly because I don't have time to post in the middle of moving ☺ and also because my Flip video cam was getting a lot of dust.

Ideally, we should be able to bend completely over at the hip joint with straight legs, but most of use have too-tight hamstrings, thus the bend in the knee (with the vertical shin, of course.)

I'm still going to post on how to stretch hamstrings later this week.

Really I am.

Stop bugging me, I'm trying to move.

---

1. youtu.be/DEMplrKpnNs About three minutes explaining and showing how to properly lift heavy objects. Also shows how messy a house gets when you're moving.

A USER'S GUIDE TO HAMSTRINGS

As promised (eight hundred car-trip miles later), The Hamstring Blog.

Some quick hamstring tidbits:

1. The "hamstrings" are really three separate muscles: Biceps femo-
ris, semimembranosus, and the semitendinosus.

2. The hamstring group passes over both the hip and the knee joint,
which means it connects the pelvis to the shins. The thighbone is
held snug in the hip socket by this muscle group. Sometime too
snug (see #8).

3. When at its correct length, the hamstring group allows the pelvis,
thigh, and lower leg to be positioned in "neutral," which means
positioned optimally for force production and joint salvation.
Like this. (Image at right.)

4. The hamstrings are commonly listed as knee flexors (muscles that
bend the knee) but this muscle group should ideally be used as a
hip extensor (muscles that pull the thigh backward relative to the
pelvis), when walking.

5. The backward motion of hip extension *should* be the primary motion
used to propel the body forward
when walking (think of an ice skat-
er pushing back against the ice to
move forward), but instead, modern living has reversed our natural
gait pattern and now we just lift a
leg out in front and fall forward.
Ow.

6. Chronic sitting and/or chronic
pelvis-tucking move the hamstring

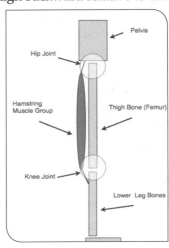

muscles' attachment sites toward each other, which requires the muscle to tighten. Why? Muscles are not allowed to slop around in the body—if your postural habits put slack in the line, the muscles respond accordingly and reduce their length. (Image at right.)

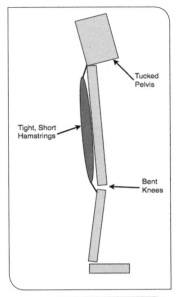

7. When the hamstrings are really (really) tight, they can prevent the knees from straightening or the pelvis from untucking. A tucked pelvis is the primary cause of pelvic floor disorders. For strong pelvic floors, lengthening the hamstrings is a requirement! You need a neutral pelvis to prevent organ prolapse!

8. When the hamstrings are really tight, joint space in the knees and hips is reduced and becomes asymmetrical front to back, which increases friction. Which makes heat. Which makes inflammation. Which prevents cells in the cartilage from regenerating. Boo.

If you don't have time to read all of the above points, they basically say that tight hamstrings are the root problem for knee, hip, low back, and pelvic floor issues and that muscles become tight by postural habits.

Why are my hamstrings so tight?

While it seems like there could be a hundred different situations that can affect human tissue, it is probably one of the four reasons listed here:

1. You sit a billion hours a day.

2. You tuck your pelvis because you think you should.

3. You wear shoes with a heel (any heel!), which causes your knees to bend or pelvis to tuck to balance the geometry, which causes the hamstrings to shorten.

4. Your psoas* tucks your pelvis, which causes the hamstrings to shorten (this is a result of high levels of psychological stress, lots of athletics or fitness, or certain postural habits).

5. You don't have a natural gait pattern (and no one does any

more, really), so these muscles are fairly inert and tighten due to underuse.

*Psoas tension is a common culprit in hamstring issues. The mechanics are a bit complicated, but if you are interested in the physics of the psoas, how it affects the spine, pelvis, and legs, and the exercise program to address it, check out the Science of the Psoas online course I created.

And now, what you've all been waiting for...

How to stretch your hamstrings

There are basically just a couple of guidelines to getting the hams to their optimal length (the one that allows the knee to straighten and the pelvis to rest in neutral on the thigh).

DON'T BEND YOUR KNEES while stretching.

DON'T TUCK THE PELVIS while stretching.

Here is a picture of my favorite hamstring Rx...ercise™, the Strap Stretch with What To Look Out For:

The strap should be on the front of the foot to pull it toward you on the floor. This helps target the calf muscle that may be preventing your hamstring.

Surround your stretching area with wispy white curtains to enhance the benefits.

This knee should not bend at all. Bending it is a way to put "slack in the line" - which keeps your hamstings at the same length despite the exercise.

You should have your hair and makeup done ahead of time in case anyone takes your picture while stretching.

The back of the thigh should be resting (completely) on the ground. If it lifts, it means that you tucked your pelvis to eliminate the hamsting stretch. Thank you, try again.

Because the muscle passes over the knee and hip joint, moving the pelvis or changing the knee joint angle during Strap Stretch is a "cheat" and leaves the hamstrings the same old length as when you started.

If you ever go to a yoga class, there are always a ton of people with the leg all the way up, blocking the sun out of their eyes. Try this: place a hand towel under the mid-hamstring point, and make sure you can't pull it out when doing this stretch. You may find that you need to lower the stretching leg almost all the way to the floor to find your true hamstring length.

Forward bends—flexibility vs. mobility

Here's a little anatomy tidbit: The term *flexibility* applies to muscle. The term *mobility* applies to joints.

The moving of the pelvis during the hamstring stretch means that you aren't stretching your hamstrings at all but are distorting the lower back joints (lumbar vertebrae) instead. People who are "hypermobile" typically rearrange their body so they don't stretch their muscles, but instead stress the joints above or below that muscle group.

In a nutshell: People who are hypermobile actually have poor muscle flexibility. And stretching is difficult because these folks don't *feel* anything in a stretch because they open the joints—not the muscles.

Example: The forward bend. Forward bends should help the user find the hamstrings, but most people bend from the spine, not from the hips, or bend the knees to get to the floor. Here is my best attempt at drawing the difference…

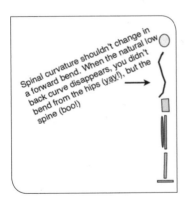

Spinal curvature shouldn't change in a forward bend. When the natural low back curve disappears, you didn't bend from the hips (yay!), but the spine (boo!)

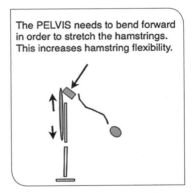

The PELVIS needs to bend forward in order to stretch the hamstrings. This increases hamstring flexibility.

Moral of the forward-bend story is this: Once the pelvis stops moving on a forward bend, so should the rest of your torso. Especially if you have any issues in the hips, knees, or lower back.

Common question: Should I bend my knees to go farther, then?

Common Katy answer: If you bend your knees, then 1) you're not stretching your hamstrings and 2) you are making your low back hold the load of your upper body. If you are using forward bends to improve muscle length and strength, then it is better to stay within your range of motion—even if it means you lose the beautiful sweeping motions you used to do in class. Replace them while mindfully honoring your current muscular boundary. You will make much more progress and do a better service to your body.

Note: I am an excellent biomechanist. Really, I am. I am NOT a very good graphic artist. So let's hear it for some lukewarm graphic artistry skills! (Crowd responds with a loud cheer!) My favorite is the squiggly spinal column.

I am awesome.

Sept 2 2011 — YOU'RE NOT STARVING, BUT YOUR TISSUES MIGHT BE!

I love to eat, and I hope that you do, too. Nutritional requirements are a biological imperative, meaning, without the intake of nutrients, we die. Everyone knows that, right? What you might not have known is, the same nutritional intake is required on a cellular level. Your cells need to be "fed" in order to keep going, and without cellular food, tissues die.

Which brings me to a little thing called degenerative disc disease. Have you heard of it? Back pain is a huge worldwide problem, and chances are, by the time your back pain has progressed to the level of medical treatment, you will see this term somewhere on your chart:

Degenerative disc disease (DDD). What is it? If I didn't know much about anatomy or physiology, it seems like I might have something, some bacteria or fungus that is eating away at my disks until the bones of my vertebrae are rubbing together, placing my spinal cord in peril. Maybe my disks are just old and are starting to break down.

Well, it turns out that DDD is just the result of years of extreme dieting—on the cellular level, that is. DDD is often linked to age, but more correctly, should be linked to years spent in a particular posture that has decreased the amount of nutrients to the area.

A little trivia Q: What do spinal discs eat?

A: Oxygen and glucose.

How to feed your discs

The flow of your blood delivers nutrition everywhere in the body, and the intervertebral discs are no exception. What is most pertinent to back health is how nutrition reaches tissues. When we think of blood moving, we picture the heart pumping away, but it is really the action of skeletal muscle—contracting and relaxing—that creates the flow of nutrients.

The spinal discs between the vertebrae are essentially like little sponges—squeezing out waste (lactate or lactic acid) and creating a vacuum-like suck when released. Each vertebra has a set of muscles that connect it to the next bone above and below. It is imperative to the health of the disk that your spine be able to articulate at each point.

Alignment and spinal health

Your habitual posture is a particular set of joint positions that you have made your default. And like most defaults, it is easy to return to this posture over and over again, until the muscles have stiffened, creating a fusion-like situation. Our modern daily living does not require us to move in unique and novel ways. It hardly requires us to move at all! Over time, the muscles, in their stiff positions, fail to assist in nutrient delivery, resulting in cellular death and eventual disc collapse. This is what we call degenerative disk disease.

Bummer, right? Well, the good news is, you can start the process of nutrition delivery simply by stretching. This creates a pull and then a release of tissues that begin the gentle massage that creates nutritional flow. And, **bonus**, it also aids in waste removal from the deepest part of the spinal disc.

This is great news! For many years, researchers believed that movement had nothing to do with spinal disc health. Now they know better, as better-designed research has shown that mechanical loading of the spine is what maximizes nutritional flow!

Some guidelines

For best back health, it's not just the spinal muscles that need to be stretched but any that affect the position of the spine. This includes calves, hamstrings, groin, and psoas muscles.

Stretches or positions that are easy for you are probably not very beneficial. It is much more important to stretch the muscles that are difficult to get to.

When stretching your spine, move slowly and be gentle with yourself! Also, pay attention to stretching guidelines. The reason I spend so much time giving parameters of safety is to increase the effectiveness of the stretch while keeping you safe and sound!

PIRIFORMIS HOLIDAY

Today is "Stretch Your Piriformis Day." I just decided it. Make sure you tell all of your friends.

So first off, how do you stretch this little bugger?

Try this: (see below image at left).

Only, by now you should notice my pelvic positioning is not good in this pic. So after you get into this position, untuck your pelvis until you feel your piriformis begin to stretch. If your hip is very tight, roll up a towel and sit on it to help you "untuck" your pelvis.

And then you should look like this: (see image, above right.)

Sitting on the edge of a chair helps.

Also, make sure you look very serious, like you have no sense of humor and aren't funny at all. That really helps the muscles relax.

Now do this at least FIFTEEN TIMES today, holding each side for at least a minute. And keep your ribs down while you do it. And let your head drop forward too and see if you can feel your back expand while you breathe.

The piriformis muscle connects to your sacrum and your femur (thigh bone) but also blends into the pelvic floor. When the piriformis is tight, it affects the strength of your pelvic floor. The more you work to lengthen this guy, the more the sacrum is free to move back into a position that optimizes pelvic floor strength and birthing space (especially you moms-to-be!).

Be gentle and yet consistent with your hips today. It's Piriformis Day!

And make sure you pick up a Happy Piriformis Day card for your dad. He'll love it. And, to confuse your friends, put HAPPY PIRIFORMIS DAY on your Facebook page with this article attached and demand a present. I dare you.

Sept 28 2011   5 TIPS ON SAVING YOUR KNEES

Today is my mom's birthday. She is sixty. Happy Birthday, Mom!! I love you. Thank you for everything you have ever done. It got me where I am today. Which is up at 5:30 a.m. writing this blog. Thanks a lot.

This is a picture of my sixty-year-old mom when she was living in some European communist country about seventy-five years ago.

I'm not sure why people used all-metal strollers or let their children pet wild animals back in the day. I love her shoes. I love that the monkey is on her stroller. I love that everything happening in this picture is highly illegal today.

Anyhow, when she was giving birth to me, my mom broke her tailbone. I guess *I* broke her tailbone, although not on purpose. I didn't *mean* to break her tailbone, in the same way I didn't *mean* to break her brand new tile-top kitchen table when I was twelve. I must have accidentally bonked my head on her coccyx on the way out just like I accidentally threw my baseball mitt with the ball still in it onto the two-day-old tile-top table.

What was not an accident was my setting the now-broken table in the middle of the afternoon to hide the cracked tile. I mean the whole enchilada, too—place mats, cloth napkins, silverware, plates, glasses, and a filled water pitcher. At 3:30 in the afternoon. Also not an accident was my exclaiming "Wow! Did you hear that?" when my mom placed the casserole from the oven onto the table about seven inches from the covered crack. Another non-accident was me exclaiming "What'd you do?" when pulling back the tablecloth to reveal the crack in the table.

She thought she cracked it until I came clean about twenty years later. Just kidding, I'm coming clean now. Just kidding, I already told her, but she couldn't hear me over the broken tailbone. Just kidding, she doesn't have a tail so how could she have a tailbone? Just kidding, she has a tailbone and it's called a coccyx, which is why I wasn't allowed to call it that when I was young. Just kidding, I was never young. I was born this age, thirty-five. Just kidding, I was born young, but I was thirty-five inches tall. Just kidding, I was born regular-sized, only my mother must have a gigantic tailbone that got in the way. Just kidding. Just kidding. Just kidding.

But you're reading this for knee tips, right?

So anyway, her broken tailbone caused some tucking that she only just began to work on about a year ago. So what happens after years of tucking, pushing the pelvis forward, and holding the body with the psoas and quads? Knee osteoarthritis.

In honor of mom, here are five things that you should do starting today, as often as you can, to save those knees!

1. Stretch, stretch, stretch your calves! This might be the single-best habit you could cultivate for healthy knees. And why the calves?

   Why is it bad to have the quads tense all of the time? See #5. (I've always wanted to write a choose-your-own adventure book. Remember those? I loved them!) (Image top next page.)

2. Stop wearing heeled shoes. The research linking positive heeled shoes to increased risk for knee osteoarthritis is abundant. Here is a picture from my new book, *Every Woman's Guide to Foot Pain Relief.* (Image next page, center.)

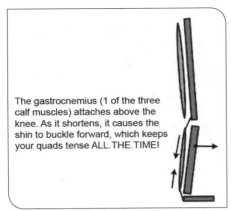

The gastrocnemius (1 of the three calf muscles) attaches above the knee. As it shortens, it causes the shin to buckle forward, which keeps your quads tense ALL.THE.TIME!

Your knees and pelvis have to balance out the angle changes at the feet. Even a "small" heel, as found on many a comfort footwear brand, can change the loading angles at the joints by 20° to 30°. Boo.

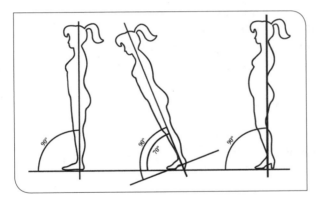

3. Get off the couch and onto the floor. There are a couple of reasons for this. The first is, you need to decrease the amount of time you spend in the 90° hip/90° knee position. Seriously, you're in it for hours a day. Sitting on the floor helps the joints articulate in new positions, which provides relief from old ones and stimulates different muscle groups. Also, when you're on the floor more, you just start stretching more. See if you can go all-floor after dinnertime. Try it. It's awesome.

4. Don't use this machine at the gym...ever. (Image top next page.)

Last year there was an article including this piece of machinery— What NOT to use at the gym. The article had great references, yet the on-line commentary was ridiculous. Personal trainers spouting off: "Well I do

it all of the time and I have great knees." Wow. This kind of commentary makes me want to poke stuff in my eye. Hey, you, don't argue with joint kinetics, okay? This machine causes very high and inappropriate forces within the knee. It should NOT be on the floor of a gym nor should it be found in any sort of knee-rehab program. There are much safer and better ways to strengthen your knees.

5. Relax your quads. The tighter your quads, the more their upward tension pulls the patella (knee cap) back into the joint space, increasing friction and then inflammation. See? (Image at right.)

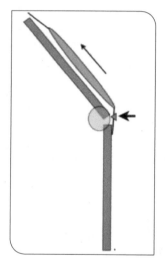

It is important to learn how to relax the quads while you are standing. It is not enough to stretch them—you need to learn how to turn them off at will. Not thrusting your pelvis is the first step.

If you want additional exercises, I recommend the Knees and Hips DVD for the super-gentle basics, and my Save Your Knees, Build a Butt webinar for the next level!

Biomechanical considerations in patellofemoral joint rehabilitation. *American Journal of Sports Medicine*, June 1993 vol. 21 no. 3 438–444.

Sept 30 2011 — SNAPPING PSOAS, HIDDEN TIGER

Have you ever gotten up from a sitting position and felt a little tweaky in the hip? Hear an audible click when taking your first few steps after being seated, or a pop when kicking your leg out to the side?

This is a condition cleverly called "snapping hip." Here's the basic lowdown. There can be two things going on. Either your iliotibial (IT) band is snapping in and out of its groove (called external snapping hip) or it's the tendon of the psoas muscle snapping over a bony protrusion (internal snapping hip).

What makes this snapping happen in the body is the same mechanics that make the noise when you snap your fingers. The amount of *snap* you get when you snap your fingers is based on how much you push them together. If you barely push your thumb and middle finger together, there's not much noise. There's no traction. But if you push them together with greater force, you get the snappy noise.

(Did you just snap your fingers? Isn't it cool to think that, somewhere else in the world, someone is reading this right now and you probably just snapped at the same time?)

Healthy muscle can stretch longer than its resting position length. But years of poor alignment, chronic stress, and other poor physiological habits all create changes within muscle that keeps it from budging. It just stays short and tight all of the time. This isn't because the muscle is stubborn, but because if it relaxed, you would fall down. You've been using the psoas muscles to hold your body upright because of forward thrusting habits and very weak posteriors.

So, tight muscle then transfers its tension to the tendons. In the case of snapping hip, the tendon of the psoas, which should be gently gliding over the pelvis as you move your leg, is pulled tight by the tense muscle and gets hung up on a small bony protrusion called the iliopectineal eminence. The noise is the sound the rubbing makes, as the tendon passes over.

This is the same mechanical thing happening in shoulder impingement syndrome, FYI.

Quick anatomy math lesson:

Tight muscles + tight tendons + bony prominences = tendonitis, then a bone spur, and then a separation (tear) in the tendon itself.

It's not fun, and if it is happening to your hip, you need to take care of it ASAP. It takes years and years to damage a hip enough for a hip replacement. The snapping noise in your pelvis and hip is your red flag.

Many people with issues like this will be told that theirs is a genetic issue. "You see ma'am, your bony prominence is just too big." Um, there is a ton of anatomical variability. All bones come with tiny variations to prominence size, lengths, and groove depths. Your ailments are not because of these anthropometric dimensions. It is your alignment and muscle lengths that create ailments. It is in the tension in the psoas muscle (not a genetic quality) that makes snapping hip. Not the size of the prominence. Makes sense?

Now here is the really fun part. There is now a surgery for dealing with chronic tension in the psoas. It's called an arthroscopic iliopsoas tendon release. Another catchy name. Still, that's better than what it should be called: cutting holes in your tendon.

Now, you may wonder (at least I'm hoping you wonder—because I'm going to tell you anyway), how does cutting a hole in something make it longer? Well, in a moment of personal science-teacher genius, I came up with a way to explain how. Kind of. Using pictures, really.

Introducing a shirt. An old shirt, not belonging to me. (Image below left.)

Add man with hairy arms, pulling as hard as possible (without pulling shirt of line). The (huge) muscular force of this man's arms is equal to what the psoas muscle is doing. Pulling. The shirt, the non-force generating material represents the tendon. Stretched to the max, it is ready to snap over a bony prominence at any moment. (Image above centre.)

Enter master surgeon, played by me and my kitchen shears. Which my husband didn't know were kitchen shears and has been using them for the last year as regular scissors, which he uses for everything, mostly cutting boxes, which is why I could barely cut the shirt. I mean tendon.

Comparing the pictures side by side, you can see the difference in length. And note: he didn't change his arm or foot position.

The structural integrity of a material is set by the continuity of its fibers. When you take a material under tension and cut a hole, the cut ends move away from each other in all directions, kind of "thinning" the tissue. Cutting into the shirt made it longer, no question. Now there will be no more snapping. Of course, there's also no ability for the psoas to be a primary-acting muscle in the body either, because it now connects to a structure that looks like this. (Image bottom of previous page)

I wouldn't do this to my own shirt, not to mention my body.

This procedure gets rid of the sound, but allows the real issue—the too-short psoas—to continue on. It's the equivalent of disconnecting the fire alarm instead of putting out the fire. And, P.S., getting your psoas to release is different than simply stretching it. You have to learn to stop using it to do everything (walk, hold you upright, etc.).

Try this. Lie your head and shoulders on a bolster or stack of pillows, making sure nothing is underneath your ribs.

If your ribs are up in the air when your thighs are down on the ground, then your psoas muscle is not relaxed to the correct length. This is a result of a chronic pelvis tucking and rib thrusting habit. And stress. Did I mention stress?

Your body should look like this:

but you shouldn't have to contract any muscles to get it there. This should be your relaxed position. It does not count if you have to work your muscles against each other.

Try this assessment, and, if you find your ribs up in the air, just sit with it. There is nothing to "do" to fix it. There is something you can "stop doing," though, and that is holding tension in the trunk.

Close your eyes, and open your mind—going in search of deep-seated tension. You're like an onion, so you'll need to switch off unnecessary muscle in waves, but it's so worth it.

If you are interested in exploring the amazing psoas more, you can learn a lot from this audio interview I did with Liz Koch, fellow psoas-lovah.[1] In fact, you can listen to it while you're relaxing on the bolster.

And, if you really want to know more or have more tools like the one presented today, I can't recommend my Science of Psoas course enough. It is totally worth it, and that's not just my opinion. My mom thinks so too.

Peace out!

---

1. coreawareness.com/podcasts/bio-mechanics-bowman/

chapter 4

PELVIC FLOOR

(and what gets stacked onto it)

SNEEZE LEAKS

Q: I am only thirty-seven years old, but over the last few months I have been "leaking" when I sneeze and I can't ever seem to get to the bathroom in time. Is this normal so young?

A: Well, incontinence is not normal. It is a failure of a muscular system, and should not be classified as normal. Common, perhaps, but not normal. According to the National Institute of Health (NIH), eighty percent of women will experience some sort of pelvic floor disorder (PFD) in their lifetime—with many seeking surgery (and then a second and third) as treatment.

Fortunately, clinical research is heavy in conservative treatments such as exercise. These exercise programs focus on Kegel exercises and biofeedback systems to train individuals to locate and strengthen the ability to contract these muscles.

Often neglected in these exercise programs, however, is the role that spinal curvature (your posture) plays in creating PFD. Tucking the pelvis under (especially while sitting) decreases the ability for the muscles "down there" to contract.

Try this: Sit on a rolled towel when you have to be in a chair for longer than two hours (like at work). Roll your hipbones forward (toward your desk). Once you are in that position, imagine squeezing a tampon. Practice all day if you like…no one will know![1]

---

1. Dear Reader, I wrote this before studying pelvic floor disorder extensively in grad school. I include this post to highlight how time and exposure to new information have developed my thinking.

AUSTRALIA

Today's blog is dedicated to the last hundred women I saw with Pelvic Floor Disorder. To the women who needed to whisper their symptoms to me, should anyone overhear them say the word "hoo hoo." Or "my area." Or "down there."

The inability to even utter the clinical term for your anatomy is indeed a reflection of how you think about this part of your body. If you think the word VAGINA is inappropriate to say, chances are you feel the VAGINA is inappropriate. Just so you know, it is not. Your VAGINA is just like your ELBOW...only different. If you are having problems there, please don't hesitate to mention it to friends and family or seek medical or theraputic attention. Keeping it a secret delays repair.

If your knee buckled every time you sneezed, you would know that there was something wrong with the knee machinery. You might even call your friends over..."Hey, look at this! ACHOOOO...See what that does to my knee?" If your pelvic floor fails when you sneeze, or if it hurts or burns, this is your body's way of letting you know there's a problem.

With your VAGINA.

Today's homework: Say the word VAGINA out loud three times. It's okay if no one is in the room, but check if you looked around first. Second: Work the word VAGINA into a conversation. Ok, maybe that's a stretch. But how about when you go to your doctor, or therapist, or pelvic floor specialist, you feel okay about articulating the problem clearly and without embarassment.

FYI: I typed the word VAGINA six times and I'll admit...it IS a little embarassing. But I'll get over it.

 ## April 28 2010 LOW HANGING FRUIT

Guess what? It's time to talk testicles. You didn't think I'd do it, but here I go.

P.S. I decided against a video blog.

If you've ever done the weight-machine circuit at the gym or followed the calisthenic stations at the park, chances are you may have noticed a lack of program development for the cremaster muscle. A gross oversight for sure, as this muscle is very important to the healthy function of the testicles. The word *cremaster* comes from the Greek verb meaning "I hang." Clever. The cremaster is responsible for keeping the sperm-generating process underway by regulating temperature in *le sac* as well as keeping the boys protected in dangerous situations. Like tennis.

Spermatogenesis is the process of generating mature sperm. This process happens inside the testicles and not only requires about sixty-four days, but also needs the perfect environment. Immature sperm cells don't thrive in heat. Depending on outside temperature, body temperature, and the tightness of one's pants, the cremaster is busy lifting or lowering the testicles to find an amount of heat that's just right. The cremaster is like the Goldilocks of the scrotum. If you jump into a cold body of water, it's the cremaster that yanks the boys towards the heat (your body) in order to protect the growing sperm.

The cremaster works automatically (without you having to think about it), and in addition to temperature, this muscle also fires under stress. In the hunter-gathering days, stressful situations usually meant a threat to physical harm. The cremaster gets busy tucking the sensitive bits out of the way. The stress-cremaster response is very old in our DNA, so it hasn't quite gotten the message that the stress you are feeling these days is more likely coming from driving in bad traffic or reading an email from the boss. You end up with testes close to the body more often than optimal, which can create over-heating of these precious tissues.

The testicles can also be lifted into the body by using the pelvic floor's "don't wet/soil yourself" muscles (pubococcygeal) or by sucking in your stomach. Sucking the belly IN is really sucking the belly UP. "Sucking it in" creates a vacuum that pulls the contents of your abdomen up into the dome of your diaphragm and raises the pelvic floor with it. The waistline flattens out, but now your testicles are perpetually heating up and your diaphragm isn't free to move when you're breathing. Now there's a hefty price for vanity.

Scientists have been wondering for years why the left testicle hangs lower in most men (about seventy to eighty-five percent, depending on the source). Researchers have even created studies examining all ancient statues of nude men, cataloging the percentage of right vs. left. Now there's a job. "Hey, get off my back, it's for research, man!" I tend to think that right/left testicle altitude might have something to do with one psoas being tighter than the other, resulting from the use of a dominant leg. Other fun testicle facts: The word "avocado" comes from the Aztec word *ahuacatl*, meaning "testicle tree." Avocado fruit hangs in pairs, with one slightly below the other. I wonder what the Aztec word for guacamole would be, then. ☺

For optimal sperm and organ (in this case, testicle) health, it is best to avoid interfering with the health-regulating processes your body uses on a regular basis (say Bon Voyage to your leather pants, unless it's a very special occasion). Stress plays a factor in many diseases, and pelvic floor and testicular health are no different. To check and see if you have your testicles in a (stress) vice, stand up and see if you can relax the pelvic floor just a bit. Most of us, men and women, think that we have to not only shut off our urine, we have to *super* shut it off. It's like we get points for really not wetting ourselves. Guess what, no prizes today. In fact, if you're going to get any prize (really healthy testes) use the minimal amount of force to keep your bladder closed, not the maximum.

I wonder, if I made an infomercial for the Cre-Master, would it sell as well as the Thigh-Master?

1,2,3,4, WE LIKE OUR PELVIC FLOOR!

Kara Thom, author of See Mom Run and an expert on keeping fit during pregnancy and motherhood (sounds easy) interviewed me on the exciting mechanics of the pelvic floor. I love the pelvic floor (kinda creepy, right?). And, as Kara reports, my position on the Kegel as a pelvic floor–weakening exercise is 180° from widespread beliefs, but what can I say? This has happened before.

The PF muscles run between the sacrum (the triangular bone at the base of the spine) and the pubic bone (the bottom-front of the pelvis). Ideally you want the PF to be long, supple, and taut, to generate long-term forces that hold up your organs, as well as have enough motor skill to open and close your bathroom muscles as needed. When the PF is too tight, it can pull the sacrum out of alignment, bringing it forward into the bowl of the pelvis. Bye-bye, strong PF muscles; hello, PF hammock. Hammocks are for vacation and have not promoted the notion of long-term force generation for some time.

A nice, taught (but supple!) pelvic floor. (Image above left.)

When the sacrum moves, it puts slack (and creates weakness) in the PF. (Image above centre.)

The red arrows are the Kegel contraction. See how the long-term result will always be slack? Eventually there is no more contraction to get and stuff starts falling down (and out!) (Image above right.)

The gluteal muscles (your butt!) keep the PF in check, preventing the sacrum from collapsing and PF from weakening. Of course, if the glutes get too tight, the same thing can happen in reverse. The secret to perfect pelvic floor tone is supple and strong glutes and PF muscles. Oh, and this goes for you too, dudes. You have a pelvic floor too, and although your organs won't prolapse or peek out of your vagina (vagina. vagina. vagina. Anyone uncomfortable?), they will settle down onto your prostate. For your prostate, this is about as comfortable as being trapped underneath a large bookcase.

The squatting exercise progression from You Don't Know Squat (below) should be done a few times a day, even in a modified form, until the tension in the knees, hips, back, and feet allow you to settle in. Then, keep doing this bathroom squat on a "regular" basis. Get it?

To read Kara's interview and my follow up (if you leave an awesome comment, you might win my "Down There" Pelvic Floor DVD!) click here.[1]

Follow up questions.[2]

---

1. mamasweat.blogspot.com/2010/05/pelvic-floor-party-kegels-are-not.html

2. mamasweat.blogspot.com/2010/05/pelvic-floor-encore.html

 June 2 2010 YOU DON'T KNOW SQUAT

Are you ready, world? In the next few years, you are going to be hearing more and more about the health benefits of the squat. Just like the barefoot movement I blogged about yesterday, the "squat movement" is going to happen in a big way, once everyone realizes that your pelvic floor, hip, and knee health require regular squatting. If you haven't squatted in the last million years (besides the two times you went camping and peed on your shoe), it's going to take awhile to prepare your joints. Be patient, it's worth it. Those of you with knee and hip replacements are going to have to stick to the first few exercises and avoid the weight-bearing squat. The artificial equipment is not designed to have the same range of motion as real joints. Bummer, I know. (To stave off surgeries of the hips and knees, start this program now!)

I've created a step-by-step squat program from my Aligned and Well program you can begin right now. All you need is a yoga mat, thick towel, or blanket. At the institute, we have begun a "Twenty Hours in Thirty Days" squat challenge. Join us in reaping the benefits of deep (deep) core muscle conditioning, and sit back (or squat) in disbelief as you realize how tight you have become! Let's go...

Place your one foot up on a rolled yoga mat, keeping your heel on the ground. Step forward with the opposite foot to stretch the back calf. (image above left)

Now step up with both feet and try to lift your tailbone until it looks like...(image below)...this! See the little curve at my low back? This is an indication of an un-tucked, healthy pelvis. If your tailbone slopes down, your hamstrings are too tight for pelvic floor and glute strength!

Before you do any deep squatting, spend five to ten minutes running through the first two exercises. They are great for preparing your joints for full flexion (bending all the way). After you've warmed up the back of the calves and hamstrings with a little stretching, it's time to get down onto your hands and knees.

It is very important that your lower legs and feet track straight back, and are parallel to each other. (Image below left.)

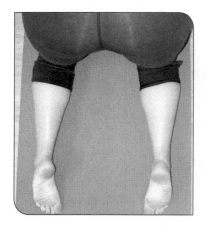 

Now, from the hands and knees position, drop the hips back as far as you can, without allowing the feet to move closer to each other or allowing the tailbone to tuck under.

Ooops! See how the pelvis tucked? (Image above right.) Don't sit back any farther if your tailbone tucks. This forces your lower back into flexion, stressing the lumbar disks. Instead, come forward, re-lift your bum, and hang out in this position.

When your tailbone tucks, this clearly shows how TIGHT your hips have become! If you can't sit back without tucking, it means that your hips are so stiff, they are preventing the regular activity of the pelvic floor and gluteal muscle groups. Do this a few times a day until you can get back with your tailbone up.

You may also see how we are getting the body in the same position as a "bathroom squat" without loading it. Better to improve range of motion before you throw all your body weight onto tight joints.

Other fun squatting tips...

If your tailbone is tucking when you sit back, it means that instead of peeing (or other things) in a downward motion, you are actually doing them in a forward motion. So that's why you pee all over your shoes. Maybe you shouldn't join the barefoot movement until you've mastered the squatting one. Just sayin'.

Now that you've been sitting back, it's time to make your feet move into their squatting position. Tuck your toes under and try to get your feet perpendicular to the ground. (Image below left.)

Spend some time stretching your feet while sitting back (aaaaand lifting your tailbone!).

Now it's time to start bearing your body weight. The strength needed in a squat is not only getting down and lifting up, but also the strength in the lower legs to stabilize the ankles. Your lower-leg muscles will usually fatigue first! To keep squats safe for your knees and hips, you must keep the alignment of the lower leg and feet. Your feet should point forward, they should be placed just slightly wider than the pelvis, and the knees should not be wider or more narrow that the feet. Your thigh, lower leg, and foot bones should be parallel to each other the entire time.

When you first squat, super tight quads and psoas can increase the pressure in the knees. Give yourself more joint space by placing the rolled yoga mat where the knee bends. (Image above right.)

The "goal" is to keep the pelvis from tucking, as well as try to stretch the heels toward the ground. Hold on to something when you first start, if you're feeling wobbly!

Also check that your feet and knees are still aligned well and they haven't twisted (image above left), and your weight hasn't shifted to favor one side. Not good for your joints! (Image above right.)

After the ankles and lower leg have had a chance to stretch and strengthen, prop your feet up with the rolled yoga mat and gently allow your knees to bend their full range. Again, it's very important that your lumbar spine (the concavity at the lower back) maintain its curve. Now you know your deeper pelvic floor muscles are toning, your glutes strengthening, and your hips opening!

I'm ready to go camping now! (Image below left.)

Don't let this person borrow your shoes. (Image above right.)

Years of shoe-wearing and chair-sitting can lead to a squat that uses more quad than glute. The best indicator of which muscles you are using is the position of the shin. The more the knees are in front of the ankle, the less glute muscle you use when coming up. Doing squats with a vertical shin is nearly impossible with our modern-living strength patterns, so to help you along, use a door knob and walk your feet back until your arms are long. When you sink down, untuck your pelvis until your shins are closer to vertical. Vertical shins are most important when you come up, so doing a few with external support will not only give you a deeper stretch but stronger buns.

Eventually you will be able to maintain the curve in your low back (pelvis position...check!) and get your heels on the ground (foot health... check!). Until then, use this program as the ultimate total-leg, pelvic floor workout. Progress through these exercises as you feel comfortable, giving yourself plenty of time (which can be weeks and months even!). I also like to turn these moves into an hour-long lower-body conditioning session when I'm feeling spunky.

And if you're really bored, you can print out all of these pictures and make a flip book of me squatting. Now there's fun for the whole family...

June 14 2010 FIXED MY PFD ASAP

I just took the weekend off to visit my family and returned to this freaky statement: You Have 1048 Emails.

I'm sorry, but, *what*?

I know, I've got a few friends, and I've got a city full of people who come to the institute, and I even have nine brothers and sisters. But even if all those people sent me an email, I still would not have 1048 emails.

In what can only be called a Pelvic Floor Frenzy, my inbox may actually be smoking. And, all I can say is: this is an exciting time. In graduate school, I focused on pelvic floor function with a passion that made me look, well, like a crazy person. My first paper was researching the "Biomechanics of Birth" (which I'm reworking for publishing purposes... yay!). My next paper was on the "Prevalence of and Effectiveness of Therapy Procedures for Low Back Pain during Pregnancy."

My graduate thesis was "Decreasing Risk Factors of PFD Through Exercise," during which the treatment group was able to restore their lumbar curvature (missing curvature being a risk factor for PFD) in four weeks (wow!). It was a fun thesis to defend because it was blowing the minds of my department professors (all men, of course) who were like: "Yeah, that makes sense. Someone should tell women. Yeah, duh."

So, when I say it's an exciting time, I really mean it's about time, because I have been working on optimizing pelvic floor function (well, the whole body, really) based on the position of the bones for about TEN YEARS. The biggest issue in getting the information out has been the fact that the current treatment for PFD is held so dear to many and is woven so deeply into the fibers of women's health, that the resistance is higher than it should be for a scientific community. But, as more and more women (at least 1048 of them) following the current model of treatment realize it's not working for them, now seems like a good time to teach the biomechanics of women's health. You with me?

So, how do I, whose work schedule barely allows me time to get my car smogged (the notice has been sitting in my kitchen for three months...), answer 1048 emails to women who are ready to hear what they need to do? Women who are giving birth in the next few months who want the event to be even more magical? Women who have organs dropping down right now who want to get better *right now*?

Hmm.

Well, here's what I am thinking. I'm thinking I can write a quick list of things to start doing right now that will stop creating forces that make PFD worse or pregnancy harder. I don't have time to explain in depth *why* these things make a difference, but if you'll take my word for it for the time being (and it's stuff like what shoes do to the pelvis, sucking in your stomach, etc.) you can start them ASAP.

I understand completely, that most of you are unable to squat because of poor knee health. And here's the kicker: Women tend to have much more ACL and patella tracking issues for the very same reasons they have PFD. That's why the two tend to go together. When I told everyone they needed to squat, it's really they need to work up to doing a squat, which could take you a month or two. My Down There for Women DVD is really the most introductory of the pelvic alignment exercises—no squatting required yet. If you do the preparatory exercises and then add in the squats, you should be on your way to a really great pelvic floor, deep hip, deep abdominal, knee, and foot health program.

And then finally, I am thinking that I could teach a five- or six-week video webinar that would teach you all the amazing stuff that's going on with your pelvis and how to reverse the downward forces you are creating. It's a posture, musculo-skeletal, and neurological (brain and nerve) issue, but you would be amazed how easily you understand what it is you are doing that creates PFD and how to stop. You could take it for yourself, or, if you are a birthing professional, you can take it to distribute the information to your lovely moms-to-be. I'm thinking low cost. I'm thinking video, so you have about a forty-five-minute exercise session, fully modified for all participants. I'm thinking if you're interested, raise your hands. If there's interest, I'll see if I can pull it together quickly.[1]

I'll add your first assignment tomorrow, instead of smogging my car of course. What's one more month?

---

1. I did it! The No More Kegels course is online.

June 14 2010 4 FAST FIXES FOR PELVIC FLOOR DISORDER

I was just interviewed for an article on Lifescript.com, and in looking for some research to share, I came upon this from the *Journal of Obstetrics and Gynecology*:

> The number of American women with at least one pelvic floor disorder will increase from 28.1 million in 2010 to 43.8 million in 2050. During this time period, the number of women with UI will increase 55% from 18.3 million to 28.4 million. For fecal incontinence, the number of affected women will increase 59% from 10.6 to 16.8 million, and the number of women with POP will increase 46% from 3.3 to 4.9 million. The highest projections for 2050 estimate that 58.2 million women will have at least one pelvic floor disorder, with 41.3 million with UI, 25.3 million with fecal incontinence, and 9.2 million with POP.[1]

In case you were wondering, POP stands for pelvic organ prolapse (lesser known than the more commonly used acronym for organ prolapse, WTF!). Holy cow, ladies. It's time we change course of this slow-moving ship called Women's Health. If the existing treatments are "scientifically proven to work," then how do you explain the doubling of incidence in the next forty years? Let's get off this crazy ride and try something new. As promised, I am giving you some homework to start right away.

1. If you have an organ prolapsing, you must reduce your impact activities until you have better muscular strength. Once the organ is prolapsing, it is clear that you don't have the strength to hold up your organs. True, your ligaments can carry the load for a while, but here's the thing about ligaments:

   Your ligaments cannot **ever** shrink back to their original position. They are not like muscles and tendons in that way. If you use your ligaments to hold up your organs (or stabilize your knees or sacrum, for that matter) instead of using your muscles, the ligaments will stretch out like the elastic on an old pair of knee socks. You know what happens to the top of socks after you've stretched out the elastic? How they kind of sadly swamp around your calf? I call socks at this stage "quitters." Don't let your ligaments become quitters! (Oh, and your ligaments don't sag because they age, but rather because they have been subjected to your bad habits for longer!)

2. Get out of your heels ASAP. This one is better explained with a picture taken from *Podiatry Management*, written by genius podiatrist William A. Rossi.

   As a biomechanical scientist, I have been trying to educate everyone (including the most educated of the medical community) about the damage heeled footwear brings to the entire body. A positive heel instantly impacts the biomechanics of ALL of the joints, which includes the sacroiliac joints and the hips. In fact, my pelvic floor thesis and corrective exercise program dealt

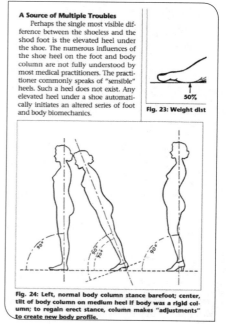

A Source of Multiple Troubles
Perhaps the single most visible difference between the shoeless and the shod foot is the elevated heel under the shoe. The numerous influences of the shoe heel on the foot and body column are not fully understood by most medical practitioners. The practitioner commonly speaks of "sensible" heels. Such a heel does not exist. Any elevated heel under a shoe automatically initiates an altered series of foot and body biomechanics.

Fig. 23: Weight dist

Fig. 24: Left, normal body column stance barefoot; center, tilt of body column on medium heel if body was a rigid column; to regain erect stance, column makes "adjustments" to create new body profile.

primarily with the legs when treating PFD. You can't correct the smaller problem until you deal with what created it.

From Dr. Rossi's article:

Perhaps the single most visible difference between the shoeless and the shod foot is the elevated heel under the shoe. The numerous influences of the shoe heel on the foot and body column are not fully understood by most medical practitioners. The practitioner commonly speaks of "sensible" heels. Such a heel does not exist. Any elevated heel under a shoe automatically initiates an altered series of foot and body biomechanics.[2]

Can we please have three cheers for Dr. Rossi? Hip, hip...!

3. Walk, but STAY OFF THE TREADMILL! The treadmill is a no-no for anyone with a hip, pelvic floor, psoas, or knee issues. Sounds weird, right? You thought walking was walking? Nope. Okay, fast physics lesson. On this planet, things move forward by:

A) Pushing backward

B) Using a credit card

C) Leaning forward, and then falling.

If you guessed A and C, then you are correct. If you guessed B, then you're probably a large bank. To truly move forward using your muscles, then you have to generate an opposing force. You understand this when swimming, or when dipping an oar into the water. Even your tires push back to move the car forward. It's how things are done. If you are falling forward, then not only are you NOT using your posterior, you are using your joint cartilage to cushion the fall. And there's not an unlimited supply. Walking correctly means you get to use your BUTT muscles. The gluteal muscles extend the hip (fancy words for lifting the leg out behind you), but on a treadmill, because the belt is moving toward you, you don't get to push back. Instead, your lift your leg out (hip flexion) and fall forward. Congratulations, you burned up a lot of calories but weakened your pelvic strength. Try taking your act on the road (as in, not on the treadmill) and practice the "ice skating" feel of a pushing-back gait pattern.

4. Stop sitting on your sacrum! In addition to your pelvic floor muscles, your organs are also held in place by ligaments. One major ligament to the uterus is attached to the sacrum, so if the sacrum is pushed into your pelvic bowl, the uterus moves down.

Even a diligent exercise program can't override the constant and displacing physical pressure.

Make sure that you're not thrusting your ribs (anatomical experts say that your ribs are different than your pelvis...) but are rolling your pelvis forward. Sit on a rolled towel to help you find the correct pelvic tilt!

All right, you've got your tips. Let's go to it!

---

1. "Forecasting the prevalence of pelvic floor disorders in U.S. Women: 2010 to 2050." *Journal of Obstetrics and Gynecology.* December 2009. Wu JM, Hundley AF, Fulton RG, Myers ER. ncbi.nlm.nih.gov/pubmed/19935030,

2. "Footwear: The Primary Cause of Foot Disorders. A continuation of the scientific review of the failings of modern shoes." *Podiatry Management.* February 2001. William A. Rossi, DPM.

June 17 2010 CREEP

I am still reading through Mama Sweat's blog posts and today I came across this gem of a comment:

> ...I think pelvic strength is great. But, as a scientist myself, I know that everything has an elasticity limit, even muscle. Pregnancy and birth can provide enough force for some muscle to never go back to its strength or length. Look at your stretchmarks, ladies—some women have more elastic skin. Nothing you can do to change it. Bottom line: don't blame yourself for your pelvic floor problems.

Signed, of course, Anonymous.

I responded with:

Hi, Anonymous,

The mechanical failure you are speaking of is called CREEP (really, that's what it's called!) and has nothing to do with PFD (this is well researched, so with a little leg work anyone can see the extreme difference between the failure of skin vs. muscle). In other words, stretch marks have no scientific purpose in a discussion about PFD.

The incorrect notion that the structure is inherently weak and is the cause of PFD is really the problem in a nutshell (thanks for bringing it out to discuss, though!), and is really undermining women's health.

And then I signed my real name. Just because. And actually, to be a bit of a butthead, I added M.S. after it, because, well, the "M" stands for Master and the "S" for Science, so I was just pointing it out. And it's been fatiguing answering questions from "scientists" who don't know their biomechanics (or tissue properties) that well. And I may have PMS, but these days, who can tell.

Right-o. So, I just wanted to bring up this post (and am I wrong to think this wasn't posted by a woman, because I can't think of any woman who would say "Look at your stretchmarks, ladies"?) to highlight the "scientific dismissal" issue. To further stray off topic, comedian Demetri Martin says that to make any sentence sound creepy, add the word "ladies" to the end of it. And frankly, I agree.

Focus. Okay, so, mechanical creep is best defined as "the elongation of tissue beyond its intrinsic extensibility resulting from a constant load over time." What makes skin and ligaments stretch beyond their ability to return is, believe it or not, not that well understood. It has a lot to do with the fact that creep is affected by so many different things like temperature, rate of loading, size of loading, hydration, and probably, hormones too. Which is why hormonal changes often get blamed for PF issues. But, to remind you about ligaments, they are the backup plan. You've always got the muscles there on the front line, so we need to get them to the right length to maximize force and minimize the demands on the supporting ligaments.

Finally, a note on scientists. There are scientists of many fields: physiotherapy, geology, chemistry, pharmacology, psychology, astronomy, etymology, etc. And while the education varies in each of these fields, a scientist will understand that there is information they don't know, and will consider that before making statements like: "Nothing you can do to change it" or "don't blame yourself for your pelvic floor problems." I for one, hope Anonymous Scientist isn't in the women's health field. Don't you agree...ladies?

And, for all my stay-at-home scientists, here's a fun journal abstract that shows the research on long-term effectiveness of traditional pelvic floor muscle strengthening.[1] (And note, when PF exercises are shown *not* be effective eight years later, the conclusion is "we need to figure out how to get these women to do their exercises," despite the fact that they reported they did.) "Scientific Conclusion": Kegels work, and women are liars. Nice ☺

1. ncbi.nlm.nih.gov/pubmed/18651881

When we think of "good posture," we often think "shoulders back," but when we *do* "good posture," we're not really pulling our arms back as much as we are pushing our pelvis forward. Our "shoulders back" has become "pelvis forward." This doesn't seem like a big deal, I know, but from a mechanical perspective, your alignment can be significantly altered (up to 30° in some places!), which means the performance of your body is significantly altered. The habit of "pelvis forward" is a major contributor to musculoskeletal issues like foot pain, chondromalacia of the knees, low bone density, pelvic floor disorder, and sacroiliac pain…just to name five of the most common ailments in the United States.

Ideally, the major axis of the body should form a plumb line, to minimize torsional forces that cause joint degeneration (in the ankles, knees, hips, and lumbar and thoracic spine) and maximize structural support (weight over the dense heel bones instead of plantar fascia and forefoot). Vertical positioning also places the weight of the upper body directly over the pelvis, which makes the hip joints weight bearing.

Let me say that again. In order to have weight-bearing hips, you must have vertical alignment. No vertical alignment, no weight-bearing hips. Hmmm. Where have we heard we need for "weight-bearing exercise" before? Why do we need weight-bearing exercise? Oh, right. Because without it, our bone density decreases. But here's the thing. Weight-bearing exercise does not mean using weights or resistance training, but carrying your body weight in a weight-bearing fashion. You want to build bone density? Get your hips over your ankles and your torso over your hips. And then walk around a lot. *Walking in alignment is the most naturally osteogenic (bone-building) thing you can do!*

Once you learn to evaluate posture using objective markers (like a vertical line), physical forces like torque are easier to understand. The non-vertical line allows anyone with a mechanical eye to "see" the forward lean, which causes damage to structure (just like mal-aligned wheels on a car.)

How does an anterior pelvis contribute to foot health?

Drawing a vertical line from the center of mass (located in the pelvis) shows, pretty clearly, the difference in weight placement over the foot. On the left, the burden is great on the forefoot (front of the foot), overloading tiny muscles that should be concentrating on supporting the arch of the

Non-vertical   *Vertical*

foot—not the weight of the body. This weight on the front of the foot can contribute to plantar fasciitis, hammertoes, metatarsalgia (pain in the base of the toes), and neuropathy.

A quick fix: Get your weight back where it belongs! It's easy, fast, and free. You may also notice that you can't do this with a positive heeled shoe on. Shoes with any heel automatically force the pelvis forward. Barefoot, try getting your weight far enough back to lift your toes off the ground. Now you have a sense of your weight being over your heels!

What about bone density and low back pain?

The lines (image next page) should help you see the "physics" of this posture. When we think of weight we usually think of how much we weigh (which seems right, right?), but weight is really how much of your mass is being pulled on by gravity. The gravitational force works in a vertical fashion, and points straight down to the ground. So, if a vertical line from your torso to the ground passes through a lot of air (like the picture on the left), then the weight of the torso is not felt by the bones of the legs. If a vertical line travels from the torso through the legs, that means the leg and pelvic bones experience the full weight of the torso. The mass is the same, but because weight is a vertical force, the less vertical you are, the less you "weigh," from the standpoint of your legs, hips, and pelvis.

What does this have to do with you? If you have decreasing bone density in your hips, getting your body vertical should be a main priority. There is no drug or exercise that can compete with actually loading

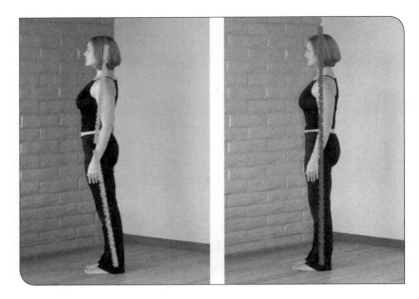

the bones correctly to send the mechanical "build bone" signal. That's the way bones work! (Good idea alert: Put a slanted floor in your bathroom and place your bathroom scale on said floor for optimal results. Your mass will be the same, but boy will you be in a better mood!)

Back pain? The constant forward motion of the pelvis relative to the backward motion of the torso create lumbar compression, which increases friction, heat, osteoarthritis, and ultimately disk degeneration. Solution: Vertical Alignment. Got it? If you're a pelvis thruster, backing your hips up (so they are in a plumb over the knees and hips) will feel like you're sticking your bum out. Get objective about it first. What you "feel" and what "is" are two different things. I just read a great quote from yoga teacher Jeffrey Lang. "Your Natural may not be your Neutral." Brilliant.

The purpose of my biomechanical model of preventive medicine is to help you restore your proprioception (knowing where you are in space) to a time before you had thoughts and beliefs and habits. Feel like your booty's sticking out? Get used to it. One more thing, the reason you have moved the pelvis forward in space has more to do with the tension in the muscle and myofascial patterns of the upper body, but we can't adjust the "vase" until we have the table straightened out!

Tomorrow, I'll continue with how a forward-thrusting pelvis in pregnancy can decrease your birthing mechanics. Back that thing up, will ya?

KEGEL QUEEN

Has anyone out there heard of The Kegel Queen? Well I hadn't, until someone forwarded me this YouTube video where KQ counters my Mama Sweat interview and takes me on. You can tell she's the queen because she's wearing a crown. No, really. She wears a Kegel crown. It is still unclear if she was born into pelvic floor royalty or elected, but I swore I'd find out.

I can, of course, understand why the Kegel Queen might be upset. I'm sure the article made her feel like someone was storming her castle, and who was I? Some squatting court jester with a large derriere! Take a look (her video is great!).[1]

I'm not sure if any of you out there have ever had someone "take you on" in a YouTube video, but it can be quite startling to hear someone you don't know say your name on film. Deemed "Katy Bowman, Kegel Critic" by the KQ, I want to point out that while my position on Kegels is they are a short-term band-aid for PF disorder, I am actually *not* in favor of weak pelvic floors or organs falling out of women. Katy Bowman = Women's Health Advocate, not The Grinch Who Stole Bladder Control. After watching the video, I did what any person mentioned in a Kegel Queen video would do. I emailed her. And then I waited for her return email. What was the KQ going to be like? Heavily guarded? A royal pain? Did she even return her own calls (does the queen of England?) or would I have to deal with her court? So many questions!

Finally, we connected for a nice long chat, and who knew how awesome the KQ was going to be? (Actually, I did. No one stutters "bogus" *a la* Max Headroom and wears a crown unless they are awesome...) We talked over many things and I wanted to share a bit of my interview with you.

1. *Was your mother also Pelvic Floor Royalty, or did you marry in?*

    My reign is actually the result of a coup. With an army of like-minded women, I've been working for years to end the rule of high-tech, dangerous options like unnecessary hysterectomies and cesareans. My rise to power is a victory for every woman who wants to keep her body, and her power, intact and use the wisdom of her own body to heal herself.

2. *When did you first start using the word "bogus"?*

I certainly have been inspired to use that word whenever I see the made-up "information" that is commonly presented as kegel instructions. Actually, the word I'm inspired to use is best avoided in polite company, but "bogus" is an acceptable substitute to describe virtually all the kegel instructions you'll find online, and in medical offices, childbirth classes, magazines, and on and on.

3. *What is the difference between Your Kegels and Other Kegel guidelines and programs?*

First: In the Kegel Queen Program we use No Devices, Ever. Women looking for kegel instructions online will be barraged with ads for all manner of things you can put in your vagina and squeeze. But some types of devices can damage the pelvic floor, and inevitably they end up gathering dust at the back of the underwear drawer when women get sick of taking their pants off to do kegels, washing the device, packing it in their overnight bag and having to explain to airport security that it's not a bomb.

Another major difference is that my program is based on scientific studies, and designed to be done in minutes a day. People who recommend hundreds of kegels or hour-long kegel sessions haven't studied the research, and they also must not know any actual women if they think we are all going to quit our jobs and do kegels all day! I also emphasize the importance of breathing and correct positioning, and teach how to successfully make kegels a habit—all of which are generally overlooked in most kegel instructions.

4. *Do you understand a bit more about the squat–kegel relationship after speaking with me?*

Of course, I was already familiar with the way the sacrum moves with squatting or pelvic floor contraction. But your perspective on muscle physiology is new and fascinating to me, and something I'm very excited to learn more about.

Because of the way kegels have profoundly benefited me and my students—eliminating prolapse symptoms, curing incontinence, and leading to radically better sex—the idea that you can squat INSTEAD of doing kegels will be a very hard sell for me. But working with positioning and other aspects of body mechanics to optimize pelvic health is a brilliant approach, and something women badly need. Kegels AND squatting? I'm all over it.

5. *Where should people go to learn more about your program?*

Go to KegelQueen.com, and watch a short video in which I explain the Kegel Queen's Top Three Reasons You Should Never Do Kegels while Driving Your Car. I'll send you a free PDF with even more detail about that too. You can also order my free Kegel Myths & Facts DVD.

6. And if you want to learn all about the correct way to do kegels, you can order my two-hour audio course, which comes with a study guide, two months' free access to my members-only website, and live Q & A calls, and other awesome goodies available only with the course, including my book, *Keep It Simple Kegels: The Complete, No-Devices, Fast and Easy Guide That Shows You How to do Safe, Effective Kegels so You, Too, Can Stop Peeing Your Pants, Get Help with Prolapse, and Have the Most Incredible Sex of Your Life.*

Now, my two cents. I think it's pretty revolutionary to interview oneself, don't you think?

1. *Why do you hate the kegel? Are just you a bad person?*

    I don't hate the kegel at all, but what people need to know is *there is a poorly understood, much larger whole-body issue going on in those with PFD that the Kegel doesn't even touch.* If this larger whole-body issue were addressed, you would never need to kegel to keep your junk in place.

2. *What about all the research that says kegels work?*

    I am assuming that most people who say "the research shows that kegels work" are people who have actually not looked at the research beyond the abstract or know the difference between what the research finds (the numbers and measurements) and what the researcher extrapolates (their thoughts on what those numbers and measurements mean). There is actually very little research showing the long-term ability for a PF "strengthening" program to continue to keep women from surgery. (And FYI, having the surgery doesn't seem to keep you from needing more surgeries either, so there.) Research showing that "kegels work" typically show that kegels done under the supervision of someone (with something possibly inserted into your vagina) register an electrical or mechanical measurement. Which is different than showing they have kept someone from needing surgery over a lifetime.

    What a statistically larger number of people experience is a short-term (and I'm talking years) positive experience, that, over time, begins to decrease. The positive benefits decrease because the pelvic floor, like any muscle in the body, cannot contract indefinitely. For long term PF function (which means continuous force generation), eccentric (muscle-lengthening) function is needed, which requires the muscles on the other side of the sacrum to be working at the same rate that the PF is working.

3. *What is a kegel, actually?*

As a scientist, this is another bee in my bonnet. The research conducted to PF muscle training uses different protocols (aka definitions) for the kegel. Some use devices. Some use long holds. Some use flicks. Or flickers (which is flashing the lights as you rapidly clench and unclench your PF. Just kidding). Some teach the relaxation portion. Some don't. Some say do it using a trigger, like a stoplight. Other evidence shows that using a trigger (or, say, stopping and then releasing your urine stream) creates an incorrect neurological dependence on these items or situations, interfering with the natural motor programming. Who knows if the method by which you are executing a kegel (that your doctor or midwife or best friend or latest issue of *Fitness Magazine* describes) was even the method researched for health benefits? The whole science/research thing has become extremely (and ironically) subjective and has resulted in poor exercise prescription.

4. *Why do kegels work for some people and not for others?*

PFD is absolutely multi-factorial, meaning there are a lot of things that happen to result in the situation at hand. To increase your short-term PF strength benefits, you probably received in-depth instruction and followed a correct, well-thought-out protocol (like the KQ has done). The person for whom it works is also someone who did not have excessive PF displacement during labor, may have paid more attention to alignment and movement (exercise) before they had an issue, and isn't a runner (A shortened PF cannot handle the excessive G-forces from the impact. Pelvic floor length may be your salvation!)

5. *Is there any time a kegel is a good thing to do?*

Of course! If you have lost your somatic ability to find your body parts (i.e. finding and moving your body parts with your brain), it's helpful to learn how to engage them. So if you have no idea how to feel your PF, doing a kegel is a first step to building a brain-body relationship. The original use of the kegel was for women just after delivery—probably the best time to begin a somatic lesson in finding your pelvic floor through good guidance. Also, as people decrease sexual activity, the muscular walls of the PF may have also lost "contact" with their brain. If you find yourself needing to sweep away the cobwebs "down there," an introduction to pelvic floor somatics may also be in your best interest.

6. *What is the "bigger issue"?*

The pelvic floor is not supposed to be a muscle you "train." It is a muscle designed to have continuous electrical flow based on the correct use of the lower limbs and muscles in the trunk: diaphragm, psoas, TVA and TVT, lumbar extensors, multifidus, intercostals, latissimus, gluteus maximus, TFL, etc. Because of years of mal-alignment and incorrect lower leg muscle development, your larger muscles are not bestowing the PF with the space and support it needs to maintain a healthy tension. That tension is necessary to its ability to generate force. A kegel is a good way to fake it, but the PF is too small to do the work for the larger, slacking muscle groups.

PFD is a sign that your body is collapsing into itself. While the sneeze-pee and organ prolapse may seem like a huge deal, they are nothing compared to the neurological damage to the spine, nerves, and tissues that those nerves supply (especially the nerves running to the lower legs). The kegel is a drop in the bucket to what someone with PFD needs to do to heal all their structures. PFD is a whole-body situation and for optimal, whole-body outcome, the body has to be treated holistically. Otherwise it's like putting a new roof on a termite-ridden frame.

Where do the KQ and I stand (or squat)? Pretty much on the same page. She's for optimizing women's health, as am I. She's taken lots of time to develop a detailed, easy-to-use program with great results. I like to think I've done the same. It is essential that women realize normal aging does not require organ removal or surgeries to accomplish what their muscles should be doing. PFD is simply a weakness of the muscles that attach to the pelvis and the pelvic floor. This situation is easily changeable. The conversation between the KQ and me should prompt you to look at the entire problem thoroughly, know your equipment (all of it, not just where your trouble seems to be located), and choose the correct prescription based on lots of information. Visit KQ's site and you can sign up for some free PF info, and maybe order her kegel course. I'm going to be working with her on the bigger alignment portion, so stay tuned to the KQ's email.

(And of course you can register for my course, No More Kegels.)

The Kegel Queen and I have decided to become allies, which makes me some country that I could think of had I paid more attention in World History. How about I be Australia, then I can be Queen of Down Under, which is kind of like "down there." How do you like my crown?

---

1. youtu.be/fjbfXzOIoCY Seven minutes of the Kegel Queen dissing the interview I did on Mama Sweat. I mean, talking about how great kegels are.

October 29 2010 TAIL OF TWO PUMPKINS

This is a true story. Really.

The other day, I was wandering through the pumpkin patch when I stumbled upon these two jack-o'-lanterns. They were having coffee and talking about their day (you know, like girlfriends do), when I heard one start to talk about some pelvic floor issues she was having.

Here's a picture of the two of them from behind:

Which one do you think was having the pelvic floor issue?

If you guessed this one:

then you guessed right!

Did you notice the lack of lumbar curvature on her as noted by her flat pumpkin back and pumpkin butt?

As I got closer, I saw the problem

And then, I got even closer, which I'll admit is pretty rude. I became sure of her problem.

Her jack-o'-sacrum, a movable bone in her jack-o'-pelvis, was sinking forward into her body. A healthy and strong pelvic floor requires the sacrum to be nutated, where the lowest part of the sacrum is farther outside the pelvis than the top part. What little Ms. Pumpkin has here is a counternutated sacrum; the coccyx (tip of the tailbone triangle) is deeper in the body than the top of the sacrum.

This poor pumpkin's sacral position had given her a short, tight, and weak pelvic floor. This is a mechanical cause of pelvic floor disorder and is actually worsened, ironically, by a common pelvic floor "strengthening" exercise called the kegel—especially when done by someone with weak and/or tight gluteal muscles.

So, I told all of this to the kegel-o'-lantern. She laughed at my ridiculous explanation to her issue.

"Kegels *causing* pelvic floor issues? What are you, some sort of crazy women's health witch??"

Her laughing caused an unfortunate accident. (Image at right.) She decided to hear me out.

I said, "I don't have time to tell you everything right now, but read my blog" (I had my computer in the pumpkin patch) "and it'll get you up to speed with the latest pelvic floor information."

And, to make things even more creepy, I checked out her friend:

Nice behind-rind. No problems there at all. Looks like this squat-o'-lantern has things under control...including her bladder. Turns out she had been following my squatting instructions for the last three months. Nice job!

My work there was done so I went on my way! Happy Halloween!!!

Feb 17 2011 SUPER KEGEL

Dear Dr. Oz,

I let it go (kind of) when you brought up "beautiful women in spandex" as a motivating factor to take yoga.

> Yoga is the most important exercise of my daily routine. Being surround-
> ed by beautiful women in spandex should be reason enough to join a
> class, but if you need more motivation, consider this: Yoga eases stress,
> lowers blood pressure, slows heart rate and increases flexibility.

—Dr. Oz (From the June 2010 issue of *Natural Awakenings Magazine*—
"Everyman's Rx from Dr. Oz.")

Ick. Don't you have an editor? You can't go around saying creepy stuff
like that, dude.

Anyhow, what I cannot let go of (without at least one blog-lashing)
is your SUPER KEGEL recommendation. I was excited to see pelvic floor
disorder (PFD) making its way onto daytime TV. That's great. Really great.
It helps get the word out that we are dealing with a super-epidemic of falling organs. Note: The show would have been even better if it included the
fact that MEN have their organs falling down too, only they don't fall out,
but down...right onto the prostate.

The topic was timely and important. The episode was, however, lack-
luster in the areas of causes, treatment options, and, of course, the SUPER
KEGEL. I figured there would be the standard "your pregnancy caused
your prolapse" presentation (yep!) and some inaccurate content about age
and lost muscle mass, blah blah.

My (first) favorite part was (and I am paraphrasing): Pelvic organ prolapse occurs because your pelvic floor is weak due to age or pregnancy
damage (never mind that women who've never had children, men, and
men who have never birthed children have a high incidence of PFD), and
while the pelvic floor organs don't weigh very much, these muscles fail
under the weight of your entire body. HUH?

Hello...Your pelvic floor organs are not holding the weight of your
body. That's what you have your skeleton for, as well as the muscles in
your legs, hips, and trunk. The pressure in your pelvis does not skyrocket
when you carry an extra twenty, thirty, or seventy pounds on your BODY.
This "logical" deduction is clearly derived by those who don't seem to understand geometry, physics, or the actual definition of weight very well.

Message to all: Extra weight on your body does not especially place
extra weight on your pelvic floor. (What DOES place large amounts of
strain of pelvic floor organs is downward pressure, which we'll discuss
later...)

Attention all "scientists" out there, or people who review literature for fun. One cannot deduce causation from epidemiological studies. You are not researching cause. You are researching characteristics. CORRELATION DOES NOT IMPLY CAUSATION! If you don't know what the previous statement means, then you should never use the terms "there is research that shows…" or "according to the literature…"

My (second) favorite part was the recommendation that, "if you have a prolapse and it doesn't bother you, you can go to your grave knowing you had a bit of prolapse." Ummmm, I guess that's OK advice if you notice your prolapse at age eighty-five. If you are less than eighty-five (say, twenty-eight, thirty-eight, forty-eight, fifty-eight, sixty-eight, or seventy-eight), I suggest you pay attention to the fact that your organs are thinking of vacationing south this year and take care of the problem before the seriousness advances to another stage or before they think about inviting their friends to travel with them.

My (third) favorite part was the recommended exercise (of course) the SUPER KEGEL.

Is it a bird? Is it a plane? No, it's the newest member of the comic book characters found on Daytime Medical Advice.

Seriously, if you didn't watch the episode, see if you have a friend who Ti-Voed it.

Evidently, when normal pelvic floor tone is not happening in your tired, old, over-birthed, over-stretched, overloaded (from an extra thirty pounds) pelvic floor, then do this:

Pretend you are a little kid trying to "hold it in." Cross one leg over the other and sit back by bending your knees (like in a little chair). Then, squeeze your butt and your knees together, and, of course, tighten your va-jay-jay (dudes, you can do this too—just tighten your pro-tay-tay).

Perfecto! You now have strengthened all of the motor programs for incorrect pelvic floor function.

Quick anatomy/exercise prescription lesson, Dr. Oz. This exercise is not a kegel. A kegel exercise is a very specific, clinical thing, that, when blended with other motor programs, changes into something entirely. In fact, people doing kegel exercises incorrectly exacerbates PFD, which is why physical therapists have to spend so much time helping people FIND their pelvic floor muscles instead of simply squeezing (guess what?) their butt and thighs at the same time. Nice one.

Now, I don't want to confuse my readers. I am not a fan of kegel exercises as a thorough, correctly designed program for PFD because kegels treat the symptom and not the problem. There is much (much) more to understand about your pelvic floor (i.e. how the PF is involved in gait, for example) than how to contract and release these muscles. There are more mechanics involved with what keeps items up and closed in your body than what is taught even in college-level anatomy and physiology. That all being said, you still need to be able to FIND your pelvic floor and know HOW to signal it to do what you want (i.e. relaxing your PF during vaginal delivery is essential!).

Do you need SUPER STRENGTH to have a healthy pelvic floor? Absolutely not. The MORE IS BETTER argument doesn't hold up with muscle tone. There is simply the right amount to optimize physiological function. MORE IS BETTER only applies to BBQ beef tri-tip sandwiches when one is thirty-one weeks pregnant. Or, is that just me?

I got a lot of emails about the SUPER KEGEL, which makes me happy, because it means that you are paying attention. And I also got to hear from my favorite Pelvic Floor Royalty, the Kegel Queen. As you can imagine, she wasn't that thrilled with Dr. Oz's prescription, either.

Next blog: Let's talk about downward pressure and how it pushes the PF organs out.

Next meal: Tri-tip sandwich.

The end.

Feb 22 2011   ACHING FOR AN ANSWER

Friends!

I was working on my downward pressure post when this email came through. I will still be posting my pressure post later today or early tomorrow, but I wanted to throw this post in while it was fresh in my mind. Sorry to clutter your inbox. Kind of.

Back in November, I received an email from a young woman suffering from bad menstrual cycles. I get this type of email often and I responded quickly for two reasons. First, she already had a long-term, structured approach to how she was going to solve a problem with a tool. Second, her note about monthly pain medication quantities gave me a tug. No one should be on regular pain medication, especially for a natural human function. Because many of her questions are similar to many of your questions, I decided to post both our email exchange and her results. NOTE: The emails are undoctored (with the exception of some formatting and correction of MY spelling ☺) and I was given Stephanie's permission to post.

November 26, 2010

Dear Katy,

I am going to do your five exercises in Down There For Women three times a day for three months (basically until Valentine's day) as an experiment to try and get rid of menstrual cramps. I am twenty-four and have regular cycles, no migraines, but if I didn't take triple doses of pain-killers, I would be debilitated for four days. I also switched the way that I sit and drive by tilting my pelvis forward and sitting squarly on the bones instead of tilting back on the tailbone.

What I want to know is, first, I saw on your blog that you had zero menstrual cramps after lots of hiking. Do you know any other success stories of people with zero menstrual cramps?

Also, do I have to do these five exercises three times a day for the rest of my life or, could I eventually maintain flexibility and strength by doing them once a day?

Thanks so much.

—Stephanie

November 27, 2010

Hi Stephanie,

Thanks for your email. We have had many people reduce or eliminate their cramps by doing these exercises, especially those like yourself who are young. Severe cramps are usually caused by extremely tight posterior leg muscles (everything from the hamstrings down to the soles of the feet.) I personally took myself from a two-day monthly debilitation to needing nothing but one hot bath the day I started. One year later I have taken no medications at all, and I used to be a two-to-five-Tylenol-a-month girl.

I think the barefoot hiking finally got into the tight lower leg issue I was having. The DVD is a great place to start, and you should notice a difference your first cycle. Please keep me posted!

The exercises are fairly simple. The bigger issue is WHY your muscles are tight. If you change the way you sit and stand (by thinking about what you learn on the DVD and applying it to "all the time"), you won't have such a chronic issue. The exercises to really learn to love are the single and double calf stretch. Those are going to be your new best friends and I do them two or three times a day (just a few minutes) because sitting and driving and shoe wearing just tighten everything back up again...

Hope this helps! Good luck...

Katy

February 19, 2011

Hi Katy,

You were right, the stretches for eliminating menstrual cramps really work! I tried a three-month experiment. I did notice a difference my first cycle: instead of taking a triple dose of painkillers for four days, I only took pain meds for thirty hours. Second cycle: fifteen hours. Third and most recent cycle, I never felt any pain at all. I took one narproxen the first day and the second day just because I was out and about and couldn't believe how successful this was—I thought it was too good to be true. YAY!!!!!!!!!!!!!!!!!!!!!!!!!!!!!!!!!!!!!!

Thank you so much for making that information available, I am so grateful!

—Stephanie

Why is this email so great? A dependency on painkillers to get through a day without pain is not a healthy model. The long-term cardiovascular and intestinal risks of taking painkillers and/or anti-inflammatories are well known, and if you are taking these medications for anything other than an irregular, acute issue, your body is sending you a message—SOMETHING IS NOT RIGHT. Most of the time, however, that SOMETHING is a slow accumulation of tissue damage due to inappropriate mechanical function, whether at the joint (big) or cellular (small) level.

Do you want to know the super-cool, amazing, wonderful thing no one is telling you? Joint issues are easily fixable (with diligence!). And that cellular-level ailment is being affected by the joint thing. Fix one, improve the other. Sounds too simple, I know, but, that's how the body works. The big muscles not only move the big joints around, but they also get the microscopic stuff to move around as well. Who knew? (ME, ME, ME, and you, you, and YOU!!!)

If you are having menstrual issues, you (like Stephanie) can optimize the mechanics of your natural pelvic functions with the Down There for Women DVD.

Why/how does it work? The posterior leg muscle and fascial system, when at the incorrect length (SHORT), keep the pelvis tucked posteriorly. This makes your "tubes" no longer oriented in the direction their muscles systems require to generate optimal force. Now these tissues have to work more for something that should be relatively easy. That tucking position also places your weight onto your sacrum through your sitting hours, which, over time, causes an increase in tension on the muscle and fascial systems of the pelvic floor. Now things have to pass through an extra-tight space, which, of course, means you feel everything more. The exercises are VERY simple yet most people will struggle with the extreme tension (positive heeled shoes, sitting in chairs, poor walking patterns) down the backs and inner portion of their legs. Open these tissues up, open your biological functions up!

Job well done, Stephanie, in searching for a solution and, more importantly, doing the work it takes to be well. Let me know if you'd like another DVD to try. This one is on me!

Now, stop bugging me with all of your emails, everyone. I have to write another post!!!!!!!!!

 PELVIC FLOOR PARTY:
ANNIVERSARY SPECIAL

June 27 2011

Once upon a time, there was a tiny little post called Pelvic Floor Party[1] on a blog called Mama Sweat. Did you hear of it? Well, a LOT of people did—it was kind of freaky—and it added an element to pelvic floor exercise routines across the globe. Because it has been a year, Kara Thom, the author of the original blog, and I got together (in cyberspace, because I have a newborn and she has, like, forty-seven kids) and interviewed each other.

We split up the interview, half on my site and half on her site. Here's one part below. Follow the link at the bottom to read the other part (including my favorite changing-table shoulder stretch.) I hope you enjoy Kara as much as I do. She's hilarious. You know, for a mom.

KATY: Okay, super-popular blogger-fit-mom. Just HOW MANY PEOPLE read Pelvic Floor Party, Kegels Not Invited? What was it like to have that many people read your blog? That's more people than have read anything, ever. Except for the Andrew Weiner Wikipedia entry. Tell me, did it make you make a lot of money?

KARA: I went from a little-known blog with around fifteen thousand visitors in my first year, to more than seventy thousand my second year, and the Pelvic Floor Party posts were responsible for that. Finding fitness in the chaos of motherhood is all well and good, but apparently, much better when done without peeing your pants. And, one year later—happy anniversary, Katy, darling!—it remains my most popular post. And since I just checked the stats, here's a few interesting factoids: the last three people who read Pelvic Floor Party were from Paris, Egypt, and Ireland respectively. We jokingly referred to it as the "post heard round the world," but it definitely is a message spanning the globe. It's not trending on Twitter yet, but it's hard to compete with the wave of political sex scandals.

Money? Have I made a lot of money? [She laughs heartily.] I have made exactly 0 dollars and 0 cents. In fact [she crawls onto her high horse], a large, prominent women's health website asked to buy the post for their site, but we couldn't come to agreeable contract terms (they wanted all rights and I wanted one-time rights) so I turned down the offer. I've also been approached by advertisers, but that just doesn't feel right for Mama Sweat right now (you're welcome, dear readers).

KATY: How did the squat/kegel information change you, personally? Did you really start squatting?

KARA: Oh goodness. Is that my pelvic floor conscience? I squat all the time and I shamelessly pee in the shower. I have not gone camping since that post, but I'm certain there would be no peeing on the shoes. That post also helped me put two and two together because I had been doing Cross-Fit workouts (famous for deep squats) for about eight months by then and had already noticed that my usual nagging hip pain/low back issues had disappeared. I knew I was stronger and it was helping but didn't understand until you 'splained it to me: a strong booty is the backup power for the pelvic floor and in my case, with hip flexors like clam shells, I need good glutes to pry them back open. But what I love about the information is that squatting, while good to include in a workout, should also be part of the way you move in general. If whatever I'm doing requires me to be low to the ground I rarely bend over now; I squat.

One thing, though, worth mentioning: once you get pelvic floor strength you still have to maintain it, as I learned when allergy season hit. I could safely get through one sneeze, but the double sneeze reminded me that my pelvic floor needed more attention. And, with my increase in triathlon training this summer, guess what I've done less of? Doing less strength training and yoga has also had ramifications for my afore-mentioned hip and low back. At my age, foregoing strength training and stretching is not an option.

KATY: What have you been doing since writing that blog? Lying around eating bonbons and stuff, or…?

KARA: What exactly is a bonbon? If it's dark chocolate with a touch of sea salt, then yes, I'll admit, that sort of bonbon crosses my lips occasionally. But the lying around part…hardly. Last summer my book *Hot (Sweaty) Mamas: Five Secrets to Life as a Fit Mom* went to press, and before it did you better believe I included this all-important info in the book. That post generated a few other exciting opportunities for me, including a video for *Experience Life* magazine, where I got to demonstrate the Bowman Squat for all to see.

I also wrote a yet-to-be-published article for *Health* magazine, that included squatting for pelvic floor health, as you know because I interviewed you! [Katy's note on Kara's squat: First of all, I remember typing Kara an email on my iPhone at 4:00 a.m. while on vacation in Hawaii to give her tips to make the squat "better" in terms of alignment. Note her bolstered knees and ankles. To work the glutes even more, the shins should be more vertical and less angled forward. Someday, I will get my hands on Kara Thom's psoas. Then she'll REALLY have something to post about!]

KATY: What is doing a book, I mean after writing it, like? I ask because I have one coming out in November and I thought all the work was done but then I see all the stuff you've been doing to follow up. What is it like having a book out? (Here's the cover. Cute, right?)

And, FYI, here's a picture of me finding your book in a California Barnes and Noble—and, guess what—my name is in it! Sweet.

You are awesome. What's it like to be awesome? That's a lot of questions, isn't it…

KARA: Let's start with, "What's it like to be awesome?" Here's the visual: I'm wearing sweatpants and a race t-shirt. It's early in the morning so no one has come in yet to do my hair and makeup, and I'll need a little makeup because I just popped a zit on my chin (I thought being awesome would give me clear skin.) Oh, what was that? No one is coming in to do my hair and makeup today? There are still toilets to clean and laundry to fold, noses to wipe and underappreciated meals to make, but I can't help but feel awesome anyway. I get enormous satisfaction that I'm writing about topics that interest me and matter to others, but mostly I get to do it while watching my foursome grow. One of my daughters asked me the other day: "Mom, will your book still be alive when I grow up because I want to read it when I'm a mom." So yeah, the post-book-birth process is as hectic, if not more, than the time spent writing it, but I love that it's out there and I'm privileged to have books to promote and children I adore, and often at the same time, like today when two childcare options have fallen through.

KATY: What's next for you? More kids? More marathons? More squatting? I want to know how being a super-fit mom and a super-fun blogger plans for her future.

KARA: More books, marathons and squatting for sure, but I've maxed out on kids. I love pregnancy, birth, and babies, but I have officially moved out of that stage and am focused on helping them become the people they're meant to be (I'm hoping at least one of the four likes running and triathlon as much as me, is that too much to ask?). This is just as fun and rewarding and I'm sleeping more at night (most nights) so that's a perk (except when I get sucked into twitter—@mama_sweat if you want to keep me up too).

To read the other part of the Kara/Katy coffeetalk, click here.[2]

---

1. mamasweat.blogspot.com/2010/05/pelvic-floor-party-kegels-are-not.html

2. mamasweat.blogspot.com/2011/06/hows-your-pelvic-floor.html

HIGH HEELS, PELVIC FLOOR, AND BAD SCIENCE

I have a twitter account and I'm not afraid to use it.

Last week I read this:

"@fitnessleak: study shows 3" high heels worn by women created stronger pelvic muscles lead 2 better sex @AlignedandWell what do u think?

For those of you who don't know twitter-speak, this means "Hey, Katy, what do you think of what Fitnessleak wrote—that a study shows that three-inch heels worn by women created stronger pelvic muscles that lead to better sex?

What do I think? What do I think? I think I want to put a hot poker in my eye.

This is not the first time someone has asked me my opinion about this article. I thought I'd answer it here, so you can all enjoy my answer. ☺ Once upon a time there was a high-heeled wearing urologist that stumbled across an article linking high heel use to schizophrenia (you should read the article, it is fascinating and just one more reason that I never wear anything with a positive heel!). This caused her great alarm as she loved her heels, even saying "although they are sometimes uncomfortable, I continue to wear them in an effort to appear more slender and taller." Because she could not refute the article, she decided to prove a healthy aspect of heels.

She and her team created a study looking at changes in ankle positions that change pelvic tilt (that's tucking or untucking your pelvis from a neutral position) and changes in pelvic floor activity. She measured this in fifteen women. It is important to note that she did not have any of her subjects in shoes. That's right. No shoes were worn by the people in the study. They just stood on wedges of different angles.

Here's where it gets a little nerdy. She based her study on a previous study looking at the same thing: how does ankle position affect pelvic position and how does pelvic position affect pelvic floor contraction. I'll call this the Chen study, after the lead researcher. The Chen study found that each ankle/pelvic angle brought about a different PF activity. They measured both resting and maximal (trying to squeeze it the tightest) force production.

The results? The greatest PF change came with DORSIflexion, which is the *opposite* of being on your toes.

Dorsiflexion looks like this (image top right):

All measured dorsiflexion angles had higher PF activity than plantarflexion.

Plantarflexion looks like this (image bottom right):

Dr. Highheel's study also measured both resting and maximal contractions and here are her results (see table below):

The left column shows ankle position. HS is flat, the P stands for plantarflex (what happens in a high heel) and D stands for dorsiflex (what happens in a negative-heeled shoe).

What do you see, my little research students?

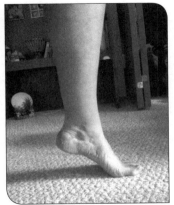

1. The greatest resting activity in the PF is, like the Chen study, higher with heels down, not heels up.

2. The maximal contraction (squeezing it hard) comes from five degrees of plantar flexion (less than a half-inch of heel), not those ankle positions simulating two- or three-inch heels.

Position	Resting contraction (mV) (mean and IR)	Maximal Contraction (mV) (mean and IR)	P value
HS	32 (28-40)	278.5 (154.4-459	<0.001
5P	35 (28-41)	322 (154-446)	<0.001
10P	35 (28-45)	320 (133-436)	<0.001
15P	40 (33-49)	316 (186.5-386)	<0.001
5D	45 (34.57)	283.5 (157-358.5)	<0.001
10D	47 (41-59)	267 (242-360.5)	<0.001
15D	58 (43-68)	233.5 (163.5-329)	<0.001

The conclusion of the study: Ten degrees of dorsiflexion (opposite of heels, remember?) might be just as effective as the fifteen degrees reported by Chen, but more comfortable. Low heels—not high heels.

How does one walk around in dorsiflexion? I don't know. The researchers don't know. It's just a study, not a practical guide on how to use the information! And remember, they aren't measuring walking—just standing on an angled board for ten seconds in each position.

So we went from studies by Dr. Highheel and Dr. Chen, showing that dorsiflexion in the ankle gives greater resting PF readings than plantar flexion, to the headline:

Three-inch heels increase pelvic floor strength and better ooooh la la!

This is how: Dr. Highheel writes a letter to the editor to the *Journal of European Urology* called: "Women, Pay Attention to Shoe Heels: Besides Causing Schizophrenia, They Might Affect Your Pelvic Floor Muscle Activity!!"

(Catchy title, right?)

In this letter she says that she has done research (the research we just looked at) and that despite being linked to schizophrenia, heels might be beneficial as noted by her data. She fails to mention that dorsiflexion has been noted to have greater impact in the two conducted studies. Fails to mention that of the positive ankle positions measured, the five-degree (the lowest) measurement of heel was "the best."

Boo. Bad scientist. Wanting heels to be healthy (as she stated she does) skewed how she presented the material to the public. That's a huge (HUGE!) no-no.

Of course, the media picked up on her letter to the editor and didn't follow up with reading the research portion.

To clarify:

No shoes were used in the study. High heels (heels at all) were never measured.

Negative ankle position measured greater PF activity than positive ankle positions, not the other way around.

And, I have a bigger issue with both the Chen and the Cerruto (Dr. Highheel) studies.

How is pelvic floor contraction measured? With EMG (electromyography). An electrode picks up on the quantity of electricity flowing through a muscle and converts it to forces the muscle generated. Ideally, EMG is measured by an electrode placed into the measured muscle with a tiny needle. But not many people want a needle jammed in their pelvic floor, myself included. So researchers (and therapists) use a patch that reads electricity around the area of a muscle, through the skin.

The problem with this method is that the patch picks up every other muscle contraction in the area as well. It's called crosstalk. And what is the nice neighbor of the PF? The lower glutes. These are the muscles you use when you've had five cups of coffee and three bran muffins and you're sitting in traffic. They are also the muscles that you use to tuck the pelvis under. Go ahead and try it. Pretend your bowels are going to explode and you'll feel those low, low glutes squeeze your cheeks together.

Any change in ankle position, up or down, causes the pelvis to tilt from neutral, which causes the low glutes to fire. This means your PF data is automatically tainted (hee hee) with glute crosstalk. Dear researchers: You're not even measuring what you think you are. You're not isolating the variable!

Muscle geometry and biomechanical science demonstrates that neutral pelvis will give you the optimal pelvic strength (not too much tension and not too little) every time. Any heel, high or low, causes your pelvis to tilt from neutral. I'm not a fan of walking around with my toes higher than my heels, or my heels higher than my toes. I like to be more grounded than that. Think flats or minimal footwear. But if you love LOVE shoes, go with Earth Kalso bottom. The negative is at least a little better than the positive.

And, take note on what you read. The story—"High Heels Increase Pelvic Floor Strength"—was reported in just about every prominent health magazine, newspaper, and television network. No one bothered to check the research.

When I moved about three weeks ago I finally chucked my last pair of heeled shoes. Note the cobwebs...

 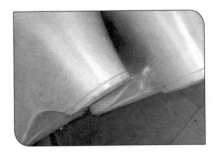

Schizophrenia? Really? What more do you need to hear?

C.H. Chen, M.H. Huang, T.W. Chen, et al. Relationship between ankle position and pelvic floor muscle activity in female stress urinary incontinence. Urology 66 (2005) (288–292)

M.A. Cerruto, E. Vedovi, S. Dalla Riva, et al. The effect of ankle inclination in upright position on the electromyographic activity of pelvic floor muscles in women with stress urinary incontinence. Eur Urol Suppl 6 (2007) (102)

Flensmark. Is there an association between the use of heeled footwear and schizophrenia?. Med Hypotheses 63 (2004) (740–747)

Women Pay Attention to Shoe Heels: Besides Causing Schizophrenia They Might Affect Your Pelvic Floor Muscle Activity!! Maria Angela Cerruto, Ermes Vedovi, William Mantovani. Accepted 15 January 2008, Published online 24 January 2008, pages 1094–1095

Surface EMG crosstalk evaluated from experimental recordings and simulated signals. Reflections on crosstalk interpretation, quantification, and reduction. Farina, D. Merletti, R. Indino, B. Graven-Nielsen, T. Dipartimento di Elettronica, Politecnico di Torino, Corso Duca degli Abruzzi 24, Torino, 10129, Italy.

chapter 5

GETTING TO THE
CORE OF IT

DOES "CORE STRENGTH" MATTER?

Q: I am so sick of everyone talking about core strength. I don't even really know what that means! Why can't I just run? Do I really need to vary my workout? I have never had a problem.

A: That's a lot of questions! Where to begin…

You are right, core strength is everywhere right now, and rightly so. It is probably the most critical part of an exercise program if you are looking to continue moving without pain well into your older years. Due to the complexity of the mechanical function of the trunk, it is also the most poorly understood component of exercise instruction. Core strength does not mean abdominal exercises! It is the ability to stabilize the bones in the upper body, rotate the torso with proper spinal curvature, and maintain pelvic position while sitting, standing, and exercising! It's the ability to control the bladder, stabilize the ligaments of the knees with the lower abdominal wall, and breathe correctly while doing all of these! Quite a lot of movement skills fall under the category of "core strength," all of which you need both now and in the future.

You also ask why you can't run. Or "just run," actually. While there are many people who use running as a fitness program, it is usually not conducive to longevity, meaning you won't be able to do it for the rest of your life. And the forces generated by running cause damage to the renewable resources of your body—often causing irreversible damage.

By not varying your program, you will use the same motor pattern over and over again, creating overuse injuries in the muscles you do use, and weakness in the muscles you don't use. If you round out your program a bit by strengthening the muscles not utilized in running and flexibility/mobility exercises for maintaining normal range of motion of all joints, you are much more likely to be healthy and well long into your advanced years.

AN OMNIVORE'S DECISION

I love food. Not just any food, but really fine-quality ingredients. And I love health. Some people (my mom) would even say that I am a health *expert*. And because I speak and write regularly on topics of human science,

I always seem to get the question: "What should I be eating?"

And I say this. The biochemistry of food and its effect on the insanely complex functions of metabolism and human anatomy are so complicated, most diets are approximations and theories that seem to work well for some and not as well for others. Those with the most pristine diet can still suffer from the same diseases as those that eat McDonald's every week. The guy that eats bacon for breakfast may not have any cardiovascular disease, while the vegan can still battle high cholesterol. There are factors beyond the calorie, beyond the carbohydrate quantity, and beyond the grams of fat. In addition to the tough science, politics have always played a role in food prescription, and once you've got profits coming from food recommendations, you should know that the bias may not be in your favor.

I don't usually answer the What To Eat question beyond the basics—avoid processed foods, stick to fresh ingredients, and cook for yourself! And now, thankfully, I don't have to do anything but point inquiring minds to *Food Rules*, a new book by Michael Pollan (*The Omnivore's Dilemma*), who was a featured expert in the documentary *Food, Inc. Food Rules* is a five-dollar pocket book of guidelines on what to eat. Not only are the guidelines brilliant, the simple information may be the best weight-loss program for anyone trying to shed a little body fat and increase overall body performance.

Perfect for vegetarians or Argentinians (those beef-eating South Americans!), *Food Rules* will tell you how to shop (around the edges of a grocery store ONLY!) and what to buy (avoid eating anything with any ingredient your sixth-grader can't pronounce). I won't give too much away (you can read the entire thing standing in Barnes and Noble), but I will say that yes, Pollan suggests having a glass of wine a day. An entire bottle of wine, however, would not be part of the perfect dinner for one.

This week I have added rice and salad every day (I am shooting for foods that have fewer than five ingredients...and rice only has one!) and will attempt small fishes (sardines) later this week. Next week I am attacking bread. Try to find a bread with only whole ingredients (stuff you can grow in your yard) and you will find yourself at a bakery three times a week. Too expensive.

You know what's not expensive? The ingredients found in bread! Who knew?!

This weekend I will get the electric bread maker (still in its box!) I got as a gift a few years back and join my father, who bakes his bread fresh twice a week. Am I really sooo busy that I can't throw some flour, yeast, salt, and grains into an electrical appliance and let it mix, knead, rise, bake, and cool bread for me? (What, this machine won't cut it, too? I want my money back!)

Food, coupled with our beliefs and preferences, is a huge issue for most people. Until we have scientists dedicated to researching food from the perspective of pure biochemistry (and not which food additives won't kill you) and time dedicated to personal gardens that generate even ten percent of our daily caloric intake, this issue will continue to weave itself heavily into our current health care issues (mark my words). Do you want to reform health care? Start with the personal responsibility of becoming a conscious (food) consumer! Join me, and take the thirty-day, eat-by-the-*Food Rules* challenge. You can do it!

P.S. Does anyone know what to do with sardines???

If you have some time, a DVD player, and a good library card, check out these movies and books:

The Omnivore's Dilemma, by Michael Pollan

Animal, Vegetable, Miracle by Barbara Kingsolver

Food, Inc. (2008)

King Corn (2007)

March 19 2010 THE BEST ABDOMINAL EXERCISE YOU'RE NOT DOING

I'm betting there are plenty of core exercises you aren't doing, but today we're going to talk about the BEST one you're not doing.

I've taught exercise for fifteen years now, and I have seen people crunch, flex, strain, bulge, and jut just trying to get their stomachs in shape. And most of that effort was in vain. The natural tone of the abdominal muscles will kick in when the pelvis and ribs are in the right position. It's true! Here is another fun fact: The strength of the abdomen always matches the strength of the back. In trying to maintain "good posture" we tend to pull the shoulders back and slightly arch the lower mid back, creating tension and stiffness in the spine just above the waist. Now you have a tight and weak back, with over-stretched and weak abdominals on the other side, or what I like to call "a back spasm in the making."

When we think of a strong stomach, we usually picture the long, up-

and-down "washboard" abdominals (rectus abdominus), but it's actually your oblique and transverse abdominal muscles (they run right to left and on diagonals) that give you a waist. If you don't have good waist definition, a crunch can actually widen your middle. Not what you had in mind when you bought that last "Three Minutes to Better Abs" VHS, I know.

Looking trim in the mid-section is a great bonus, but the true benefit to a waist that is equal to or less than your hip measurement is the stress it removes from your cardiovascular system. The smaller arterioles in *all* your muscles should be holding their share of blood, reducing the pressure in the larger arteries. The tighter the muscles, the less blood they hold, and the higher your arterial blood pressure. The abdominal aorta is most susceptible to plaque accumulation because the pressure there tends to be very high. It tends to be high because most of us have a very tight mid-section as a result of the excessive time we spend in hip flexion (sitting), the lack of rotational movements we have in our daily lives, and the tension we have in the back (holding ourselves upright with the spine instead of the glutes and hamstrings). Tight, larger abdominal muscles (obliques and transverse) can't aid your cardiovascular system in the way they should—which is why a larger waist size is a risk factor for coronary heart disease.

To start whittling your waist and decreasing the pressure in your arteries, try a simple spinal twist.

Lie on your back and bring one knee towards your chest. Try to take (don't force it!) your knee to the ground, leaving the rest of the body where it started. If you waist is very tight, you won't twist so much as roll to get the knee to floor. You can tell you rolled if your spine and shoulders all rotated over with your hips. Ideally, your shoulders and chest all stay planted on the ground and your waist muscles are long enough to allow your pelvis to move. If you are a "roller," do this exercise every day until you start to twist, decreasing waist size, increasing the strength of the stomach and back musculature, and decreasing your blood pressure. Make sure to do both sides. (This is also a great exercise to improve your golf game, too!)

Today let's travel down biomechanics' sister path, biochemistry.

For the last five days I've had a bad taste in my mouth. Not in a meta-phorical way, but in an actual way. It isn't an overall bad taste, but a spe-cific, metallic taste along the most posterior (back) and sides of my tongue. It started off as a little metallic, but just kept increasing. Now I can barely taste regular food—my tongue is "sweating" this bad taste.

It took me five days until I decided to do something about it. My water intake is good, but I drank four extra glasses today. I scraped my tongue, as recommended by Asian and Ayurvedic medicines. I brushed and flossed as recommended by the American Dental Association. Nada. So then I did the next logical step. I searched my symptom on the internet.

When I search for physical issues on the computer, I like to play a little game called "how long before my Mac tells me I have cancer?" It's pretty fun. See how many pages you have to look at before you get that diagno-sis. (Which reminds me of the time last week when I found tons of blood in my stool and looked up "tons of blood in the stool" on the internet. I was knee-deep in colon and rectal cancer before I remembered I had eaten five whole beets the day before. My bad.)

Well, this time my search turned up surprising results. Metallic taste is the mouth was ALL OVER the computer, linked to pine nuts. *Quoi*? And the weird thing was, I had just purchased an entire bag the week before and proceeded to make pine nut and pesto (made with pine nuts) pizza. I ate pine nuts in my oatmeal all week. I was eating them out of the bag. I had never eaten so many pine nuts! It was true, I was having a physiological event stimulated by pine nuts.

A little note about me: I am like a two-year-old when it comes to asking "Why?" You will notice kids have a tendency to continue to ask "Why?" because you haven't actually given a satisfactory answer. They want to know the mechanics of it all…and so do I.

Q: Why do I have this taste in my mouth?

A. Some people get this taste in their mouth when they eat pine nuts.

For some, this is satisfactory, but for me, I am very rarely satisfied with that type of answer. What about the mechanics of it? One explanation for this pine nut phenomenon is that the nuts may be rancid or tainted. But there were at least two of us who ate the implicated nuts, and I was the only one with a mouth full of metal. Rancid? Not apparently. At this point, my tongue is still sweating, but I no longer care. I am on a mission to know what are the chemical properties of the pine nut.

Pinolenic acid in pine nut oil stimulates the production of cholecystokinin (CCK), a digestive hormone that helps break down fats and proteins. Cholecystokinin, literally translating to "move [*kinin*] the bile [*chol*] sac [*systo*]" (from Greek), stimulates the contraction of the gall bladder and helps it open, to release bile. If you don't release bile, or an optimal amount, your body is unable to digest and process dietary fat. Woo hoo! Eat fat and not have it process? Sounds good, right? Well, not really. If you aren't processing the fat, your body doesn't register that you have eaten, meaning your appetite will continue to rage on. Probably why CCK is beginning to be researched as an appetite supressant, and why low-fat diets are unsustainable without eating huge quantities of sugar (and lots of cardio exercise). Another bummer related to not processing your fat? Dermatitis, or skin inflammation. Dermatitis can range from uncomfortable to a big deal. Until the 1950s, the lineolic acid found in nut seeds was the primary treatment for dermatitis. Neat.

Additional info:

- Pine oil is used regularly in treatments of gastritis and diabetes in Russia and China with double-blind, clinical results published in their medical journals.
- Increased fat metabolism causes the taste buds on the back of the tongue (Ah-HA!) to increase the secretion of lingual lipase, a digestive enzyme that helps increase the breakdown of fats. I wonder if it tastes like metal?

According to the literature, I have "pine mouth," an unknown reaction to pine nuts caused by rancid or bad nuts. I say hooey. I think I am being alerted to biochemical processes going on in my gallbladder, small intestine, and evidently, my mouth. Why me and not anyone else who ate this bag of nuts? I have a history of dermatitis that has been resistant to every treatment under the sun. I have been known to shoot oil straight through my digestive track, where it comes out unprocessed on the other side (a lot of bathroom talk today, yah?). I'm not processing oil very well. My delta-6-desaturation process (the breaking of the fats) seems to be impaired. Could it be genetic? I do know this desaturation process requires magnesium and zinc (metals!!!), of which I tend to run a bit low. Maybe the sudden increase in fat metabolism has the temporary effect of metal mouth, or, maybe I'll go into anaphylatic shock this week. Do I understand it all? Nope. But the science seems stacked in my favor.

PS. If I don't blog tomorrow…send help!

March 27 2010 IT'S A GAS

Can I write a blog post on farts?

I think I can.

Are you feeling a bit uncomfortable right now?

Good. That's kind of my point.

Gas is a natural phenomenon experienced by every person on the planet, although the quantities can vary between "a little bit" and "my Dad." A little known fact about your digestive system: Your digestive tract, from beginning to end, is considered to be outside your body. Essentially a long tube, anything you put into your mouth needs to be broken down into tiny particles (first with your teeth, then with chemical enzymes, then with the washing machine–like agitation of the stomach) so the energy units, minerals, and vitamins can move across the tubing walls and into the body. Food with very low nutritional quality basically stays outside your body. Even though you eat it, very little energy can be extracted. The result of regular, low-quality food consumption is an increased need to consume more. Calories don't really make a difference when it comes to satiety (feeling biologically satisfied), as artificial fillers don't even make it inside your body. They just shoot out the other side. (Diet tip: Try eating only whole foods for a week and notice the decrease in appetite and the surge of energy!)

Gas is created in a few ways. It is the sum total of the air you accidentally pull into your stomach by mouth or nose, the by-product of digesting certain foods, and what is created as a result of incomplete digestion. Normally, oxygen you inhale should be funneled into your lungs. Gulping air while eating, however, can cause air to move into your stomach instead. Any air you take into your mouth will have to get out somehow, and a fart is a natural equalizing process. If the air *doesn't* leave your intestines, this trapped air does what air tends to do to baked goods and wet towels. It dries them out! Healthy intestinal walls are moist. Diffusion of nutrients and energy need the water to move across the membrane. Dry intestines mean poor nutrition, even if you are eating the perfect diet.

Certain foods can make gas, as can food that takes too long to digest. Because food is trying to move down, slouching or reclining while you eat decreases the ease of downward movement. The longer food takes to digest, the more it festers in your tract. The more it festers, the greater the gas. Keep your spine up tall when eating and take an easy walk around the block just after to help food move down quickly.

Here's an algebra problem you've never seen before: GAS + TIGHT MUSCLES = A FART (See, math is fun!)

If you have audible farts (which means you can hear them), you can bet that your pelvic floor muscles are too tight. Passing gas is A GOOD THING. For the health of your intestines and to optimize energy pathways, you need to get gas out ASAP. Because gas is perceived as rude or embarrassing, we tend to tighten our anal sphincter or butt muscles (glutes) to keep it in. This is not only a bad idea from a nutritional perspective, but also from a musculoskeletal one. The gas-holding-in muscles are also the pelvic floor muscles. If you have been holding gas in, you may have noticed gas worsening as you get older, which means you have probably been tightening the muscles even more. Although we tend to think tight muscles are strong muscles, this is incorrect. Tight muscle fibers are just as weak as over-long muscle fibers. I can also tell you that the more you can hear your gas escaping, the tighter (and weaker) your pelvic floor muscles are.

Tips:

1. Don't talk while your mouth is full. This increases the intake of gas into your stomach.

2. Eat dinner with your torso upright and avoid tucking your tail under or rounding the back forward.

3. Farts are a natural biofeedback exercise. Work on developing control over your PF muscles. Can you relax the gas-stopping muscles and decrease the sound of air passing?

4. Do not hold gas in! Get up and excuse yourself until your PF muscles are relaxed enough to allow gas to leave silently.

5. Share this blog with your family, friends, and coworkers. That way you can all fart away and feel good about it!

WARNING: Silent gas can be a potential weapon when you use a little technique called "crop-dusting." Best used while walking down the aisle of a boarding airplane.

 ## WHEN THE POOP HURTS

When you are constipated, what does it mean? Not enough water? Not enough fiber? Not enough exercise? While these things can definitely impact your, uh, impact, many people are still plugged up after addressing these items.

The question we need to ask is, what else is going on down there? What other things could be affecting the downward movement of food? The answer: You may be creating an upward force that is preventing the natural movement of your food. What feels like constipation may be YOU, holding your food in place. Now I know you aren't doing this on purpose, but you still may be doing it.

Upward Force 1: Are you holding your stomach in? There is a definite difference between contracting your transverse abdominals (the corset-like muscles under your belly button) and sucking in your stomach. The telltale sign of "sucking it in" is a tucked-under pelvis. If you don't have butt muscles or a natural curve to your low back, you may be mistaking abdominal work for diaphragm contraction. Sucking in your stomach will give you the appearance of a flat belly, but really you are pulling all your festiveness upward. Great for the waistline, but not so good for the waste line (hee hee). This upward action is a force that works against the digestive process. You are basically pushing food back into your body, so when you go to the bathroom there is no relaxed pathway out of the body. You have to strain (against yourself) to get it out.

TRY: Standing and relaxing your belly, letting it drop all the way out and down. If you do this in front of a mirror, you will see that your stomach moves down lower than the level of the belly button. This is a great self-check to see if you are, in fact, pulling it *up* and not *in*.

Upward Force 2: The psoas is a very large muscle that attaches from the spine to the front of the leg. Research has shown this muscle shortens in response to stressful or anxious situations (ever been in one of those?). If you sit a lot or have spent many hours performing athletic activities such as cycling or using non-natural movements on treadmills and other cardio machines, the psoas can become very tight. Because the psoas runs along the entire digestive system, the tension created will also create a "standstill" in the intestines. To evaluate the length of your psoas, try this exercise.

Begin with one leg bent behind you. Just getting into this beginning position may reveal how your psoas tension is affecting your knee flexion (bending). Make sure to keep the knees squeezing together. (Image below left.)

Because the psoas also crosses the hip joint, we need to add hip flexion to fully test your range of motion. Bend forward to place your palm on the seat of a stool or chair. Work on holding one minute. If your psoas is tight, your spine will bend forward, but your back leg will not move. (Image above right.)

Ideally, the thigh of the bent leg should stay in line with the torso (see how the thigh has lifted equal to the distance the torso has travelled forward?) and lift up as you move forward. Eventually your psoas should yield (release) enough for you to touch the floor (!), but the chair is a nice place to start. (Image below.)

SO! Let me know, HOW TIGHT IS YOUR PSOAS?? If you are having constipation issues, know that nutrition is important, but so are the mechanics of what MOVES the food down!

WHAT A WAIST!

All right, all you biomechanists-to-be. Here's your first test. Can you tell the difference between Picture One (image below left) and Picture Two (image below right)? (Don't cheat…really take a look☺.)

Take a close look at my low back and on the left you will see a lumbar curve (a "sway" in my back) and on the right, nothing but a straight line. Now let me tell you what's going in the muscles beneath the skin.

In the picture on the left, I've sent a signal to my transverse abdominals (TVA), which are muscles that, when contracting (shortening), lift the abdominal wall up toward my spine (or if standing, back toward the spine). Awesome. Now I have increased space in between my lumbar vertebrae, my glutes are in a position to be active (especially if I was upright), and I've flattened my abdominal wall, not only to look fantastic by the pool, but also to start a series of microbiomechanical changes that actually have a positive effect on plaque accumulation in the abdominal aorta (but that's for another time!). TVA activation is a win-win situation for all systems and your body really needs this muscle firing constantly for optimal function.

The picture on the right, however, is what most people do when they *think* they are using their stabilizing abdominal muscles. The beauty of biomechanics is how clear it is to see what's going on. If you see zero lumbar curve, then you know, for sure, the TVA was not used (there is no physical way it can tilt the pelvis). In the picture on the right, I see (and hopefully you do too) a pelvis that has moved into what is called a posterior tilt (tuck). The evidence is in, and the abs are out.

Most people have replaced deep abdominal activity with "sucking their stomach in," and think they are constantly using their muscles. In actuality, the sucking in motion is a pressure (like creating a vacuum) that pulls the abdomen's contents up (not in), displacing the guts up against the diaphragm (hiatal hernia, anyone??). You get a flat stomach that looks fantastic by the pool, but you also get a tucked pelvis (do we need to talk more about where pelvic floor disorder comes from?), no butt muscles used when walking, no real work done in the abdomen, and excessive friction in the lumbar spine, hips, and knees. That friction is called osteo-arthritis.

Here's another side effect of sucking it in: The constant, upward generation of movement caused by sucking it in makes downward-moving processes like digestion, lower-leg (and pelvic organ) circulation, and monthly menses more difficult for the body. For example, many people strain on the toilet because they don't realize they are keeping these processes at a standstill. Prolapsing ladies, Stop Sucking It Up, and learn how to muscle it in instead. There's nothing that will push an organ out more than straining to bathroom. Check in with your abdomen and make sure it's relaaaaxxxxeeeed.

You're probably not going to enjoy the reality of letting your belly go. If you've got an extra twenty or more pounds on your body that you've been sucking up, letting all of that mass out of your trunk (do you hear your organs breathing a sigh of relief?), will show you what you've go to work with. And, it's going to be ok. Here's the cool thing. That mass that you've been sucking in will go away now that you're allowing the muscles underneath it to work, increasing your metabolism with trunk, butt, and hamstring muscles. It will get better and you will be healthier, right away!

Now that you're no longer sucking in, it's time to practice actual TVA activation.

The Exercise: (Use a mirror if you've got one handy...)

Start on your hands and knees, with your hands and knees below your shoulders and hips (image at right). Let your head relax completely and breathe in a relaxed manner. Now, let the wall of your belly completely drop toward the floor (you get extra points if it touches!)

and let your spine unfurl, making sure that you release your pelvis completely, lifting your sitting bones up toward the ceiling. Hang out there for thirty minutes. No, just kidding. But hang out for a little bit playing between where you feel most comfortable (your habitual position—watch what you do with your pelvis) and totally relaxed.

Now, exhale and pull your belly button up toward your spine, but do not let your pelvis move. See how long you can keep your belly button up, without holding your breath, without moving your pelvis, and without sucking it in. Can you feel the deep abdominal contraction? Do this a few times every day in this exaggerated position, and then, take it to the real world. When walking and standing, do the same thing. When squatting with your tailbone out, you can also add TVA activation.

Moms-to-be, this is a great exercise to practice while you're pregnant as it can prevent (or repair) diastasis recti. Another fun birthing fact: The TVA is the best birthing muscle as the TVA's fibers works in the same direction as the uterus does during the expulsion phase of delivery, only with a lot more force. So many moms-to-be are never taught about relaxing their abdomen and have deeply ingrained Sucking It In habits that get in the way of trying to push a baby out. Practice releasing now and make delivery that much easier! (That's my baby shower gift to you all ☺. I can't afford a thousand pairs of booties and onesies and other things that end in "-ies.")

 July 10 2010 GUT INSTINCT

Today, a rare treat, is an entire day off. My choice for the day was to sleep as much as I'd like (til 7:30!), get up with a cup of coffee and a mystery set in an Amish town (I have NO idea how this book happened into my house), and spend a couple of hours dedicated to my hunting and gathering lifestyle-in-the-making.

I hitched up my backpack (no asymmetrical shoulder loading for me), my iPod Nano (well, it *is* four years old, so that makes it kind of primitive AND I was listening to songs from the eighties, so there…), and I set out for the 2.5-mile walk to the farmer's market downtown.

Not shopping for anything in particular, I allowed myself the luxury of purchasing whatever I fancied, without thinking "what will I cook this with" or "what will this item go with," etc. Walking back, smelling the aromatic hodgepodge of ingredients, I took a mental look at what I had selected. And then I took an actual picture to show you what I bought:

Cherries, plums, pomegranate juice, olives, tomatoes, basil, cilantro, onions, avocado, and hummus.

Walking home, I was wondering what made me choose these particular items. I love *all* fruits and veggies and am especially fond of the peaches, nectarines, artichokes, and asparagus that are in season now, but they didn't make it in the bag. Why not?

A quick (okay, honestly, two-hour) research session into the biochemistry of my selected groceries showed a heavy, *heavy* concentration of potassium, vitamin A, lutein, anthocyanins, and melatonin. Why had I gravitated toward these items?

This last month has been pretty busy and, well, constant. I haven't rested well and have been suffering a low-grade and continuous ache behind my eyes. In search of a solution this week, I had a great massage, eliminated all sugary items (desserts, not fruit), and kept myself to one cup of coffee or tea a day, but still, the eyes! I was pretty surprised that, when following my gut instinct, I had managed to pick the top two foods recommended for macular degeneration prevention (lutein), one of the few food sources for melatonin (cherries), and a selection of anti-inflammatory foods and compounds that have been shown to relax the blood vessels. And after drinking half my juice, and eating half a container of olives (okay, really, I've almost eaten all of the olives), my week-long headache is gone. Really. (Note to self: Study "Food as Medicine" in free time.)

I had also, according to registered dietician Marilyn Sterling's December 2001 article on anthocyanins in *Nutritional Science News*,[1] selected plant pigments (deep red/purples) that were "eaten in large amounts by primitive humans." Nice addition to my hunter/gathering-type day. Wow, I'm totally retro. Forget the seventies, I'm more like the -100,000s. With an iPod. Wonder what they wore, and if BC fashion is indeed going to come back into style. Bare feet and squats certainly have, so stay tuned.

More from the article: "Anthocyanins are the active component in several herbal folk medicines such as bilberry (Vaccinium myrtillus), which was used in the 12th century to induce menstruation and during World War II to improve British pilots' night vision." All right, then. Anthocyanins are apparently multi-purpose.

When looking up nutritional information on these foods, I had to dig pretty deep into some biochemical research (not my favorite science!). Through the digging it became pretty clear that our Western "nutritional science" has a very limited way of classifying and quantifying foods (macronutrients, vitamins, and calories), which may be quite elementary when compared to food classification systems in the East, which really examine the effects of the entire food on the entire person. Most scientists seem to be examining the result of food when burned by fire, not quite an *in vivo* study, as each individual has unique chemistry (and further unique chemistry in that moment) that will influence the effects of food.

For example, the avocado, rich in potassium, gives me the cellular-balancing (to oversimplify quite a bit) and optimal muscular or action-potential function I need. So why not pick a banana? (I mean, besides the fact that it had to be flown here from another country, causing way more pollution than its potassium may be worth.) Bananas are astringent, which means the tannins found in them can shrink and contract mucous membranes. I am sure there is a time and a place for the astringent banana, in the case of ulcers or diarrhea (I never spell that word right, so thank you, Spellchecker Fairy). But since I don't have those things, the banana would not be the better choice for potassium, *for me*. Which doesn't not mean that the banana is not right for you. Which starts to make one think that diet should not be prescribed for large groups of people, but for the individual, in a particular place, in a particular time in their life, in a particular situation (disease, fatigue, pregnancy, etc.). Even general macronutrient guidelines can be significantly "off" (do we have to talk about the whole "fat-free" fiasco?).

There are so many more components of diet than the protein/carb/fat/calorie model. It's amazing! And even more amazing to me is that when listening to my body, my hunting-and-gathering gut *knew* what I needed, and I bet yours does too. Next time you're out at the market, see what you gravitate toward, and notice what your kids or spouse would prefer. They've got unique chemistry as well, and their choices may provide you with insight to their "mechanics."

Next week: Stalking Cookies, The Video. (Warning: Carnage maybe too graphic for young Vanilla Wafers.)

---

1. newhope.com/nutritionsciencenews/NSN_backs/Dec_01/antho.cfm

THE PSOAS, LIZ, AND ME, THE QUEEN OF DOWN UNDER

A while back I received a call from Liz Koch, who runs a website called coreawareness.com, wanting to know a bit more of my thoughts on the pelvis. She had read the Mama Sweat article on squats instead of kegels and had long been teaching how the tailbone-out position (which is called anterior pelvis or bootylicious, depending if you're talking to an anatomy professor or Beyoncé) is required for correct psoas function.

Psoa muscle is one of my favorite topics to teach. This muscle has such a rich history in anatomy, and when working correctly gives the body tons of stability. When not working correctly, it can wreak havoc on the knees and hip joints, degenerate the lumbar spine, and increase fracture risk of the thoracic vertebrae. Each psoa also attaches to the intervertebral disks of T-12 and the lumbar spine. If your disks are prolapsing into the body, constant psoai tension can be the primary cause.

The word *psoas* is interesting, even before getting to what it does. In modern Greek, the word *psoa* means "pertaining to the loin." In the 1500s, French anatomist (but obviously not Latin language expert) Riolanus began to refer to the pair of psoa as the *musculus psoas*. Because there were two. Technically, two psoa make a psoai, which was the more common form used by Hippocrates and other ancient Greek physicians. That's why you will hear me refer to the psoa or psoai when I do a podcast-tele-interview with Liz Koch next week. But wait, there's more. Before the word psoa was used, ancient Greek anatomists referred to this muscle as the origin or womb of the kidney.

νεφρομητρς = origin of kidney

νεφρο = *kidney*
μητρς = *womb, origin, source*

This was a compound between "kidney" and the second part, which means *womb, origin, source*. FYI, trying to get your computer to type in ancient Greek is a time-consuming task, but I thought you'd appreciate it. And no, this information did not come from Wikipedia (ew), but from (NERD ALERT!), the *Lexicon of Orthopaedic Etymology*, which, yes, sits by my bed. Thanks for asking.

Issues of the psoas are often considered emotional in root by almost every somatic therapist and non–Western trained medical professional. The adrenals, or organs of reaction, as I like to call them, are located at the top of the kidneys (*ad* = above, *renal* = kidney). One of the psoai attachments is located just at this location. Anecdotal data shows that the curling tendency (spinal flexion) of the body occurs when stressed, just as a cat (and many other animals) flexes its spine to appear menacing when threatened. Is it to appear larger? Is it to protect the vitals? No one knows for sure why it happens, but still the reflex is in our biology. Turns out those ancient Greeks were pretty smart. Just watch *Clash of the Titans*. Not the new one, but the old one. Harry Hamlin. Meow! It's chock full of history.

There are also structural reasons the psoai do not yield properly, like if you use them to hold yourself up all the time. Forward pelvis, anyone? If the psoai released you would fall over, so it turns out your psoa tension may be a self-induced stability mechanism. The length and therefore electrical flow of the psoai are absolutely dependent on skeletal position. The more you know about where your body is in space, the better you can determine if you've got your muscles at the right length.

Anyhow, my point. Liz Koch will be interviewing me in a format you can all join to listen and ask questions![1]

---

1. coreawareness.com/podcasts/bio-mechanics-bowman/

 ## UNDER PRESSURE (PART 1)

Sept 19 2011

I'm having a hard time writing a post on diastasis recti (DR). Why? Because it is a musculoskeletal issue that has various components. The average woman seeking treatment in physical or non-supervised treatment programs spends hours learning about form and exercise. And you want it in one thousand words.

I think it would be easier for me to explain how a diastasis recti happens in the first place. If you are reading this and don't have this condition—DON'T HANG UP! The info in this article is about how your body should be all of the time so that you don't develop this, or other pressure-based conditions like pelvic organ prolapse, varicose veins, constipation, hemorrhoids, hernias, and high blood pressure, to name a few.

I know you want me to tell you how to fix it. And I will. But my answer is not going to be "Do these four exercises for five minutes three times a day for ten weeks and then call me in the morning." One of the reasons people have a difficult time correcting their own ailments without surgery is that they are not changing what they did that caused it in the first place—and I don't mean what they were doing the moment before the ailment happened. Fixing DR is the same as anything else. It is not a situation caused by pregnancy, or a giant baby. It is caused by carrying high pressure in the abdomen. Trying to do exercises to fix a pressure-induced issue while continuing to hold your body in a way that continues to create high pressure will never work.

If a pressure-based ailment has developed in your body, then your entire body needs to be aligned so the pressure goes back to the correct homeostatic values. We have become a spot-treatment kind of culture. An "I'll give my health at least sixty minutes a day" kind of population. An "I'll spend more money on my clothes than on what's under them" kind of population. But if you are really interested in fixing your DR for good, you need to learn what caused it and what you are doing, right now, that is keeping the split in place.

So, are you ready?

Diastasis recti: A musculoskeletal injury, where the rectus abdominus tears at the connective tissue, separating it from the linea alba—a collagen cord that runs from the bottom of your sternum to the front of your pelvis.

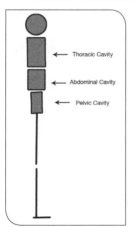

Ok, science class! There are three main cavities in the body and probably more than that if you loved Now & Laters as a kid.

The cavities include the thoracic cavity, which houses your heart and lungs; your abdominal cavity, housing your guts and possibly a baby-to-be if you're lucky; and finally, your pelvic cavity. If you grew up in a more natural way, walking long distances per year, squatting to bathroom and birth, barefoot, and were well rested, the position of these cavities would be aligned like this (image on previous page):

(Don't make fun of my pictures. Even if the feet are tiny and the femurs excessively long.)

Alignment is different than posture. Posture is about how something looks. Alignment is about how something works. So, to work on your alignment is to pick a position that improves performance, not necessarily aesthetics. The alignment of the cavities (image previous page) is very important because the pressures within the cavities are affected by various positions.

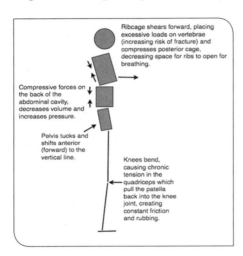

Ribcage shears forward, placing excessive loads on vertebrae (increasing risk of fracture) and compresses posterior cage, decreasing space for ribs to open for breathing.

Compressive forces on the back of the abdominal cavity, decreases volume and increases pressure.

Pelvis tucks and shifts anterior (forward) to the vertical line.

Knees bend, causing chronic tension in the quadriceps which pull the patella back into the knee joint, creating constant friction and rubbing.

So let's talk about pressure.

The best way I can explain pressure is like this. If you have twenty small children and put them in a small room, the pressure will be very high. If you wanted to increase the pressure you could either add more kids or make the room smaller. You could also give all the kids candy.

The only way to decrease the pressure would be to 1) remove some of the children or *stuff* from the space or 2) get a bigger room. Preferably at someone else's house.

Now back to body alignment, or, more importantly, how alignment affects inter-cavity pressure.

After decades and decades of sitting, shoe-wearing, sucking in the stomach, and various postures that we have selected for aesthetics or cultural reasons, this has become the common stance for most Western populations (image above left).

The most important thing about this picture, at least for our discussion today, is what is happening to the abdominal cavity. When the ribs lift in front, they drop in the back. Pinching the back of this "abdominal balloon" places the contents under greater pressure. And like things under great pressure, they try to escape. Because the spine bone makes a strong boundary in the back, your guts (or your baby, if there's one there) have to go…somewhere. The choices are up, down, or forward. Or any combination of the three. In diastasis recti, the forward pressure on the uterus causes the uterus to push forward into the muscle. And it's not just babies. High-pressure abdomens cause all sorts of ruckus.

Here is a fun pic showing how things escape the abdominal cavity: (image right)

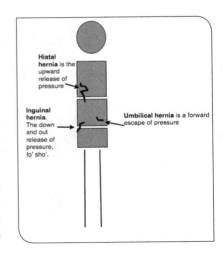

A hiatal hernia is when the organs move up through the diaphragm and party with the heart and lungs. Hint: It's no party. And, FYI, neither are the other two.

The pic might be a little silly, but these hernias (especially the hiatal) can be life-threatening situations. And if you happen to have one of the more minor ones, then take it as a red flag. Pressures are not as they should be in your abdominal cavity (and neither is the alignment).

If you are ready to say buh-bye to your DR, or your particular pressure ailment, the solution starts with whole-body posture. Before you do a stomach exercise, a pilates roll-up, or try any other type of solution, you need to change your pressure gradients. And before you can do that, you need to see how to recognize the body positions that make IAP (intra-abdominal pressure) worse. So, students, what do you see here (don't cheat—see what you see first!):

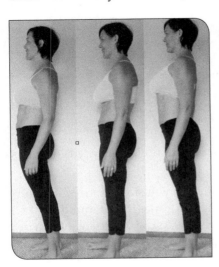

And once you've checked out my super-duper postures and bedhead (yes, I took these about ten minutes after rolling out of bed after sleeping for about two hours less than I needed...please don't zoom in), let me know if you saw this (image top of next page):

You probably noticed that the title of today's post has the ominous words "Part 1" in the title. So this is where we end today's presentation.

Your homework, my dear students, is to study these pictures—and

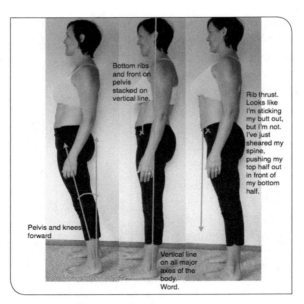

Bottom ribs and front on pelvis stacked on vertical line.

Rib thrust. Looks like I'm sticking my butt out, but I'm not. I've just sheared my spine, pushing my top half out in front of my bottom half.

Pelvis and knees forward

Vertical line on all major axes of the body. Word.

understand the differences between them. Focus particularly on the rib shearing and pelvis thrusting, and not so much on the bedhead, which technically isn't an alignment issue.

The second part is to start to look for and recognize these postures as you are out and about. Once you learn how to see these alignment snafus, you'll start noticing them everywhere and then eventually in yourself.

As for Part 2, tune in again to learn other high-pressure makers, some exercises, and which body parts play a significant role in the chronic knee-bending, hip-flexing, rib-shearing posture that so many of us are so fond of.

Till then, ta-ta. I'm going to go brush my hair now.

 ## Sept 22 2011 UNDER PRESSURE (PART 2)

Continuing on from the last post, let's talk a little about what increases intra-abdominal pressure.

1. Sucking in your stomach. Seriously. If you stand in front of a mirror and "let it all hang out" (just make sure it's after midnight...) you are going to see some stuff. Now I know that many of you out there aren't engineers, but I'm going to ask you something. Where

do you think all of that stuff is going? I'll give you a hint. Or better than that, I'll give you the answer.

You are displacing your organs upward (or downward) to make room for the stuff, while also increasing the pressure in the abdomen. What gives? Eventually what gives is the tissue (fascia) that connects one discrete muscle to the next. When the pressure gets very high, the guts push into the muscular wall, which gives out, thinning and eventually tearing the fascia between the muscles.

If you've got stuff, then you need to deal with it—not hide it at the expense of your organs.

2. Breathing in a non-optimal way. I tried to write this out a million times and eventually I just ate a PB&J sandwich and made this video instead. I hope it makes sense.[1]

   Many of you have learned belly breathing as a relaxation tool because it helps the diaphragm relax from all the hours spent sucking it in. Babies breathe a belly breath, but only when they are lying down. That's when you should be belly breathing too—when your spine does not need transverse support to keep the vertebrae from smashing the discs in between.

3. Holding in gas. Has the following ever happened to you?

   You dress up for a party in your oh-my-gosh-I-can't-believe-these-fit jeans. You swear you won't eat anything. You look great. You feel great. But then, a couple of hours into the party you do eat something because looking this good makes you work up an appetite. And on an empty stomach, that something you ate caused some gas. You can't fart though, not in this smokin' hot outfit. So you hold your stomach and gas in, for the rest of the party. When it is time to leave, you can't leave fast enough because now you have a stabbing pain in your lower abdomen and the desire to fart one thousand farts. But when you try, they don't come out with the vigor you were hoping. Once at home you put on your super-stretchy, comfy pajamas only to find that the stomach that was flat earlier at the party, now resembles a stomach in the sixth month of pregnancy.

Be honest. You know what I'm talking about, right? And my over-the-shoulder-reading husband pointed out that this would never happen to a man. It is simply not in his cultural programming to suffer painfully as far as clothes or farts go. But more on that later.

So how does all of this relate to diastasis recti and other pressure-related ailments? The above list is part of the contributing mechanism of these ailments. These are habits that must start to change for the healing to begin. These are habits that, in conjunction with the extra loading from pregnancy, can result in "pregnancy-related" ailments. It's not the pregnancy; it's what your body brought to the pregnancy table that creates the ailments.

Don't worry. There's more, my chest-breathing, rib-thrusting friends...

---

2. youtu.be/F3j_gQdicWl A seven-minute video explaining the mechanics of breathing. With help from a very bony assistant.

October 17 2011 ROLLING OR THUNKING LIKE A BALL?

I'm trying to post every Monday.

Today, I am still in my pajamas at 11:30 a.m. I haven't a shower in the last thirty-six hours. We've eaten breakfast and I've played with my baby and answered emails the rest of the morning, and that is all. This is my life. So, the post you get is this:

Rolling or thunking like a ball. What is it about? It's about the exercise Rolling Like a Ball. Great name. At least, it *would* be a great name if the exercise were more rolling than it was plopping on the ground and then straining your neck and psoas to get back up.

People always ask me which exercise is good. And I hesitate to answer that because it's not the exercise that matters, but how you use your body while doing it. Rolling Like a Ball is a great example of how an exercise can be executed (not killed, but completed) with your neck and your psoas (not the optimal muscles) or with your abdominals. Are you using momentum to get the exercise to look right, even though it makes things worse? Can you use your muscles in a beneficial way?

Try this out.[1]

Also note: If you are straining in any way during this exercise, with special awareness to any downward force you may be creating (did you pee your pants or pass gas when you did it?), these are signs that you are not using your abdominals but are wrestling your own pressure systems to attempt an exercise you're not strong enough for.

Start with just lying on your back, knees held firmly into your body with fixed arms, and work tiny movements from your abdominal muscles only. If you only move one or two inches , that's great. Only progress when you feel your belly driving the motion. Leg swinging, head swinging, gritting your teeth, and grunting are all distractions from the facts, man.

Let's stick with the facts, man.

---

1. youtu.be/RL99Ih57qh0 A four-and-a-half-minute video demonstrating the correct and incorrect way to do the "Rolling Like a Ball" exercise.

 ## HAVE PSOAS, WILL TRAVEL

I'm currently on my book tour, which started in California, because I started in California, both in alignment work and in life. The thing is, I just moved away from California, which means that to get to my book tour I had to:

- Drive sixty miles in my car.
- Take a ferry to Seattle.
- Take a cab from the ferry station to the airport.
- Take a 2.5-hour plane ride to California.
- Drive from the airport to my family's house in Orange County.
- Eat a bunch of food that wasn't Greek.
- Drive sixty miles to the closest and bestest Greek restaurant in Los Angeles.
- Ride a train for two hours to Ventura.
- Drive from train station to bestest friend's house.
- Eat leftover Greek food.

I've done more sitting in the last six days than I have in the last three months combined. My body is letting me know that this is not okay.

The psoas is a muscle that, when functioning optimally, makes the world glow just a little brighter. Sitting makes the psoas about three inches shorter than it should be. And when the psoas ain't happy, the pelvic floor, the hips, and the spine aren't happy either.

The psoas (a muscle running between the spine and the femur bone) is often clumsily lumped in anatomy books with the iliacus (a muscle running between the pelvis and the femur) and referred to as a "hip flexor" or a "pelvic tilter." And while the psoas can do these things, anatomy books fail to mention that the psoas also moves the ribs forward into a thrust, lifting the chest.

When checking out your body's profile, if you see a prominently lifted chest cavity or the bony bottom ribs thrust forward of the pelvis, you need some psoas relief info A.S.A.P.

Psoas issues are no joke. A psoas that won't release can affect baby position in utero. It can prevent the hips from extending and the glutes from building. It can compress the disks in the spinal column. It can keep the hamstrings and the calf muscles short. The psoas, like all muscles, responds to the position you spend the greatest amount of time in. Most of us sit—at work, in the car, in front of the computer, in front of the TV, at meals—all the time. And when we stand, we thrust our pelves forward and lift our ribs. We suck in our stomachs and walk on gym machines. All of these things make the psoas muscles a fraction of the length they should be.

In light of all my additional sitting, I've had to dust off my psoas protocol to try to find some relief from my new traveling tightness. If you want to work on restoring your psoas with me, try my favorite, easygoing low lunge.

Note these things about this exercise:

1. The front shin should be vertical.

2. The pelvis should be neutral, which means that the pelvis does not tip forward to sink deeper—only go as far forward as you can while keeping the front plane of the pelvis vertical.

3. The ribs do not thrust and the back does not arch.

4. Make sure you have Wonder Woman arms. This stretch won't work unless your hands are on your hips and you look kind of stern. And have an invisible jet.

5. Do not do this stretch unless you need a haircut. And have a headband.

The effects of a home psoas-stretching program have shown to have positive impacts on improving gait patterns[1], so get on it! Doesn't matter how tight you are when you start—you will notice quick results with regular practice! I find the human body's ability to transform mindblowing. Simply amazing.

A funny tidbit from my real life: I just learned that A.S.A.P. is called an acronym and not an anacronym, the term I've always used. Because I think it sounds better. But that's because I'm a verbal moron, evidently.

For in-depth anatomical and biomechanical info on the psoas, consider my three-hour lecture and exercise presentation, Science of Psoas.

---

1. ncbi.nlm.nih.gov/pubmed/21497319

Dec 12 2011 DASHING THROUGH THE... HOUSE FOR THE BATHROOM?

The word "holiday" has become synonymous with the word "indulge." At no other time is it considered a good decision to have oysters, two glasses of wine, three bites of your aunt's peppermint fudge, and a turkey leg before noon. Are you sick yet? I feel bloated just typing it out!

Combining rich and/or heavily sugared foods with the stress of extra to-dos, less exercise than normal, and more time in a seated traveling posture, you're likely to end up with some gut issues. I would type out all of the issues that can arise (or explode) as a result from your holiday habits, but I'll refrain out of politeness.

Many digestive issues, at the holidays and otherwise, are a result of not only what you put into your mouth, but also how the food is shuttled through your body. The intestines move food with smooth muscle action (called *peristaltic activity*) through the mechanical pathways your posture creates.

Here are a few tips to keep you at the party and out of the bathroom:

1. Mind your posture while you're eating. Slouching at the table is not great for the colon. To keep the geometry of your tubes maximized, keep your pelvis neutral and your torso upright.

2. Swallowing your food can be more or less difficult based on head position. Keep your ears stacked over your shoulders. If you have any issues with acid reflux, don't take your meals in a recliner! Even small amounts of angle mess up the natural mechanics of digestion.

3. Take a short, easygoing walk after each meal. This will help the natural movement of food through your abdomen.

4. Relax your belly! Sucking in your stomach or holding in gas can increase the pressure of digesting contents, increasing the time it takes for food to break down.

chapter 6

FOR THE LADIES

I consider myself a very "healthy" person. I eat well and walk and stretch every day. I challenge my balance to make sure I still can. I look both ways before crossing the street. But even with this preventative mindset, I still manage to end up wounded every month like clockwork. Don't worry, it's nothing my body isn't designed to handle, and it is a wound borne by half of the people on the planet at about the same frequency.

Still haven't guessed yet?

The term "menstruation" comes from the Latin term *mensis*—which means "month." The word "mensis" was itself derived from the Greek word for moon, *mene*. Why were these lunar/calendar terms selected? Maybe it was due to the typical menstrual cycle of twenty-eight days being equal to the moon's orbit around the Earth (about twenty-eight days), or perhaps it was reference to the animal species whose menstrual or estrous cycles coordinated with the cycle of the moon.

"What's all this nonsense about the moon?!" you ask.

I didn't make it up, really. If you are reading this on the internet, then you are part of a "nightlighting" culture. Nightlighting means we use artificial light in order to stay up far beyond the hours of the sun and in doing so have to bear the effects.

"Exposure to light at night can inhibit the pineal gland's production of melatonin. The pineal gland directs your body's rhythmic activities—including sleep, appetite, and the onset of puberty—through its production of melatonin. This hormone is primarily secreted at night, and it requires darkness to be produced. Bright light suppresses melatonin secretion."[1]

Data collected on non-nightlighting cultures[2] as well as research designed to measure the affect of nightlighting on the hormonal cycles of mice have shown that perhaps your menstrual cycle could be more in line with that of the moon if you didn't have so much exposure to light. That's pretty cool!

Getting back to my womb wound.

It turns out that the monthly shed of the endometrium from the inner lining of the uterus is just like falling and skinning your knees. Similar to any skin wound, there is usually blood and then these tissues need time and rest to heal.

Last month Jacqueline Maybin won the Max Perutz science-writing prize with her essay "Secrets of the Womb." It is fabulous, much more well-written than THIS blog, and I strongly suggest you read it. Maybin is a PhD student at the University of Edinburgh and has discovered a link between the quantity of "healing proteins" (HIF factor) a woman secretes and the quantity of blood flow she will experience during her cycle. Maybin also points out that the health of your other tissues dictates how well your uterus heals.

So what does this mean for you at home? If you are constantly stressing your tissues through alcohol, caffeine, or nicotine intake, garbage food, caffeine, excessive exposure to environmental toxins, overly vigorous exercise, not enough exercise, caffeine, extreme stress, lack of sleep, or caffeine, you may have a tough period because you will not be able to heal as quickly. Catch my drift?

I wanted to compare these "new facts" about the menstrual process with my trusty 1950s *Modern Medical Counselor*. Oh boy!

Dysmenorrhea (Painful or Difficult Menstruation)

Two days before the expected time for the flow to begin, reduce the amount of work done and increase the amount of rest. Take a warm tub bath each evening for thirty minutes.

When the flow starts, go to bed and keep hot-water bottles to the feet and lower abdomen...

To help prevent future attacks of dysmenorrhea give attention to the following:

a. Regular habits of eating, sleeping, and exercise.

b. A wholesome diet, free from spices, condiments, greasy or fried food, tea and coffee, with little or no flesh food.

c. Avoid tight clothing, and see that the limbs, neck, and chest are prevented from chilling.

d. Correct constipation, if present...

Many cases of dysmenorrhea can be partly or wholly relieved without calling a doctor. Correct health habits will do much toward making the female organs function properly. Taking cold, constipation, sitting in the same position for excessive amounts of time, mental stress, late hours, and dissipation are frequent causes of pain at menstrual periods.

Once again, I am impressed with the "back to the basics" information we have here. And while I know it isn't possible for us to get out of work on such a regular basis, you can cut way back on stimulants, make better food choices, and replace the vigorous cardio workout for a nice long walk with relaxing music. If your menstrual experience isn't what you would like it to be, or even what it used to be, check in on your overall lifestyle habits and see if they could be contributing. Be good to yourselves!

Which brings me to a joke I love[3]:

Q: Why do they call it PMS?

A: Because "Mad Cow Disease" was already taken.

---

1. Singer, Katie. "Fertility Awareness, Food, and Night-Lighting." Wise Traditions in Food, Farming and the Healing Arts, Spring 2004. See section on Night-Lighting.

2. Cohen, Sari. "Melatonin, menstruation, and the moon." Townsend Letter for Doctors and Patients (February–March 2005). findarticles.com/p/articles/mi_m0ISW/is_259-260/ai_n10299307/pg_1.

3. I strongly suggest that all men refrain from telling this joke. Just trust me on this one.

 KEEPING YOU ALL ABREAST...

Oct 22 2009

Since October is Breast Cancer Awareness Month (make that National Breast *Health* month, please!) I thought I would share a magazine article I wrote a couple years back. Enjoy!

Linda G., a breast cancer survivor, is joyful every day—a feat that became more difficult as she developed a frozen shoulder and then a mild case of thoracic outlet syndrome in the year following her single mastectomy. "I know that I should be grateful that I kicked cancer, but it was really difficult to be happy when my shoulder ached on a daily basis," Linda said.

For many years, women receiving breast surgery were told to "baby" their arm to prevent lymphedema (a swelling of the armpit, arms, or hands). And while this advice seems appropriate when trying to prevent agitation of the surgery site, it has a drastic effect on the mechanics and health of the shoulder girdle. As we develop a better understanding of the role healthy muscles play in overall body health, research is starting to show that gentle use of the arm during the healing process can keep the tissues of the shoulder girdle mobile, help the scar tissue lie down in the correct direction, and actually reduce swelling in the arms.

In addition to aiding in the healing process, working on skeletal alignment can actually create a healthier environment for the breast tissue before there is an issue. In fact, many asymptomatic women are slowly collecting lymph in the armpit area due to poor tone of the rotator cuff, pectoral and latissimus muscles. This excessive swelling in the armpit and around the bra area is often accumulated waste, even though it looks and feels like fatty tissue.

How does your lymph system work?

Circulating alongside your blood vessels, our lymph system drains the cellular waste products removed from the cells. But unlike the heart in the cardiovascular system, lymph has no big pumping mechanism of its own. Lymph movement depends on regular use of the muscles in the area. And even if you work the large muscles in the gym, the smaller muscles often get neglected. Whether you are interested in preventing waste accumulation or reducing swelling due to stagnant lymph, exercises to slowly stretch and strengthen the area are greatly encouraged!

The "egg-hole" test

This is a simple way to evaluate the tone of the muscles that define your armpit. Muscles that are easy for you to innervate (which means to supply with nerves by contracting) indicate the lymph has a better chance at being removed from the area.

Stand in front of a mirror with a clear view of your armpit. The less clothing you have on, the easier it is to see. Bring your elbow out to the side and lift it until it is the same height as your shoulder. Gently pulling your shoulder blade downward should create a hole about the size and shape of an egg. (Image at right.)

No hole? If you have some lymph accumulation in the area, the wall of your armpit will be flat. If you have a *lot* of lymph lymph accumulation in the area, your armpit will bulge.

If you have a generous amount of swelling in the armpit, it may be a good time to schedule a breast examination and have your nodes checked—and start these exercises right away!

What types of movement are safe?

When you first have surgery, most movement is uncomfortable and there is a lot of fear of re-injuring or damaging the surgical site. Your surgeon should let you know when the suture has healed and you are no longer at risk of opening the site. During that time, however, you can still move the surrounding areas without impeding the healing process.

 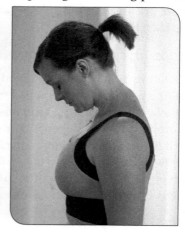

Keep your hands active to keep neurological activity moving through the arms. Do "piano player" exercises with your fingers, touching each finger to the thumb, going up and down the scales working on speed and fine motor skills. Add gentle hand and finger stretching often, to keep the tissues from atrophying while you are healing (pictured above left). Also, add the Head Hang (pictured above right) to stretch the back of your neck. Healing tissues tend to be stiff from disuse, so keep the neck supple by allowing the chin to drop until it touches the chest.

Put the swing back into your step!

Of all the exercises you could do with your shoulders, walking with natural arm swing is the most important to breast health! Many people have adopted the race-walking posture of bending the elbows when going for their daily exercise, but long-armed swings pushing behind you (imagine pushing ski poles behind you) is the more natural and calorie-burning way to walk. Remember, race walking is a sport, so these athletes are trying to conserve energy and expend less calories—why would we want to do that?

Stretching your chest with floor angels

Reclining on a bolster or stacked pillows, reach your arms out to the side, keeping the palms up. Try to get the backs of your hands to the floor, keeping your elbows slightly bent. Once your chest can handle that stretch, slowly raise your arms above your head, trying to keep them on the floor. Do this for five to ten minutes per day, and be gentle with yourself!

You may not feel up to heavy bouts of exercise right after treatment or surgery, but as you begin to feel better, small movements can aid the body in healing and have an overall energy-increasing result.

These exercises from the Aligned and Well program can be used as both preventive health measures as well as part of a recovery program. They are designed to gently restore the circulation of electricity, blood and lymph mechanics, which all lead to better breast health!

January 29 2010 DON'T BECOME A NUMBER: GET CURRENT ON WOMEN'S HEALTH

(Reprinted from my Gaiam blog)

I love the section some magazines have where they list interesting and fun statistics, like... 48: Number of bugs eaten per year in your sleep; 3,154: Number of texts sent by the average teenager (per month); 14: Number of states visited in an average lifetime.

I have some statistics for you, and I assure you, nobody's laughing.

- 80: The percentage of women who will suffer a pelvic floor disorder in their lifetime.

- 1 in 9: The number of women who will have surgery on their pelvic floor.

- 50: The percentage of women who, after having a first surgery, will have a second, third, or fourth surgery for the same issue.

- 10 million: The number of Americans estimated to have osteoporosis—8 million are women.

- 25: The percentage of us who experience debilitating foot pain, RIGHT NOW.

Weak Bones. Leaky Bladder. Fragile Feet.

Are these the burdens of being female?

The plight of the woman seems like an eternal struggle. We can't be the CEO of a corporation because we have to have a child. We can't be a firefighter because we don't have the strength. We can't pee standing up…and I really wanted to do this one as a kid. But as we make progress on most gender-related issues, we still seem to be failing in the genetics arena. Sorry about your bone density—you just don't have enough space for minerals in those tiny bones! You and your mom have bunions? Well that's just genetic. Pee when you laugh? Or sneeze? Or walk? That's just the penalty of having a baby, of getting older, and you got it, being a woman.

Now, what if I told you that your bones were actually strong and big enough to maintain your bone density, but the posture you are choosing to stand in is signaling your bones to weaken? What if I told you that your foot pain, your plantar fasciitis, your neuroma, or your bunion is caused by how you walk? And that your mother's bunion is caused by how *she* walks? And that you learned to walk by watching her. What if I told you that there was one thing, one habit you have that you do every day, that is causing your organs to move down, your bladder to fail, your lower back to hurt, and your hip joints to die?

Well, that's what I am telling you.

Women's health, while making huge progress in some areas, is progressing like a limping slug in others. As a biomechanical scientist specializing in pelvic-floor physics, I am continuously appalled by how misinformed people are regarding their own equipment. The muscles in your pelvic floor have very important jobs. First, they keep your bathroom functions running…uh, smoothly. Second, they support the weight of your pelvic floor and abdominal organs. This is a lot of work for this complex group of muscles. And if you ask most medical and exercise experts how to strengthen these muscles, they will give you the same exercise: the Kegel.

Named by OBGYN Dr. Arnold Kegel, the kegel exercise is a contraction of some of the pelvic floor muscles. I doubt that he invented it. In fact, I'm pretty sure people had been practicing kegel exercises for hundreds of years before it had a name. I want to name a muscle action after myself, too. How about lifting the second toe? Let's call that the Bowman. We'll see if it sticks.

Anyhow, the kegel seemed like a good solution to increase the tone in the pelvic floor back in 1948, following Dr. Kegel's creation of a device that measured the pressure on the pelvic floor. In the last sixty years, however, there has been much more evidence, research, and anatomical understanding of this area, which has shown that the kegel, while providing a short-term solution to pelvic floor issues, actually creates a greater problem later on, both in pelvic floor function and with sacroiliac pain. Yet I bet most of you, when asking your doctor, favorite movement teacher, or friend what to do about your incontinence, will be given this outdated, situation-worsening exercise.

In the age of information, it is time for women to understand how their bodies work and what simple things they can do to take back their health. No more hearsay! Here's the real, most up-to-date science on what pelvic floor muscles, bones, and the tissues of the feet need to stay functioning well. You might be blown away by what you didn't know and how quickly you can improve your health.

1. Osteoporosis is not an all-over bone disease, but an indication of where your bones are being loaded properly and where they are not. You need to know where in your body your mineral density is low to design an osteogenic (bone-generating) program that is specific to your bones. Foot pain can be significantly reduced through exercising the feet!

2. Twenty-five percent of the number of muscles and bones are from the ankle down. Using your feet while exercising the rest of your body doesn't do much for the health of the feet—they need their own set of exercises.

3. Pelvic Floor Disorder (PFD) is typically thought to be a result of childbearing, but the fact of the matter is, the statistics of PFD show that women who have not given birth are just as likely to develop a problem in their muscular function.

 BOOBS

October 7
2011

I had no idea how amazing boobs were. I was aware of their magical power of course, but it took calming another ranting, inconsolable human being with them for me to truly understand. And then I had a kid. October is not only National Walking Month, but also National Breast

Cancer Awareness Month, which I petition to rename National Breast Health Month, because breasts do not need more cancer awareness, they need more health.

I am also glad that National Breast Month is not National Running Month, because that would be uncomfortable. How uncomfortable? One biomechanics researcher took a look:

The study: Breast displacement in three dimensions during the walking and running gait cycles.

Methods: Fifteen D-cup runners, reflectors attached to their nipples (doesn't it pull the hair out when you take them off? What? Doesn't it?) walk and run on a treadmill while their breasts are filmed. (It's for science, honey, for SCIENCE!)

Findings: During walking, breasts move evenly in all directions. This is just one of the reasons I love walking. During running, however, "More than 50 percent of the total movement was in the up-down direction, 22 percent side-to-side and 27 percent in-and-out."[1]

Wait. What's "in-and-out?" You mean like a punching balloon? Nice.

Why the difference in walking vs. running? Because walking is entirely different from running. Your body travels much more up and down, even if that up and down movement comes from your knees bending to absorb the shock and not your feet leaving the ground much.

Boobs. Your new gait-training tool.

Your boobs are essentially a biofeedback device, visually demonstrating the inefficient movements of your whole body. Body swaying right and left? So will the girls. Breasts moving in and out? Your body is not flowing forward smoothly and has points of start-stop creating a stall in every step. Forget the hips. The boobs don't lie.

Another interesting tidbit:

Researchers found that the poorer the breast support, the greater the effect on the whole-body kinetics of the runner. Using D-cuppers again (the more boob mass you have, the more it affects your whole-body performance), this study found an increase in medial loading of the pelvis and knees as a result of the breasts' momentum.

If you're taking the ladies out for a run, invest in some nice support. Especially if you're a breastfeeding mom also working on getting your pelvis back together. That excessive medial loading is killer on the pubic symphysis. Even though I'm a walker, every girl needs to run across the street now and then. This D+ cupper would die without Athleta—for bras and for everything else. Dear Athleta, I love you. You complete me.

KB+ ATHLETA = 4EVER.[2]

1. Quote from Jeanna Brynner's article here: livescience.com/1864-bras-support-bouncing-breasts-study-finds.html

2. Dear Reader, Since the writing of this article, more data on breast health—specifically cancer—has come to light emphasizing the importance of breast tissue environment. For more information on bras (and breasts) please review more recent posts on katysays.com.

Ergonomics. 2009 Apr;52(4):492–8. The effect of breast support on kinetics during overground running performance. White JL, Scurr JC, Smith NA.

Journal of Applied Biomechanics. 2011 Feb;27(1):47–53.A comparison of three-dimensional breast displacement and breast comfort during overground and treadmill running. White J, Scurr J, Hedger W.

Journal of Applied Biomechanics. 2009 Nov;25(4):322–9. Breast displacement in three dimensions during the walking and running gait cycles. Scurr J, White J, Hedger W.

chapter 7

ARMS, ELBOWS, WRISTS, AND HANDS

Don't panic, this isn't a political blog. But it does have to do with guns...
Mrs. Obama's *super-coveted* biceps, that is.

Now, I'm not all that interested in buffing up people's bodies. I've always been much more interested in returning the human machine back to the way it functions best. And while that does involve a serious increase in lean body mass, it is much better for your body if that lean mass is everywhere, and not just looking fine in a sleeveless shirt. All these things I explained to my client Rose during her third session with me today. And she totally understood what I was saying.

But she really, *really* wants those arms.

Yeah, yeah, I know.

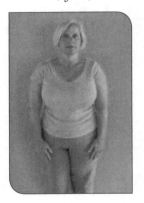

So, the good news is, I said, toned upper arms are a good indication that other functions, like lymph drainage in the chest and shoulder area and the pulley system that supports your cervical and thoracic spine (neck and mid-back), are working.

Now that I had "healthy motivation" (as opposed to the tenth commandment), we could move on.

Step one for Rose, and for all of you, is to stand in front of your mirror with your arms relaxed down by your sides. What most people see is that both arms hang slightly in front of their body, and the backs of their wrists face the mirror.

See Rose Stand (image above):

When you can see the backs of your hands while standing in front of the mirror, it is a significant implication for how things are working up in the shoulder. This posture means your arm bones are not in a place where your upper arm muscles could even *be* toned!

When your arms aren't as toned as you would like, how about putting them where they're supposed to go?! Once you get the muscles into the correct position, they will start to work (and tone) on their own.

The Soldier Stand: Super-simple exercise alert!

Move your hands down along your sides (like they would be touch-

ing the racing strips on athletic pant legs) and press the palms into the sides of the thigh. (Image at left.) Once your hands are flat, see if you can get the entire arm to touch your body—working all the little muscles in the shoulder as well as the large muscles of the latissimus!

I know it doesn't look like much, but I can assure you, it is! Especially for those who type or drive excessively, are under lots of stress, and carry tension in the shoulder area! You can do this one anywhere...all you need are your arms! Once you've done this for about a minute, try this next exercise.

Ah, the Posterior Block Hold...I know this looks SUPER EASY, but it is SUPER CHALLENGING! All you need to do is hold a yoga block behind you by *pressing* your palms into it (no gripping!)

Can you see how Rose's elbows are bent? (Image below left.)

They should be straight, like this...(Image below right.)

The surprising thing is, this is the *largest muscle* you have in the upper body, but our shoulders have gotten so tight, the "lats" no longer generate very much force. Think of what your metabolism could be if this baby was working all the time!

Time for one more?

Tension in the neck and shoulders leads to tension in the arms and hands. This wrist exercise is great to see how the mechanics are functioning in the area where many people are battling carpal tunnel syndrome.

The backs of my hands and thumbs touch...do yours??? (Image top of next page.)

SWING FORWARD?

Ahhhh, moving the clocks forward gives us that extra hour of light at the end of the day. A perfect hour for an evening walk. That's what I thought as I went for my first weeknight walk in the sunlit evening. I walked instead of blogging. You can give me grief when you see me next, but I really don't feel that badly, I have to admit. It's hard to feel bad when walking gives me all those endorphins. You should try it. So, I'm out walking and noticing many new folks out along the beach path, good for them! After witnessing this I decided on today's topic. The arm-swinging thing.

Take a closer look at the arm motion of your fellow walkers next time you have a chance. There are arms that don't move at all, and arms that swing right to left instead of front to back. There can be one tight arm that moves less than the other, and my favorite, those trying to work their arms extra hard by holding weights. Wow! It's an arm buffet, I tell you. Believe it or not, arm swing is an extremely important part of a natural gait pattern, and when digitally analyzing gait, you can tell a lot about shoulder and spine injuries-in-the-making just by watching what the arms are doing.

Why do our arms swing when we walk?

The best, most muscle-building, calorie-utilizing, metabolism-enhancing, heart-strengthening, and blood-circulating gait pattern is one that is smooth and symmetrical. All of your body mass is moving in one direction—forward. Think of the bouncy walker. It does look pretty cute, I'll admit, right up until you open up the knee and hip joints and see what that bounce does to the cartilage in the joint. Ouch.

When you walk forward, it is *supposed* to be due to the fact that one leg pushes off behind you (most people lift a leg out and fall forward, but that's a different blog). So now you have your right leg behind you, pushing back. That large quantity of mass (our legs make up quite a bit of our body weight) can twist your pelvis around, creating torque on your spine. There needs to be something to balance out the twisting tendency walking creates on the spine. The easiest thing to do is reach the arm back on the opposite side of the body, to help balance the twist. It's called reciprocal arm swing. The opposite sides of the body reach back at the same time, an arm on one side and a leg on the other. A backward-reaching arm is not only great for reducing overuse of the spine, it is a nature-designed workout for the backs of the upper arm. Awesome! Lifting the arm up behind you keeps your tricep muscles toned and the armpit lymph free. If you thought three sets of twelve tricep exercises with a five-pound weight were effective, just wait until you use your arms correctly when walking. It tones those arms right up!

When you are out walking, you are going to see another strange phenomenon. People are lifting their arms out in front of them because they think that makes means they are working harder and burning more calories. Sorry! Pumping your arms out in front creates extra tension in the front of the shoulders, muscles that are usually already super tight and fatigued from computering, driving, and sitting all day. The tighter the muscle, the fewer calories it burns. This forward motion tenses the trapezius and neck muscles too. And, if you aren't swinging your arms behind you when you walk, there is nothing to help balance the torque created by the leg moving back, either. Nothing except the muscles in the lower back. Anyone our there have weak triceps (back of the upper arm) and a tight lower back? Don't make your low back muscles do the work for your arms. You'll end up with shoulder issues and lumbar disk degeneration. Arms swing side to side? This is an indication that your chest and shoulders have gotten so tight, the bones have rotated forward. Really pay close attention to keeping your arm swing in forward-pointing parallel lines to your body.

One more thing. Bending the arms at the elbows to mimic speed walkers actually reduces the energy you expend during a bout of walking. Race walkers are trying to minimize their energy expenditure so they can go very fast for a very long time. Want to ramp up the kcals burned during your next walk? Rapunzel, Rapunzel, let down your arms.

Exercise: Stand and let your arms relax down by your sides. Lift one arm behind you, one at a time, to see how high you can get it (don't twist the hips or shoulders, that's cheating). Let it drop down and forward. It will swing out a little in front as you drop it, but don't do any extra work to get it up higher. The work of your arms while walking should always be behind you. They relax when they come forward.

Really working a fully extended (no bending elbows!) arm while walking is going to change everything about your daily walk. You'll get much more out of it, including increased metabolic and strength benefits.

P.S. You can't swing both your arms if you are talking on your cell phone, now can you? HANDS FREE!

**March 19
2010 WHY DO YOUR HANDS SWELL WHEN YOU WALK?**

When you're up and walking around with your arms fully extended, gravity will tend to pull all the blood down to your fingertips. Active arm swinging (check!) also creates an additional centrifugal force that pushes blood into the hands. With your arm fixed at the shoulder, the arm and hand have a pendulum-like motion, swinging back and forth along a curved path. If you've ever put lettuce in a salad spinner, you have seen how spinning motions cause the water to move outwards, towards the sides of the bowl. The centrifugal force can fling the water in a salad spinner in the same way that it can move fluid mass towards your fingertips when your arm is swinging.

The centrifugal force (not to be confused with the centrofrugal force, which is the force of nature that keeps you from buying stuff for yourself) is a force that really only applies to non-active tissues. If the blood is being pushed into the hands when you are walking, then I would suggest you *use* your hands while you walk—actively reach through your fingertips to lengthen your arms or make and release fists while you are walking. Keeping the muscles active gets the blood circulating through the hands instead of staying there. Stretch your hands before and even during the walk, and treat them to a cold-water rinse when you get back. The cold water causes the tissues to contact, reducing any increase in volume that may have occurred.

Don't let your hands flop around passively while you're walking.

Your arms don't end at the wrists! And waving a friendly hello now and then wouldn't hurt, either. ☺

April 6 2010 ALL THUMBS?

This weekend I watched *Gorillas in the Mist*, which led me to research gorillas and sign language, which led me to research opposable thumbs. Anthropologists are constantly saying how great these opposable thumbs are. I think we should be calling them applaudable thumbs. Interesting Fact: My spell check doesn't recognize the word "opposable." The very thing that has perhaps been the catalyst for my computer's very own existence, doesn't even compute as a word. Interesting. Verrrry, verrry interesting.

What does "opposable" mean? Technically, it means that the joint of the thumb allows the tip of the thumb to touch the tip of each of the other fingers on that same hand. It's the only finger that can do that. But here's another fun trick to try. Can you get the pinkie finger to touch each other finger, one at a time (it won't be on the tip, but can you at least make contact?) Then try to get your ring finger, and then your middle finger, and then your index finger to touch all other fingers one at a time. Easy for some, and difficult for others, depending on the tension in your hands, forearms, or even neck.

Give it a try, and make sure to do both hands!

October 19 2011 YEAR OF THE UPPER BODY

I've decided that 2012 will be the year I dedicate to unprecedented upper body strength, of course knowing that upper body strength requires middle and lower body strength, which requires a suppleness to all tissues, so I guess 2012 won't be any different than any other year. Except that it will.

A month before I got pregnant, my friend-of-all-friends got me a pass to a two-hour-long trapeze class, which I took. It was awesome and afterward I realized that swinging and hanging from your arms is the natural movement most missing from our daily lives. Swinging and hanging require more than your arms. This full-gravitational pull requires every muscle around every joint to pitch in to hold whatever is below it. It's a

full-body phenomenon that, let's face it, our puny, blistering hand-skin isn't ready to take on. Time to get over it.

Here's where I was about sixteen months ago in terms of whole-body awesomeness.[1]

Do you like the part where I didn't know how to keep my legs together while swinging? I'm thinking that adrenaline must affect hearing somehow.

Anyway, flash forward a pregnancy and a baby and breastfeeding and all-nighters and a move and some serious forgetfulness when it comes to keeping up with my upper body. But starting today (or yesterday, actually) it will be different. No more excuses. I need to hang and swing and until I have my indoor jungle gym promised to me, I will find another way. And I did.

This branch is hidden in the trees out back (image top left).

After mowing the lawn this week-end, I found this natural gym in my very own backyard. This branch is my personal trainer because it 1) cannot mock my pathetic attempts at any trapeze efforts 2) is thin enough for me to hold it comfortably while strong enough to hold me and 3) is ten steps away and available for sixty-second playdates seven or eight times a day. (Image top right.)

Also, I have a cow to cheer me on (image previous page.)

There's no huge time commitment needed for whole-body strength. You just have to decide.

1. Find a suitable tool (branch, bar, whatever).

2. Write down all the reasons you can't do it (it's too far, too hard, too long of a drive, too much time, too cold where you live, too, too, too).

3. Throw that list away.

4. Decide to do it.

After that, it's a breeze.

1. youtu.be/PCN8Xa94kPg Two minutes of me on the trapeze. SO MUCH FUN.

SHOULDERS

(and what gets stacked onto them)

MY MOUTH MARATHON

So here's my deep dark secret.

I haven't had my teeth professionally cleaned in twelve years. TWELVE YEARS! Okay, I feel better just saying it. Yes, I have had check-ups to make sure things aren't going by the wayside, but I just haven't had time. In twelve years. I haven't had one free hour.

Yeah, I'm not buying it either.

And it's not that I am ambivalent about oral hygiene. Quite the opposite. I am So Obsessed with clean teeth that I brush four times a day. Hard. You know, to make up for the fact that I haven't had my teeth cleaned in TWELVE YEARS. And I floss. Like, in the car. I get it done, man.

So today was the big day, and as the hygienist poked her rubbery hands into my mouth and jabbed me with her "what could be a weapon if she felt like it," she said:

"I see some gum receding here. It's usually caused by vigorous brushing, teeth grinding, or chewing tobacco. Could you be doing any of those?"

And I said, "Hassshhhhhp meii ffotwer."

And she said, "I thought so."

So in my failure to take preventative steps, my overcompensation for a "healthy body" was to overbrush my teeth, which winded up overstimulating the tissues until I created a problem. This process is very similar to how we think of exercise. We fail to do the required, preventative level of movement all day long, so we cram in a high-intensity session to make up for it. And while the effect can have some positive benefit (just like my brushing), it can also come with a cost. I will be entering into my future with less gum support to my tooth, just as the runner will be moving into their future with less cartilage in their knees and hips.

It's a lot of work always making the best choices for longevity!

And an interesting fact…flossing increases the tone of the muscular gums so they do a better job of keeping the pocket around your tooth smaller in size. The smaller and more toned the pocket, the less space for bacteria to become trapped! That little pack of dental floss is like weights for the gums. Who knew my dentist was also my personal trainer?

FYI, the best DDS in Ventura is Teresa Monzon, hands down. I got a FACE MASSAGE before my tooth cleaning. How classy. ☺

THIS HEAD POSITION CAN BE HARD TO SWALLOW!

If you are super-busy, chances are you are reading emails, spending a lot of time on the computer, and driving in a rush to get to all of your obligations—and spending a lot of time in the first head position shown here! During my client Rose's first appointment we covered a lot of ground, but one simple movement such as the one pictured above, can decrease—in an instant—the level of strain you are placing on your body.

Which of these pics go with the above head positions??

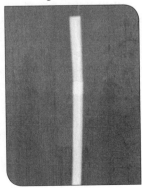

This simple straw example shows you the changes the "tubes in your neck" go through when you alter head position. What's the big deal? Well, let's look at what normal processes happen in the cervical neck. There are three main ones...blood flow, airflow, and the movement of food. Now, air is going to get through pretty well no matter the curve (just like a straw), but blood and food are a different story.

When you change the geometry of a blood vessel, you increase the risk of plaque accumulation in that area. When blood cells accelerate around a corner, the cells (which have sharp edges compared to the thin-celled walls of the blood vessels) end up hitting and injuring the wall of the curved vessel. This wounded wall now makes a scab of plaque to heal, which starts to limit the tube size and reduces the quantity of blood, and starts the injury/plaque formation cycle. Ideally, blood flow should be down fairly straight hallways. Imagine rolling a ball down a curved hallway. It's most likely to run smack into the wall in front of it, and not make the curve...especially a curve that is not supposed to be there!

Not motivated yet? Then let us consider food. The muscle action of moving food down is fairly minimal. It's really not that hard to move things down, as gravity is going that direction anyhow. But look at the bent straw. Your food is hardly moving straight down...it's moving almost across! So now you have to create a lot of excessive muscle force in the throat to contract to move food without the aid of gravity. Food that does not move quickly through this area increases the risk of choking.

According to the Agency for Health Care Policy and Research (AHCPR), 6.2 million Americans over the age of sixty are suffering from swallowing problems. The causes of this issue are thought to be neurological, or a result of stroke, but to eliminate a mechanical contribution to the problem, try the exercise Ramping Up.

Without lifting your chin, slide your face back like you are making a double chin until your ears stack up over your shoulders. You instantly increase the alignment of your cervical spine and decrease the mechanical damage to the blood vessels in the neck and muscles in the throat. It's a great movement to relieve the strain on the small muscle of the neck that can lead to headaches, disk problems, and "foggy head."

I just got back from tending to my grandmother following her shoulder surgery. I know many of you out there tend to your ailing senior parents and other family members on a daily basis, and to all of you I say BRAVO. I thoroughly enjoyed my time with Gram—the smell of her Elizabeth Arden face lotion, more episodes of *COPS* than I care to count, and the sound of my name being called…over and over.

Was I being summoned to take care of an oozing wound? Did I need to find a needle from the sewing basket in a Stitch-Popping Emergency? Nope. I had to reset the clock on the VCR. And change the numbers on the speed dial of her phone. And rub her sore…shoulders? Not a chance. It's the FEET that need rubbing. Those are pretty far from the shoulders, Gram, but any increase in circulation is a good thing, so we'll let that slide…all three times.

And I did it all happily, as this woman handmade my clothes growing up, cooked us weekly dinners, and best of all, let us each, one at a time, spend Friday nights at her house, get a DONUT, and go garage sale-ing Saturday morning. I totally owe her.

The result of this weekend was a very happy lady, a glimpse of my inevitable role to come, and some tips for any of you who may be going through a similar surgery. Shoulder surgery, especially rotator-cuff repair, can be a fairly simple outpatient procedure. The most difficult part is the recovery—or really, the one-armed living habits you must cultivate while your tissues are healing. Only using one arm is DIFFICULT! (Next time you are in the loo, try pulling your pants down and then back up with one hand. It can't be done, I tell you!!!)

If you or someone you care for is getting ready for shoulder, breast, or hand surgery, send them these tips:

1. Shirts are pretty much out of the question just after shoulder surgery. Going topless is equally out of the question. ☺ An excellent compromise: Purchase one yard of fleece and make a poncho! Just fold in half and cut a neck hole (you can also make a 6–8" vertical cut down the front to prevent throat or neck aggravation) and viola! You have full coverage and warmth with no buttons, zippers, or sleeves to deal with.

2. Ask for any braces or slings ahead of time and practice putting them on beforehand. My grandmother's sling was *very* compli-

cated, and even with my engineering background, I may have put it on backwards. I suggest putting it on with the help of your doctor's office and have them label which Velcro straps go where (a Sharpie works great).

3. Buy a set of inexpensive underwear one size larger than you normally wear. This makes bathrooming an easier task for the single-armed.

4. Pain medication can wreak havoc on your stomach and upper intestine. For most, medication is a necessary part of healing, but you need to make better dietary choices during this time. Avoid "acidic foods" such as decaf and regular coffee (that's a food, right?), sodas, chili and seasoning peppers, and limit tomatoes, citrus juices, and fried foods—especially if they upset your stomach or cause heartburn. Focus on a whole-food (no processed foods), higher-fiber diet during this time. Your body is trying to heal a surgical wound—a wound you chose to inflict because of your overwhelming desire TO FEEL BETTER! Choose well.

Finally, find multiple people to tend to you. Try not to depend solely on one person for the entire duration. This will keep everyone fresh and reduce some of the inevitable irritation that comes with the slow healing process. And find someone to rub your feet. Especially one that, as they say, owes you big time.

March 9 2010 YOUR HAIR

When I was three I drew a picture of my great-grandmother Nana Kathe in preschool. I proudly brought it home to show her. She asked in her thick German accent "Vas ist das?" as she pointed to tiny black dots I had drawn on the lower half of her crayoned face. Well, I didn't know *what* they were, only that I got zapped by them whenever she tried to kiss me. Electric eels on her face? No. Thick hairs that punctured my skin? Yep. Hair is complicated. It's never quite the style you'd like. You may have too little...or too much. Maybe you don't know even know what color it is *sans* intervention. But today, we're going to look at the facts and see if we can shed a little light on what hair is all about.

First, it's important to know that humans have three different types of hair: terminal, vellus, and lanugo. Lanugo hair was the hair you had in utero, which is all gone now. Vellus hair consists of those tiny hairs that you have all over your body, the peach fuzz. Terminal hair is the type of hair you most likely think about when you think "hair." Head hair, armpit, and pubic hair are all terminal hair. Chances are, some of that may be gone now too, with or without your consent!

The purpose of hair isn't known, although there are a few theories. In addition to making you miserable about one day a week (really…you're going to stick up like that?), hair has been touted as a heat-maintaining mechanism, friction reducer, and heat conductor. All that being said, different races and different genes give us different quantities of the stuff, so whatever hair is supposed to do, its purpose is probably not that simple. Because hair is generated from the skin it grows from, the better the circulation to the skin, the healthier the hair. How do you get healthy skin? Increase the circulation to the muscles just beneath it (you knew there was going to be a muscular solution, didn't you?!).

Head hair is a multi-million-dollar industry from both an aesthetic and medical point of view, as hair loss has become an increasing problem in the US affecting both men and women. What are some of the biomechanical factors? Friction. Excessive friction (think putting a hat on every day) can cause hair follicles to inflame and potentially decrease rate of growth. On the other hand, a lack of circulation to the muscles of the head can also decrease hair production.

Research shows that stress, tension, and anxiety are major contributors to hair loss. Why? Firstly, terminal hair development is affected by certain hormones. Stress levels can raise and drop hormone levels, often in an instant. But let us also consider the tension in the facial muscles when our stress loads increase (think Grand Canyon–sized crevice between the eyebrows). Neck, shoulder, head, and face tension will also negatively impact blood flow to the head muscles, skin, and HAIR muscles.

Wait, did I say hair muscles? You bet I did. For every hair you have on your body (vellus and terminal), there is a small muscle (arrector pili) that can move the hair as well as pump oil from the oil glands at the base of each follicle to keep hair protected, which also helps hair maintain its strength. Hair strands themselves can be pretty strong in terms of tensile strength if your daily mineral intake is correct and you aren't over-using minerals to deal with bone, stress, and muscle tension issues. Thin and unhealthy hair may, in fact, be a good indication of mineral and metabolic issues that are happening at deeper layers of the body. Ayurvedic science uses hair diagnoses to discern correct mineral, vitamin, and nutritional prescriptions. Studies measuring tensile strength (how hard you can pull before it breaks) and mineral content of horse mane hair found that dietary intake changed the mechanical and chemical properties of hair after a few months. The chemicals in shampoos can also alter the tensile strength of hair. The more frequently you wash and heat-blast your hair, the weaker it gets. Am I saying wash your hair less? Yes! We in the US wash our hair more than any other country, with the motivating reason being the smell of accumulated oil and cellular waste that comes out through our head's sweat glands. Don't like the smell? Your diet most likely contains items that your body cannot break down into usable units of energy and secreting waste through your sweat glands is a fast way to get rid of it. Ew, I know.

Quick tips for stronger and more abundant hair:

1. Clean up your diet, limiting processed and mineral-altering substances like diet sodas, and get your daily required minerals (see a nutritionist for a good understanding of your particular needs). This will get you the correct building blocks to better hair and reduce your need to wash your hair daily.

2. Gently massage your face and entire head daily to stimulate blood to the skin and muscles that support healthy hair follicles. A gentle hair brushing is a great way to relax your partner, child, or self at the end of the day (but it's way better if you can get someone else to do it).

3. Keep neck and shoulders loose and try to relax your face when you feel tension come on. Don't cut off optimal flow to the head muscles with muscular tension just below!

4. Reduce stress. (Great tip, I know. And specific, too!)

5. Don't be so sure that your hair loss is explained by your genes or your menopausal state. Until you have removed the junk (that means diet and regular soda) from your diet, aligned your head, neck, and shoulders, and increased blood flow to the skin of the head, there is no way to tell!

 **March 10 2010 EARWAX**

Today I'm going to talk about the physics of earwax.

I SAID WE'RE GOING TO TALK ABOUT EARWAX!

Clean those things out, geez.

Earwax, an oily secretion from the tiny hair follicles inside the outer ear, has many theoretical purposes, ranging from collecting dust particles to trapping bugs before they get down into the ear. I know about the latter as my father, missing an eardrum, has his earwax cleaned out from down deep by a doc a few times a year. Once the doctor pulled out a full-sized, fully intact fly...just in case you thought I was making the bug thing up. The tackiness of earwax is a good collector of foreign particles as well as a great catcher of sloughed-off dead cells from deeper places that aren't used to much friction. You probably don't exfoliate your inner ear, which is a good thing, but inner ear skin and hair follicle cells still die and drop off your body just like skin does. These cells need to clear outta there! Earwax is a great way to collect these tiny items into one heavy clump, making it easier to move up and out of the ear canal.

Of course, if the earwax *doesn't* move, now you have lumps of wax and waste material sitting in your ears, decreasing the basic functions of the ear, which can include both hearing and balance! How does this wax move out of the ear canal? The muscle action of the jaw, of course. Q-tips and cotton swabs are a recent invention, but Mother Nature's plan requires full range of motion and regular use of the temporomandibular joint (that's your jaw joint) to keep your ears functioning optimally. If you eat and talk, then chances are you are getting some good muscle action. But what if you carry your tension in your jaw? Tense jaw muscles decrease the range of motion of the joint, limiting the full movement of earwax. Also, many people talk more out of one side of their mouth as a learned speaking or accent pattern. Some prefer to chew more on one side of their mouth. In these cases, the muscle action is not symmetrical between sides, which (if you were paying attention) means one ear has more accumulation than the other.

The health of the ears is one of those things we assume will go with age, but many people are not suffering from age-related hearing loss, but creating a "conduction deafness." When sound waves enter your ear, they generate tiny vibrations in the hearing bones that your brain interprets as sound. Accumulated materials and fluids in the ear decrease the ability for waves to enter, and in the deeper part of the ears, can limit the movement of the three hearing bones. Nerve health of the ear also depends on adequate circulation to the tissues, and a head projected forward of the torso, or hours of daily jaw clenching, will reduce circulation to that area. Muscular tension in the jaw is not only related to pathologies of the ears, but can also create issues of the teeth, jawbone, and disk degeneration in the neck!

Healthy ear tips:

1. Massage the area of the neck just below the ears and pull on your ears from all different angles to stretch and increase circulation to the temporalis muscle that your ear attaches to.

2. Vigorously massage your "chewing muscles" before you go to sleep, or when you have a bit of free time in front of the TV or computer.

3. Let your head hang forward to stretch the back of your neck, and take turns dropping each ear towards its corresponding shoulder. Your chin should touch your chest and your ears should touch their corresponding shoulders. Really! Do this every day, spending a minute holding each stretch as you relax your shoulders.

4. Reduce stress levels (are you seeing a trend here?)

Many people come into the institute and report they have "age-related" hearing loss...in one ear. I always ask, "How old is your other ear, then?" We need to start thinking about these health items we accept to be true. When I point out a little item like "If you have age-related hearing loss, but only in one ear," ask yourself, does that actually make sense? Just putting a bug in your ear. ☺

 CAN YOU SEE ME NOW?

When I was in Africa, I was struck by many amazing things. But one thing in particular was the lack of eyeglasses. In the country of Zimbabwe, I would pass hundreds of children walking to school in brightly colored uniforms denoting their grade level—purple, orange, yellow. Giant smiles, agile limbs, and no glasses! It was a beautiful thing to see. Back in the US, however, The National Eye Institute (a branch of the National Institutes of Health) says that nearsightedness (called myopia) has increased from twenty-five percent to forty-one percent since the 1970s. What's up with that?

Farsightedness, or the ability to see things at a distance, is the most relaxed version of the eye. When the ciliary muscles inside the eye are relaxed, the lens in your eye flattens and allows distant images to be seen clearly. When you need to look at something up close, the ciliary muscles contract, making the lens more round, which gives more clarity to the image just in front of you. Up-close vision requires muscle tension, distance vision means muscle relaxation. Dig?

So the million hours I spent with my nose in a book as a kid probably had a lot to do with training my eye to see up close. Nice. Big brain. Big glasses. Guess who would have been lion supper back in the jungle days? The nerd in the front row (me). Cool.

The increase in myopia can probably be tracked to the hours and hours we spend training our muscles to tighten—school, computers, reading, television. What if you picked up a fifteen-pound weight and held it up with your bicep muscle group? Eventually the muscles in the arm would shorten, to make the task easier to do. This fatigues the muscles and re-programs the motor units between the brain and arm, resetting the resting length to be the shorter version. With all that up-close eye strengthening, your eye muscles may not even remember how to relax!

Parents and teachers, outside time is not only about large muscle ex-ercise, but also about the smaller ones. While tracking animals in Africa I had to be *taught* how to relax my eyes to see the giant herds of animals in front of me. I didn't know how to see in three dimensions until I real-ized that I held my eyes tense at all times! Try taking a break from your computer screen and look out a window for a minute or so every hour. You should be able to see layers of images, and not just one flat picture. Can you see which tree is the farthest? This is a great way to practice your long-distance vision. It will also keep you from wearing out your up-close vision so you have greater muscle health longevity in the eyes as you move into the later years of life.

 AND THE OSCAR GOES TO...

March 13 2010

Jeff Bridges in *Crazy Heart*! Awesome for The Dude. I really enjoyed this movie, right up until Maggie Gyllenhaal's clavicles appeared on the screen, freaking out my biomechanical sensibilities.

Clavicles (also known as collar bones) should be horizontal, but Miss Maggie has the habit of hiking hers up so they form a "V" shape. If you watch her on film, you can see that she has a tendency not only to lift her shoulders up to her ears, but also to round them forward. You may think "big deal," but the clavicle bones have muscles attached to them. And when you start moving the bones around, the attached muscles become ineffective at doing what they are supposed to do, which is hold the head directly above the shoulders. If you take a look at Maggie from the side view, you can see that her head is significantly out in front of her. If she would keep those bones down, it would help hold her head in the correct position. So much of her health could be affected because of this habit— disks in between her neck vertebrae, circulation to the head (wonder if she has headaches), thyroid health, middle back tension, and shoulder stabilization.

Where should your clavicles be? They should create a horizontal line across your chest, and not make a "V" shape on your chest. Muscles that help anchor your clavicles are the latissimus dorsi (your lats). These are large, downward-pulling muscles that are constantly fighting the upward tension in the trapezius muscles (sides of the neck). Lifting your clavicles can be a postural habit you have of communicating your mood (I lift mine when I'm feeling embarrassed), OR the result of holding a heavy bag on one shoulder (you can have one clavicle higher than the other).

Take a look in the mirror after a shower and evaluate your clavicle position. Any award-winning positions out there?

P.S. Maggie, if you're reading this, I am happy to give you a complimentary session. Call me.

March 17 2010 | IS IT FREEZING IN HERE, OR IS IT JUST YOUR SHOULDERS?

For a week now, you've been reading about the biomechanics of the body parts north of the shoulders. From a physiological standpoint, the health of this area is a huge deal. The bulk of your sensory input (seeing, hearing, smelling), the command center of your entire body (brain), and your heart (circulation) and lungs (breathing) are all wrapped in and affected by the muscles of the shoulder girdle.

One problem I see constantly at the institute is the lack of range of motion of the gleno-humeral (GH) joint (shoulder). FYI, there is no "shoulder bone." The gleno-humeral joint is a bunch of muscles and bones that come together at one point, which we all call the shoulder. Many people are dealing with a shoulder impingement issue (deltoid tendonitis, bursitis, bicep tendonitis, or a rotator cuff tear) or frozen shoulder (yuck) and it is one hundred percent due to the poor health of the muscles located here. Try this exercise to test your range of motion.

Reach your arm behind you, keeping it straight across if you are a beginner or bending the elbow more if you have more mobility. Hold here for one minute (image below left).

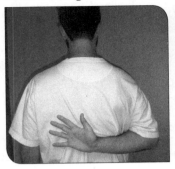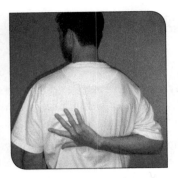

Then, to see what's *really* happening in the shoulder, flip your palm down so it is (maybe) flat against your back (image above right). Holy cow, right? Rotating your palm this way is a motion called supination. When you supinate your lower arm, it should not affect the shoulder at all, unless the muscles have gotten really tight.

How tight is your shoulder? You may also notice that one shoulder feels different than the other, yes?

The better you get, the more you can bend your elbow, until your forearm is more vertical and less horizontal. Many of you won't have any problem with this at all, due to hypermobility of the scapula (shoulder blade), so I am going to add one more instruction. When your shoulders are very VERY tight, your scapula will "wing," which means you will see the bony edge of the shoulder blade will poke out of the skin. Winging is a desired look for some, and why I don't know. The reason you see a wing is because your shoulder blade is now oriented front to back instead of right to left. This is a bad situation for the cervical spine, circulation to the head, and strength of the upper body. To each his own!

While doing this exercise, use your other hand to feel for any winging, or ask someone to help you out. Your scapula should not move when your arm does! You may have to pull your shoulder far forward to get that wing to flatten out, and when you do, the position of the arm bone (even if it is way out in front of you), is the actual length of your muscles. See if you can keep the scapula flat when flipping your palm down, too.

The harder this exercise is for you, the more you need to do it. Rotator cuff injury, neck tension, and optimal cervical–thoracic muscular function (all the muscles from the ribs to the skull) depend on this range of motion. If you don't have it, you are accepting sub-standard performance. Don't sell yourself short! Enjoy!

When I was a kid, I loved books, my Michael Jackson tape, Greek myths, and my accordion. Oh, and unicorns.

When I was a kid I hated eye drops, nose drops, eardrops, and throwing up. And rectal thermometers.

There are few therapies that I won't try, if it makes sense and I can see a reason for it. Even though I am extremely open-minded when it comes to this stuff, I still freak out if someone tries to shove something in my nose. I'm not talking about people trying to hurt me, but a nasal spray. You bring that bottle out and I turn into a tantrumy second grader. Foot stomping and the whole thing. I act like a total baby.

I seemed to have picked up a cold over the weekend, and my throat and nose were feeling kind of crummy and mucousy. I hopped into the shower and I am SHOCKED at what happened next. I cupped my hands, filled them with water, and snorted it up my nose. This is a big deal. A HUGE deal! Completely out character. I basically gave myself a water board treatment, inhaling water up into my sinus cavities to rinse them out. It's very similar to the sensation you get when you accidentally get water up your nose when swimming. And I felt better instantly.

Your sinus cavities, while not containing muscle as anatomists have defined it (smooth, skeletal, or cardiac), contains contractile tissues called cilia. Cilia have the ability to contract and generate force just as muscles do, only they do it by spinning. When the cilia spin around, they create an oscillating effect, wagging their little tails. The linings of many organs contain cilia to move particles out (dust and pollen from your sinuses and lungs) or move particles in (the cilia in the fallopian tubes move the egg from the ovary to the uterus). When you have mucus accumulating in your head, your cilia are not performing correctly.

There are two things that can decrease the function of the cilia, mucus thickness and physical resistance. Mucus thickness: The more viscous the mucus, the harder it is to move. Excessive particles will thicken clear mucus just like mixing flour into water will create a more tacky fluid. When pollens increase, this thickens the mucus, making it too heavy for the cilia "muscle." Physical resistance: Your body can actually decrease the motion of the cilia by increasing the pressure inside the cavities. Excessive tension in the head and face muscles will increase the pressure inside the sinus cavities and decrease the amount of mucus leaving the area. Squinting in the sun, stress, jaw tension, eye tension from reading or computer-ing, furrowing your brow, and even sniffing increases pressure and results in more mucus.

We are so used to thinking that "allergies" are the result of a histamine response, but sometimes mucus is less a chemical reaction and more a physical one. A couple of tips to decrease anything you may be doing that is making your situation worse:

1. Booby-trap your nose. Add a layer of tacky substance (I use my organic lip balm) to the rim of and just inside each nostril. Par-ticles will get caught on the stickiness instead of making it up into your sinuses. Be sure to change it every few hours, or more if you are walking or playing outside.

2. Closing your eyes, see if you can tap into any face tension you didn't know you were carrying. Relax each feature (each eye, nose, ears, jaw, gums, lips) and then relax the spaces in between each feature. I am always surprised to find I am tensing the back of my throat, or the muscles behind my eyes.

3. Rinse your nasal passages. If the mucus has gotten heavy, the cilia can't get any work done. Doing a rinse helps remove excess mat-ter and can give the cilia the leg up they need. If you have chronic sinusitis, avoid chlorinated water as this could further inflame the tissues. (The shower is a great place to do this. You can hock up water and spit it out, and there's no mess!) Avoid hot or cold wa-ter, warm water is best.

4. Studies show that cilia seem to move best in warm and moist en-vironments. If you live in an arid environment, or the season is cold, try keeping a teakettle boiling or invest in a humidifier. Your sinuses will thank you!

April 19 2010 CHEW YOUR FOOD!

This morning on NPR there was a very interesting segment on bone density in the face.[1] Dr. Langstein, a plastic surgeon, was talking about why facelifts don't always look right. People wanting a more youthful appearance end up looking "windswept." Turns out, windswept is a look best saved for your hair, and not your jowls. Dr. L didn't know why this was happening either, so he collected CT scans on skulls to see what was happening beneath the skin, fascia, and muscle tissue he was used to working with. The results? Turns out that shrinking bone density in the skull was contributing to skin sagging just as much as the fibers in the skin.

Best line of the article: "It's one of those things that, in retrospect, you sort of say, 'Duh, I should have known that!'" says Langstein. Yeah. That's what I want my surgeon saying.

Why is our bone density decreasing in our face? The default is to think that all tissue is aging, so it therefore it all must be shriveling up. And while aging is going to leave you with tissues different than those in your youth, the key to keeping the bone density in your face is keeping your teeth.

The signal for maintaining your bone density is a mechanical one. You can throw fancy chocolate calcium chews into a blender with milk, coral, TUMS, calcium-enhanced orange juice, and kale and, if you haven't "squeezed" the little cells inside the bone, your bones won't continue to generate. Nutrition is not enough to keep bone density high. You have to use the bones you want your body to maintain. The alveolar processes of the maxilla (also known as the mustache bone) hold the teeth. When you push against your teeth, the teeth push against the bone, signaling the bone to stay dense. Now, you can sit there pushing on your teeth all day and hope no one asks what you are doing, OR, you can chew your food.

You may not know this, but the amount of processing our food undergoes requires us to put very little force into eating. The term "mastication" simply means to break food down with the teeth. Chewing begins the digestion process by breaking food down to increase surface area, warming food up via agitation, and enhancing enzyme action. Chewing is an herbivore thing. Cows are kings when it comes to chewing food. They chew, swallow, and move food to the stomach to extract nutrients, then chew it again, and repeat the whole process a few times. (I do that sometimes too, but only by accident.) Carnivores don't have the same motor program for chewing, hence the term "wolfing it down." We come with the chewing reflex, as our diet for the last 200,000 years has been mostly plant-based.

Not consuming a plant-based diet is only part of the problem. Knives and heat have also drastically reduced the amount of masticating we do. Picture a potato. Not a baked potato with butter, sour cream, and chives, but a raw, cold, and slightly dirty potato. Imagine the gnawing you would have to do to eat it. Imagine the forces it would put on your teeth, and therefore the bones that hold the teeth. A lot more work, right? Applying heat to food reduces the amount of energy our bodies have to apply to eating. And while it may be more pleasant to eat softened foods, our bones pay the price.

Another factor in the mustache bone wasting away is the liquid-based diet. How many of your daily kcals are coming through a straw (or Starbucks cup)? No tooth work required there, which is not only damaging to the bone, but also to the teeth. The mastication process is a self-cleaning one. Chew the food and you get an increase in saliva production, which helps wash the food off the teeth. Consuming a liquid-based diet gets you sugars in your mouth with little saliva and tongue action, increasing your risk of cavities.

For better bone density in the face:

Eat firm, raw veggies throughout the day. Radishes, celery, and parsnips are a great low-sugar option. Apples and carrots are delicious and sweet.

Replace one "soft meal" for a firm one, or add raw ingredients to your soft meal to increase the chew power. I like to add raw corn to salads or warm bean entrees, as it gives me something to chew.

Chew your food! Even if you are eating cooked foods, you can still chew them down to finer pieces.

Turns out an apple a day CAN keep the doctor away. The dentist, the endocrinologist, and maybe even the plastic surgeon. Duh.

---

1. npr.org/templates/story/story.php?storyId=125387566

 July 16 2010 **$2000 FIFTEEN-MINUTE CHAIR MASSAGE**

This groundbreaking information on NPR this week: Amazing Massage Technique "fixes" mysterious throat ailment.[1]

> Julie Treible caught a common spring cold this year and lost her ability to speak—for six weeks. The cold had turned into bronchitis, then laryngitis—at least, that's what she thought.
>
> "April 1 I got a cold, and this part never got better—my voice," says the 43-year-old single mom in a forced whisper.

Treible communicated with her family using a white board, because trying to talk made her dizzy. She even passed out once and fell down the stairs, scaring her ten-year-old son and landing her in the hospital.

They put you in the hospital for scaring your kids?

Scopes of her throat, CT scans, an MRI, and visits to various specialists yielded no answers. But after an appointment at the Cleveland Clinic Head and Neck Institute, she's hopeful that will change.

Um, does anyone want to take a stab at figuring out what multiple scopes, CT scans, an MRI, and visits to various specialists cost Treible and her insurance?

Claudio Milstein, a voice specialist, is optimistic he can get Treible speaking again and send her home that day. With Treible seated in a chair, Milstein stands behind her and places his hands on her neck. He begins by manipulating the muscles in her throat with his fingers.

"I'm going to stretch the muscles around your voice box," he says. A few small pops are audible as he works. "Those cracks, you hear that? Those are normal, so don't be concerned."

Don't be concerned even though this sounds like I may be breaking your neck, okay?

Treible remains silent, wincing slightly as Milstein presses against her voice box. He turns her head from side to side and then presses down on her shoulders. To the casual observer, it looks something like a chair massage.

Dear Casual Reporter, I think it looks something like a chair massage because it *is* a chair massage.

While Milstein works, massaging her shoulders, neck and throat, Treible makes a continuous "e" sound, pausing for breaths, then switching to "oooo." Within minutes, the raspy whisper begins to disappear.

By massaging and relaxing the muscles in her throat, Milstein was able to release the vocal cords and allow Treible to speak again—all in about seven minutes.

"In order to produce voice, you need to have a very good balance of muscle tone in a lot of different muscles that are involved in voice production," Milstein says. "So an intervention like this, where basically you manipulate these muscles and try to restore the internal balance, is very effective."

The problem, it turns out, is fairly common. Milstein estimates he sees about 10 patients like Treible every month, some of whom have been without a voice for up to two to three years. One reason for the delay in treatment, Milstein says, is that things like cancer, infection, acid reflux and neurological problems need to be ruled out—and those diagnostics may involve multiple doctors, medications and expensive tests.

Wait, back up a second. The "let's check if your neck muscles are too tight to allow your vocal cords to work, and then spend seven minutes fixing that" idea is lower on the checklist than "try multiple doctors, medications and expensive tests?"

The rest of the story goes on to say that no one knows why or how this condition occurs. Really? Are we, a culture that offers the most technologically advanced medical treatments and spends the most on researching treatments, forgetting the basic laws of musculoskeletal science? Did we forget our freshman-level physiology and anatomy classes? Ever take biology? What about Physics 1?

Allow me to clear my throat before yelling YOUR MUSCLES ARE NOT JUST FOR EXERCISE (they're not even FOR exercise!!!), BUT DRIVE *EVERY* FUNCTION IN THE BODY, including your vocal cords. <Coughing fit> In every health issue, consider what the muscles in that particular area are doing. Underworking? Overworking? Are they stiff and rigid to the touch? Sore to the touch? You might have an inexpensive treatment in massage. Wait, you mean massage works? Guess what? Bodywork is one of the oldest forms of treatment for many ailments, and currently, a well-trained therapist can be a great source of help for a multitude of aches and pains.

As for the neck and throat area in general, this is a critical part of the body. Circulation to the brain, lymph node waste removal, thyroid function, and, as you now know, your vocal cords, all require supple, flexible muscles to work optimally. Stiff in the neck? Loop a warm or hot wet towel around the neck to "steam" tense muscles, especially after long bouts of talking (especially if you're ranting). Find a well-trained, licensed massage therapist to help you unwind those stubborn knots, and finally, stretch.

Stretch your neck and shoulders *every day*.

Try this: Let your right ear fall to your right shoulder (yes, it should touch) for a few one-minute bouts per day. Don't force it, but allow it to slowly relax and yield. Don't forget to do the other side. Also, let the chin drop to the chest (it should touch here as well), keeping the back of the neck long.

Okay. I'm done now, and unfortunately will *not* be getting a massage now. But a hot bath sounds good...

---

1. npr.org/templates/story/story.php?storyId=128359885

THE EYES HAVE IT

I wrote a while back about the muscles of the eye. I've said it before, but that rarely stops me from saying it again: The length of your muscles creates physiological states in the body. These physiological states of the body are given names that sound all formal and scary and disease-like. *Myopia* is the medical term for nearsightedness. Myopia could also be called *muscles in the eye that are too short*—but that doesn't sound as smartypants as myopia.

This is a creepy picture of my myopic eye:

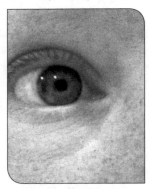

The problems with science done by one large group of people that all hold the same cultural beliefs and habits is, they rarely look at their habits as disease creators. But of course, the solutions to health problems, just like our car keys, are often hidden in an obvious place, aren't they? As I wrote last year, the rate of myopia is steadily increasing. Tension in the eye comes from the failure to use our eyes in their relaxed, long-muscle orientation. Looking at faraway places or gazing at layers of trees upon hills all utilize different muscle patterns than looking at computers and books and iPhones and Kindles. And the latest generation of kids will have worse vision than we do.

Natural movement is what our bodies require to perform optimally, and that includes our eyes. But if you can't be outdoors all of the time, shoot for a couple of hours a day. Fourteen hours of outdoor time a week is correlated to a decrease in myopia, as presented at the 115th Annual Meeting of the American Academy of Ophthalmology last week.[1]

In addition to getting your eyes outdoors (hey, I know, why don't you blend your outdoor vision relaxation with fourteen hours of weekly walking?), there are corrective exercises too. Just like you can train and restore the big muscles in the body, you can do corrective exercises for the little bitty ones. William Bates, a physician from the 1920s, came up with an entire program dedicated to restoring the proper function of the eye. I was totally stoked (totally!) to find that the content of his book *Perfect Sight Without Glasses* (1920) and years and years' worth of *Better Sight Magazine* is free online.[2] I've never read his book, but it's on my to-do list, as is trying out his program.

Another thing I do for my own personal vision issues:

I have worked on my own eye-muscle meditation for the last year. I sit quietly (without any glasses or contacts) and close my eyes. I imagine I am an inspector of my eyeballs and teeny tiny muscles. With my eyes closed, I let my brain go exploring for tension patterns that are so habitual for me, I don't even realize I am constantly gripping my eyes. Once I find tension spots, I let them go—kind of like finding out your jaw is clenched (hey! when did my jaw get all clenched!) and relaxing it.

I use this technique for headaches too and have found that all I need for eye strain and headaches are about eighteen minutes of eye-socket spelunking. Try it. It is awesome, effective, and free.

P.S. It would make a lot of sense if Bates' books were audio, right?

Then we could listen to them while walking outside.

And for the physics and anatomy nerds out there—your body, while it seems like one fixed entity, like a point mass, each individual part of your body (and then, each teeny tiny part of every teeny tiny cell) is responsible for dealing with its own excessive distortion created by the way you move. Notice, in this slow-motion film of the eye, that when the bulk of the eye has stopped, the smaller parts of the iris wobble as they catch up.[3]

Have you ever pulled a wagon with a ball in it? Ever noticed that, when you stop the wagon, the ball continues to roll? Everything has its own inertia. Even the parts of your eyeball are independent from the outer eyeball and the muscles that move it.

These micro-wobbles are happening all the time, in all the tissues of the body and to a much greater extent in bodies that are not moving in a choice-driven, stabilized, and controlled manner. These micro-movements add up over time to disease or injury. To minimize them, each of your joints needs the correct amount of mobility and stability—which, of course, requires the correct muscle length.

Love always,

A Broken Record.

---

1. sciencedaily.com/releases/2011/10/111024084639.htm

2. en.wikisource.org/wiki/Author:William_Horatio_Bates

3. youtu.be/Fmg9ZOHESgQ A three-minute video showing slow-motion eye movements. The iris REALLY WOBBLES! Amazing and weird.

chapter 9

PREGNANCY, CHILDBIRTH, BABIES, AND CHILDREN

I'M WRITING AN ARTICLE
FOR *MIDWIFERY TODAY*

I stood behind my sister as she breathed through her fourth hour of contractions. As a nulliparous woman, I was completely in awe of her natural ability to grow and then pass through her a completely formed human being. What a gift of nature. And even though I was focused on attending to her every need, the scientist in me couldn't help but watch the event through the eyes of an anthropologist. I became aware of her posture while laboring. Now, I am not a mother yet. But I am an expert on the anatomy, physiology, and physics of the female pelvis—a graduate-level biomechanist that has an entire facility dedicated to increasing the performance of the human machine through restoring the alignment of the skeleton. I teach how the natural functions of the body—like circulation, digestion, bone growth, and birth—work best. And there's no better time to have your pelvis working at its best than during the birthing process.

So here's what I was observing while massaging the lower back and pelvic area of my sister. Her feet were turned out because the muscles in her calves and hamstrings were tight. Her pelvis was tucked under following years of precedence set by the women in our family before her. I couldn't help but coach her into the postures I knew would allow her musculature to aid her in the delivery process. Me, the fool, expected her to take this advice with a cheery attitude. "Thank you, big sister, for telling me how to do it 'right!' Just like old times." But I had been with her through three of her four births, so I knew not to take it personally. She pressed on in her fashion and delivered a healthy, robust 10.4-pounder, and I knew that I would be writing this article on my flight home, using her as my muse. An article that would have given her the information on how to best prepare her body before the big game. An exercise program that could be done much in the same way as a baseball player would train before the first pitch. And a piece of essential information that birthing professionals and those responsible for women's education could disperse to their ladies-in-waiting to empower them to take control of the somatic birthing system in addition to natural instinct.

More to come… ☺

HOW TO RAISE KIDS

While I don't have any children, I feel I am perfectly qualified to write this blog as I am armed with the *The NEW MODERN MEDICAL COUNSELOR* (the 1954 edition!). I picked this gem up at a garage sale last month, and my purchase has already generated the most fun I have ever had under five bucks.

First, let's take a look at the Suggested Daily Schedule for the Child (from ages two to six):

7:00 a.m. Rising. Toilet. Wash hands and face. Brush Teeth. Dress.

7:30 a.m. Breakfast. Toilet for bowel movement. Wash hands. Out of doors when weather permits. Play in sun whenever possible.

10:00 a.m. Toilet. Wash hands. Outdoor play when weather permits.

11:45 a.m. Toilet. Wash hands and face. Rest.

12:00 p.m. Dinner, which should be the heaviest meal of the day.

12:30 p.m. Dress for nap. Toilet. Wash hands. Nap.

2:30 p.m. Toilet. Wash hands. Dress. Milk, fruit, or fruit juice if desired and if experience proves it does not impair the appetite for supper. Out of door play when weather permits. Play in sun whenever possible.

5:15 p.m. Supper. Toilet. Wash hands. Brush teeth.

6:00 p.m. Bed. Lights out, windows open, bedroom door shut.

Wow. Besides the food recommendations (not a fan) and possible bathroom-fixation issue this kid may face as an adult, check out what Child Y has learned:

Movement should make up the bulk of my day!

Rest is crucial! I sleep when I am tired (and don't reach for the cup of coffee instead!)

I go to bed when the sun goes down so I can regenerate all the used cells and allow my brain to integrate the information from this day.

Talk about solving a lot of health care issues we face as a nation! By my calculations, this child moves 5.5 hours per day, soaks in the natural vitamins provided by the sun (we may need a little sunscreen these days), and celebrates with a two-hour siesta mid-day. Um, I'd like to sign up for this suggested program, please!

Next...

> Children are normally active, and one does not often need to urge them to take more exercise. It is wise, however, to study the probable effect of the activities which they are seen to choose, and to provide a good place and suitable playthings and apparatus so they will naturally choose the types of play or exercise that will promote good posture and lead to the harmonious development of all their muscles.

Translation: If you put a computer or TV there, a kid is going to "naturally" choose to slouch in front of it all day (it's the biological law of thermodynamics!). It is best to not give them the option to choose poorly and let movement and healthy joints, dense bones, and appropriate body weight happen "organically."

On food...

> As children emerge from babyhood, great care should still be taken in educating their tastes and appetite. Often they are permitted to eat what they choose and when they choose, without reference to health. The young are led to think that the highest object in life, and that which yields the greatest amount of happiness, is to be able to indulge the appetite. The result of this training is gluttony, then comes sickness, which is usually followed by dosing with poisonous drugs.

Well, that's just a little bit of 1950s fortune telling, isn't it?

> Parents should be capable of rearing them in physical, mental, and moral health. Parents should study the laws of nature. They should become acquainted with the organism of the human body. They need to understand the functions of the various organs, and their relation and dependence. They should study the relation of the mental to the physical powers, and the conditions required for the healthy action of each. To assume the responsibilities of parenthood without such preparation is a sin.

WHOA! That is a bold paragraph, but I will have to agree, as our habits of health are developed early on. The more you can coordinate your family's biorhythms with the rhythm of nature, the more rested and relaxed the entire clan will be. Stress is a huge component of disease across the board. Take the time to teach CALM. Plan a day with no agenda but to walk around, enjoy light play, or snuggle on the couch as a family.

Finally, I'd like to end with what I feel is a great suggestion from our friends in 1950.

Being the sort of person a parent wants his child to be is the best way of helping him to become that sort of person.

(There were no women in the 1950s...obviously.)

THE HUNTING AND GATHERING **MAMA**

If you want to start a fight at the next party you go to, bring up birthing politics. Hospital or home. Drug-free or epidural. Birth tub or stirrups. But not so long ago, there was only one option. It was called, You're Doing This Now, Whether You Want To or Not.

I have no problem stating my personal preference for a natural, drug-free, non-hospitalized birth. The scientific and statistical evidence supports that, if all goes well, this is the ultimate in "healthy" experience for the mother and child.

One of the biggest supporting arguments of natural and non-hospital births is that the birthing process is an instinctual and natural event that has been happening for hundreds of thousands of years without outside aid. This is true. But there are many additional things the hunting-and-gathering mother was doing that prepared the delivering muscles much better.

Looking back a hundred years, the birth-induced mortality rates (of both baby and mother) of non-industrialized civilizations were more favorable than rates of births happening in city areas. Medical journal articles from the 1800s were looking at this fact back even back then! Why were women who lived in the larger, more industrialized areas of London having such a difficult time birthing compared to Romany women and Tinkers? The populations without medical intervention fared better (less death and cranial deformity in babies and less death or injury in the mothers) than those with the advantages of surgeons, antibiotics, and a more sterile environment. Why?

The answer back then, and still quite relevant today, had to do with movement habits, specifically the quantity of walking done per day as well as the practice of squatting to defecate and urinate. The squatting action, preferably done since birth, creates a wide pelvic outlet (the space where the baby passes out). Starting from childhood, squatting to eliminate aids in the ideal ossification (bone shaping) of both the pelvic bones and the sacrum. The wider the outlet, the safer and easier the baby passes through. Squatting also lengthens the muscles of the glutes, hamstrings, quadriceps, calves, and psoas. When these muscles are tight, they can actually reduce the movement of the pelvic bones and increase stress and pressure on the baby (and mama) during delivery. Back then and today still, the populations of people who move a lot (and I don't mean exercise an hour per day) have better, easier births.

If you want to go about a birthing process naturally, then you should add in the other "natural" processes women have been doing for years—walking five to six miles per day (this doesn't have to be done all at once), and restoring your muscles, joints, and bones to their correct function for a fluid and safe birth. Needing to "train" for delivery does not mean that the birthing process is not natural, only that the mother has let her "birthing equipment" adapt to unnatural living (sitting in chairs five to ten hours per day, not stretching the bathrooming muscles of the pelvic floor and legs) and has become out of shape when it comes to whole-body endurance.

Whole-body endurance is your ability to walk long distances using the strength of the legs. Being able to support yourself with your leg weight will come in handy when you need to labor for an extended (but hopefully not too extended) amount of time. Cycling, aerobics, running, and swimming won't get you a stronger strength-to-weight ratio. A hunting and gathering mother needs to mimic the daily required walking that keeps her birthing muscles in top form.

Delivery preparation tips:

1. If you aren't walking at all, begin with one mile per day, increasing your distance by half a mile every two to four weeks, until you hit five to six miles per day. Doing all your mileage at once will help you with endurance, but if you are feeling tired or sore, break your distance up over the course of a day.

2. Start your squatting program NOW. Hamstring and calf tension (both muscle groups down the back of the legs) tuck the tailbone and pelvis under, instantly impacting the size of your delivery space. FUNNY STORY: I made the mistake of trying to teach my pregnant sister this exercise while she *while she was giving birth.* I'm not going to write down what she said here. ☺ It's also imperative that you learn to squat. Holding onto a doorknob will help you until your legs are strong enough (and loose enough!) to hold you in this position for a minimum of one minute. A rolled towel or yoga mat under the heels will also make this exercise more easy to start. Work up to doing it with your heels on the ground!

Note: The picture on the left (above) shows a tucked tailbone and rounded spine, NOT the position you are trying to achieve. Also note how far forward the knees have moved. To really open the pelvic muscles, untuck your tailbone by sticking out your bottom (see picture on the above right and note that the knees are over the ankles). This will get your pelvic bones in the right birthing (and bathrooming) position.

Not giving birth in this lifetime? This information still applies to anyone wanting to optimize low-back, pelvic-floor, and digestive-system health. If you haven't fully flexed your knees and hips in some time, it's best to start lying on your back, pulling your knees to your chest to "warm up" the joints. All over the world, eighty-year-olds drop down into this squat like it's no big deal, because they have been doing it three times a day at minimum! It's time to get started!

The more we prepare for delivery, the more successful the outcomes. The greater the success stories of natural birthing, the more it will be seen as a viable option. Make your natural birthing story by preparing…naturally!

July 22 2010 NATURAL PREGNANCY, NATURAL BIRTH

So, you'd like to run a marathon, you say? Great. The laws of specificity state that to improve performance at a task, you must train correctly, using the muscles you'll need for the event. Swimming, while keeping you "fit," isn't going to help much. Cycling, while keeping you "fit," isn't going to help much. You must do with your body what you'd like it to do. If you want to run a marathon, you're going to need to do quite a bit of running for the best outcome.

So, am I hearing you correctly, that you'd like a natural delivery? Well then, following the science of physiological adaptation, you must train your body specifically. We need to train for delivery because, while birthing is absolutely a natural event, we have become, it seems, unnatural women. I know, I know. You eat organic food. You take yoga classes and wear Birkenstocks. You even drive a hybrid car or maybe take the bus every now and then. Maybe. And while these are all very eco-friendly things to do, they are, for the most part, completely foreign to our biological animal, which makes them completely unnatural.

Natural, in its broadest definition, means "in accordance with nature." Well, the last time I checked, nature wasn't busy designing toilets, chairs, cars, or shoes. Nature doesn't exercise four or five times per week. Nature doesn't exercise at all, but rather moves continuously throughout the day. Nature doesn't eat foods not available to the location or season, even if they are nutritious. Nature also doesn't secrete stress hormones while commuting from one part of the forest to the other, affect metabolism regulation with a flick of a thermostat, or take anti-inflammatory medications at the drop of a "my back is sore." We have, within a few thousand years, completely reduced our ability to be "natural," yet we still partake in these amazing, natural processes of digestion, sensory input, elimination, growth, and of course, birth.

The first thing to understand is, while pregnancy may feel like an unnatural state for your body, it is actually quite natural. What makes it feel so awkward and possibly uncomfortable is the extreme loading done on an unbalanced, rickety frame. I once bought a cool table from the Goodwill, even though it didn't balance quite evenly. It wasn't that big of a deal in the store, but once I brought it home and tried to put stuff on it, the lack of stability became more of a functional issue. It's the same thing with all of you out there with chronic low back and pelvic pain, feet that are flattening, birthing canals that are narrow, and abdominals that are splitting (diastasis recti). These are not issues of pregnancy, but issues of pregnancy on an unstable frame. A woman who wears her pelvis out in front of her is not a Stable Table, if you know what I mean. Loading her up with twenty-five, forty-five, or sixty-five pounds is going to increase the effects of this misplaced weight and make pregnancy more difficult than it needs to be—way more difficult than it is for other animals.

You've come with all the equipment needed for a successful, natural birth—a movable sacrum, a strong transverse muscular system that runs in series with the uterus, and thick thigh muscles that support the entire weight of the torso. But guess what? Poor alignment, especially the forward thrust of the pelvis, turns all of these things off. The sacrum becomes jammed up and the more mal-aligned the body, the weaker the abdominals. Thrusting your hips forward also pushes your belly contents right through the wall of the abdomen. Another fun fact: Diastasis Recti has nothing to do with pregnancy. It happens in men and women who habitually thrust their hips and have extra stuff in the midsection. Beer or baby belly, it doesn't matter. You want to avoid it? Stop shoving your guts through your abdomen. Stop thrusting your hips and wearing shoes with heels. If you want stronger leg, thigh, and hip muscles you have to walk...a lot, like animals do. You have to squat often, like animals do. If you want to have an optimal natural delivery, you should train with a natural pregnancy.

There are many pregnancy "myths" that have permeated their way into our cultural understanding of birth. This misinformation makes obtaining the correct birthing mechanics more difficult. In graduate school I wrote a paper outlining all of the research on what we *think* are birthing truths. My favorite study was on the pregnancy waddle. You've all seen a TV show from the fifties that showed Mom-to-Be in a flowery pregnancy frock with her hands on her back, belly shoved forward, struggling to get up off the couch and walk to the kitchen to get some pickles and ice cream (which I discovered is pretty awesome, by the way...). Well, that walk isn't a natural occurrence with pregnancy, but the walk of someone who doesn't have the strength to carry the additional weight. My grandpa walked like that too, if I recall. Yes, your midsection is growing, but if you were in the correct alignment, the glutes, hamstrings, and transverse abdominals should also be growing equal in strength, to keep you walking perfectly upright and not so much like a staggering sailor.

My paper also called for this information to be taught to birthing professionals, fitness professionals, nurses, and doctors, to pass on to moms-to-be, to optimize their mechanical ability to birth easily at home. General prenatal fitness has very little to do with real birthing mechanics, as required by the laws of specificity. It's kind of like swimming to train for a twenty-five-mile hike. The swimming isn't bad for you, but isn't the best program design.

Some training tips:

1. Get to know the geometry of the body. I'll continue to post which markers to look for.

2. Get out of positive-heeled shoes. It will make all the difference in the world!

3. Squat, a few times every day. If your body is already too damaged to squat, follow the more basic, non-squat exercises until you are strong enough to handle the full range of motion.

4. Walk, walk, walk. Work up to five miles a day, if possible, broken up throughout the day if needed.

5. Minimize sitting in chairs and change up your sitting postures often.

6. Find your transverse abdominals and see if you can fire them.

7. Stop tucking your pelvis, right now. In fact, stick your butt out while you're reading this.

Midwives: What if you could help prepare your mommy's mechanics? Moms, Midwives, and Every Woman are invited to take the No More Kegels course that will walk you through the exercises you need to know for optimal delivery, pelvic floor health, and knee, hip, & low back longevity.

And, for those of you who want to see some serious natural birthing going on, check out this elephant birth.[1]

---

1. youtu.be/lllv20mScP8 Six graphic and amazing minutes of an elephant giving birth.

August 2 2010 CAUTION: KIDS (NOT) WALKING

I just got back from vacation, visiting the Pacific North West, where my sister and her twenty-five kids live. Did I say twenty-five? I meant four. But is there really any difference? Not to me, and, probably not to her either. Speaking of fun times with kids, have you ever tried to take a fast walk with a two-, five-, and seven-year-old? Turns out you can't travel at the pace you'd originally envisioned. Turns out that walking with little kids is less like walking and more like herding cats. Have you ever tried to get three or four cats to do the same thing at once? Turns out that walking with kids is more difficult than one (without kids) can imagine.

There are many reasons why walking with little ones is challenging, but the most basic is, they aren't at the same level of physical performance as you. They get tired. Then they get pokey with their little hands and feet. Then, in a team effort, they all manage to start crying at the same time. They also seem to manage to do it when you're at the farthest point from the house. It's awesome. Turns out exercise isn't always a stress reducer. Well, maybe us exercising was relaxing for their mom, who got to stay home and take a nap.

So, where does the walking child's fatigue ("I'm tiiiiired!") come from? After all, we are beings that (at all ages) have always walked, for hundreds of thousands of years, multiple hours per day, with our very survival based on walking endurance. Simply explained, a child's fatigue is a result of poor training. Based on my own personal logbook of thousands of miles walked, I can say that I have rarely seen a child on a walk beyond returning to the car from the restaurant. Oh, I've seen kids held and packed. I've seen them on skateboards and bikes. I also see kids WAY too old to be in strollers being pushed along while mom is watching her heart rate monitor. Law of Specificity indicates: If kids don't walk, then when they do walk, it's too hard. Seems like a pretty simple answer, but really, this is the basic principle of exercise science. Of course kids are playing on jungle gyms and participating in sports—which is fine—but they aren't walking. (Don't get me wrong, I see teenagers glumly walking around wishing their parents weren't so lame, but I'm really talking about kids aged two to twelve, with developing bodies.)

In the olden days (not like two hundred years ago, but twenty thousand years ago), walking was inherent to our biological culture. From a biomechanical perspective, it is clear we need to return to walking this kind of distance for the survival of our biology. We can start with a fraction of the distance. I can't say this enough: Doing other exercise **does not replace walking**, as your physiology depends on the very particular mechanical signals found in regular, well-aligned locomotion.

How do you get your kids walking?

1. Start walking yourself. (Uh-oh.)

2. Start walking with your kids. This is going to go a lot slower than *your* walk (trust me), but it's important to walk as a family at their pace. Discuss that you will be walking "because walking is the most important form of exercise." Let them help plan the route you take. Pick a short distance and make that a habit (i.e. let's walk to the store or to the creek), so they don't feel the walk will go on indefinitely. Kids *love* things that seem to go on forever without end.

3. Understand that your kids need to develop the motor programming, strength, and endurance to walk with you. Develop your walking distances accordingly, increasing them no more than ten percent each week.

4. You probably need to carve more time out for health. If you only have forty-five minutes allotted for walking, realize that this is

inadequate time for everyone (you and your kids) to get what they need biologically. Consider replacing a planned activity with vitamin D.W. (Daily Walk). This walk is more important than just about anything…trust me.

One more thing about kids and walking. They don't like walking because it's booriiiing. But guess what? Parents probably think it's boring too, which is why music and classes and gyms and special outfits exist. I once heard "If you're bored, you're boring." Ouch! But it's kind of true. There's nothing boring about your body, in dynamic motion, with its 200 bones, 230 joints, and 600 muscles all alive with neurological connection to your brain with each step. Nothing boring about the planet you're walking on and the bugs, animals, and people you share it with. Develop a daily appreciation for your freedom to walk, and then pass it on.

Many people come to me as adults in despair, wishing that they had had one iota of health presented to them as children. In overzealous response, many of us are replacing our lack of natural movement with fitness, which can take care of one issue but creates others. Walking with your kids is free. No classes, special camps, or equipment required. No more excuses. Kids need more than "playing all day." They need to be able to walk quite a distance. Start your family training today. Their bones and brains will thank you!

March 7 2011 WHEN PUSH COMES TO SHOVE…

Last week we talked about pushing out the pelvic organs. Not the best thing.

Today we talk pushing out a baby. Much better thing.

I'm going to attempt to explain the physics of vaginal delivery to you in less than a thousand words. So far, I am already at fifty, so this should be interesting. Fifty-five now, but I digress.

Anyhow, there's this thing called the uterus. The uterus (a.k.a. The Womb) is the organ in females that wraps almost entirely around your growing, not-yet-born, baby. I say almost entirely, because the uterus is shaped like a balloon, with an opening at one end. Why the opening? Because the baby is eventually going to need to get out of the organ and conveniently, the uterus is equipped with a doorway. Equally convenient, to keep babies from falling out all over the place, there is a temporary development of tissue that plugs the door until you and the baby are ready to go.

The doorway is down at the lowest part of the balloon, because frankly, babies don't have the motor skills to climb up and out. They like to use the path of least resistance, which is going along with gravity.

So, you have this balloon-shaped organ wrapped around the baby, called the uterus.

Geek Notes: Why is it called the uterus? You'll love this. The uterus comes from the Greek word *hustera*. From the word *hustera* comes the term for a medical condition called hysteria—a disease believed (in the days of Hippocrates) to be a condition of the uterus.

> In the middle of the flanks of women lies the womb, a female viscus, closely resembling an animal; for it is moved of itself hither and thither in the flanks, also upwards in a direct line to below the cartilage of the thorax and also obliquely to the right or to the left, either to the liver or spleen; and it likewise is subject to falling downwards, and, in a word, it is altogether erratic. It delights, also, in fragrant smells, and advances towards them; and it has an aversion to fetid smells, and flees from them; and on the whole the womb is like an animal within an animal. —from Aretaeus the Cappadocian

Until the 1930s, hysteria was the psychiatric label most frequently applied to women.

Until 10:00 a.m. this morning, hysteria was the "pet name" most frequently used about my house.

Have I totally lost you? Let's get back on track.

The balloon-shaped uterus has a nice layer of muscle (myometrium) that, when the time comes, begins to contract. Actually, the myometrium has been contracting all throughout the pregnancy, but with such sporadic frequency and low force, it is fairly impossible to detect, especially when you're constantly distracted by life. And television.

Fun Party Trivia: If you're ever at a party and someone asks, "What is the strongest human muscle?" you can disagree with them when someone says the tongue. Unless it's a dude, in which case it might be, for him. But even stronger than the human tongue is the myometrium. What makes this muscle the "strongest" is the fact that it can generate an insanely high amount of force with very little tissue. Pound for pound, the uterus has got the rest of the musculature beat when it comes to force production.

There are two things the uterus has to do. First, it has to open the cervix. How does it do this? No one knows quite for sure, but one thing is clear. In order to get the cervix to open rapidly, it helps if the baby's head is "knocking at the door." This knocking is called the head-to-cervix force. There are two things that maximize the force—the first is baby head position and the second is, of course, being upright and walking during the dilation phase (increasing downward pressure). Women who progress slowly or end up needing to have a cesarean have been shown clinically have low head-to-cervix force.

The second thing the uterus has to do, after fully dilating the cervix, is expel the baby. You, Baby—to the principal's office!!!

Just kidding.

This super-strength of the uterus begs the question: If the uterus is so very strong, then why are we creating excessive downward force during labor? Why are we blowing out our pelvic floors and pushing the organs out when we've got the Arnold Schwarzenegger of female organs on the job?

This has a lot to do with the history of facilitated labor. It was believed, early in the development of medicine, that the expulsion phase should happen very quickly, yet in a controlled format. And, because mums in the hospitals were lying down to birth, the biomechanics of the uterus's forces were altered—no longer is the baby pushing down, but more horizontally. Now, to push the baby horizontally, here come the commands for Valsalva pushing and the "triple push pattern" of voluntary maternal force generation (typically bearing down just before, during, and just after peak contraction force). The result is a shorter expulsion phase, but also greater maternal fatigue (and increase in interventions), and greater pelvic floor (and fetal) stress. When left to spontaneous pushing, the expulsion phase is sometimes longer, but the natural timing of "pushing" with peak contraction seems to align itself more naturally and results maximal force generation and energy conservation—basically MOM having a greater amount of energy to get the job done while keeping her pelvic organs and muscles intact.

There is also ONE MORE THING that affects the ability for the baby to get out. And no, it's not how much you want labor to be over with. ☺ The ability for the baby to move downward depends on how little you are resisting. What do I mean? I mean that every upward force you create out of habit, and the tension you carry in your pelvic floor muscles, will make it harder for the uterus to do its job. The uterus works at a slow and steady pace. When met with chronic tension of the psoas, hips (quads, inner thighs, hamstrings), piriformis, and pelvic floor muscles directly, the uterus becomes less effective. See, the uterus can't work harder than it is biologically programmed to do. It can only generate the force it generates, which means the more relaxed and stretched your muscles are *before* you get to the maternity ward or your backyard birth tub, the more effective your natural process can be.

Now, what are you going to do with all of this info?

1. Know that, although vaginal delivery is a completely natural process, you no longer have naturally aligned equipment. Your heeled shoes, hours spent in chairs, habits of sucking in your gut, chronic stress and tension, high chest breathing, and tucked pelvis make both head-to-cervix and uterine forces lower than they could be and your resistive forces (hip and PF tension) higher than they should be.

A. You can't do much about it once you've gone into labor.

    i. You can do something about it while you are growing your baby or before you get pregnant.

Something about it:

1. Release your psoas and guts.

2. Stop tucking and thrusting your pelvis.

3. Get out of your positive heeled shoes and go for barefoot, minimalistic, Earth-brand, or super-flat shoes.

4. Get out of your chair and create a standing workstation.

5. Work (diligently) to open your hips (quads, hams, inner thighs, and IT band), making sure that your pelvis doesn't move when you do your leg stretches.

6. Walk at least a few miles EVERY DAY, while thinking about all of the things above.

7. Learn about baby positioning. You CAN work with your baby early on for more optimal baby/uterus alignment. Find more info at spinningbabies.com (and I love Gail's work so much, I'm going to interview her soon!).

One more thing. All this uterine biomechanics stuff doesn't ONLY have to do with delivery but also with conception, both natural and IVF. The fluid dynamics of the uterus are drastically affected by position—something not really known to most professionals, even though it is in the biomechanical literature.

The Gravity of the Situation

"We are all subject to the Earth's gravity forces, and all biological process must also obey Newton's basic laws of physics," says Prof. Elad, who has been studying the biomechanical engineering of pregnancy for over 15 years. "Uterine contractions push the fluid inside a woman's womb in a peristaltic fashion, which helps sperm reach the ovum in the fallopian tube. And after fertilization, this same peristalsis propels the embryo to its implantation site in the uterine wall. It's a fluid mechanics issue.

"By thinking about these biomechanical processes during IVF treatments, we can help physicians, and prospective parents, see better outcomes," he says.

See, I'm not the only Chicken Little in the room, shouting "Newtonian Physics, Newtonian Physics!"

P.S. This blog would be much more interesting with pictures, so I thought I'd include a picture of my uterus:

I really need a camera with a flash…

More reading:

E. Showalter, *Hystories: Hysterical Epidemics and Modern Culture* (1997)

Br J Obstet Gynaecol. 1996 Aug;103(8):763–8. Head-to-cervix force: an important physiological variable in labour. 1. The temporal relation between head-to-cervix force and intrauterine pressure during labour.

Br J Obstet Gynaecol. 1996 Aug;103(8):769–75. Head-to-cervix force: an important physiological variable in labour. 2. Peak active force, peak active pressure and mode of delivery.

Med Eng Phys. 1997 Jun;19(4):317–26. Simultaneous monitoring of head-to-cervix forces, intrauterine pressure and cervical dilatation during labour.

Obstet Gynecol. 2009 April; 113(4): 873–880. Biomechanical Analyses of the Efficacy of Patterns of Maternal Effort on Second-Stage Progress.

ALIGNING OR RELAXIN BEFORE PREGNANCY?

This morning on the Aligned and Well Facebook page, I got this question and thought it was important enough to share:

Does relaxin not do enough to loosen the muscles of the thighs and abdomen to facilitate labor? Will I still need to stretch my crazy tight psoas once the hormone starts working?

For those of you who don't know relaxin, she's not asking about relaxin' as in, swinging in a hammock or getting a massage. *Relaxin* is the name of a hormone that targets the collagen of certain tissues with relaxin receptors. Relaxin is often called the pregnancy hormone, but both men and women make it for various collagen-softening reasons (the most current being relaxin's role in cardiovascular function, but...for another time).

Contrary to popular belief (which in terms of scientific information is typically like one long game of Operator[1]!), relaxin does not affect muscle tension, because it does not affect muscle tissue. I'll say that again. Skeletal muscle, which is the muscle you think of when you think of your muscles (that's not confusing, right?) *is not affected by relaxin.*

Relaxin targets collagen-heavier tissues like ligaments, cartilage, and smooth muscle (like the uterus). In pregnancy, the tissues most affected by relaxin are the pelvic organ ligaments, the uterine muscle, and the pubic symphysis (which I will refer to as the PS because I have to look up how to spell symphysis every time.) The PS is the soft joint between the right and left halves of your pelvis. No, your pelvis is not one fixed bone, but the slightly malleable interaction between your right and left side. And this is a good thing, too, as you want your symphysis to be supple to allow for greater birthing space.

So, great. I've got relaxin. Awesome. Then I'll just chill on the couch until it's time to push 'em out, right?

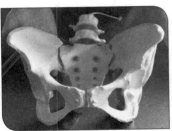

The pelvis, viewed from the front...

Not so much, because here's the thing. While relaxin secretion (when your body moves in alignment and has all the muscles at the right length) is an awesome biological occurrence, relaxin secretion (when your muscles and bones are *not* where they are supposed to be) can also result in pelvic ligament sprains, low back pain, and joint instability. Not fun, especially with an extra ___ (fill in the blank) kilos on your frame.

And here's the other thing. Nature's excellent programming of flexible parts during delivery can be rendered useless by something else—your tight muscles.There is an incorrect belief that your muscles soften when it comes time to deliver. Nope. When it's time to push, you've got what you've got—there aren't any hormones that will bail you out of muscle tension. You need to take care of that ahead of time.

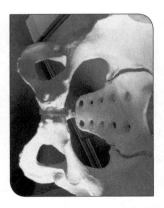

Pubic symphysis up close (the blue squishy part between the bones).

And before and after your delivery, your muscle tension can make pregnancy uncomfortable by allowing your muscle imbalances to do more damage. With softer connections, your habits place your now excessively flexible parts at risk (think sitting on that tucked pelvis with a sacrum that's a little less stable).

Or (as pictured at right) tighter hip musculature on one side that now rotates the left half of your pelvis away from the right (pubic symphysis pain, anyone?). This is NOT what you want to be dealing with before, during, or after labor.

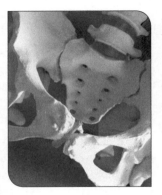

To recap: Relaxin is not your ticket to relaxin'. In fact, the fact that relaxin is there at all really requires that you to work back to your natural alignment ASAP to prevent tissue damage and have an optimal delivery.

To answer the question from above more specifically—the relaxin will do nothing for your psoas muscle. That, my friend, is your job. ☺

---

1. The game of Operator! (a.k.a. Telephone) goes like this: A group of people sit in a circle. The first person leans over and whispers in the ear of the person next door and says something like "I think that John should have to make dinner tonight," and by the time it makes it all the way around the circle, the last person says aloud what they heard it as, which is usually something like "She's in the sink with Shauna but that's all right."

A BABY STORY AND GIVEAWAY RESULTS!

Sorry to be gone for so long, peeps. I was busy, you know, having a baby. I am clearly not one of those bloggers who can post while in labor and take pictures of my placenta and put it on Facebook within an hour of delivery. Sorry.

P.S. The placenta is totally bitchin' though, so if you do have a Facebook account, troll around for someone's placenta pics. Placentas = cool.

Here's the lowdown on my experience.

1. Water breaks. On the couch. Which reminds me…
   ONE COUCH FOR SALE…CHEAP.
2. I think that heading to the Farmer's Market for a few last minute items is a good idea.
3. Water REALLY breaks at Farmer's Market. Which reminds me…
   ONE PAIR OF NAVY BLUE TRACK PANTS FOR SALE…REALLY CHEAP.
4. Then nothing much for four or five hours and then some huge[1] contractions for the next four or five hours.
   A note on contractions: Holy $#(@! What the &*@#?? Seriously? &*(#^$(@#Q@#@(%!!@&$)(#%. Okay. I'm okay.
5. Enter midwife.
   M: Your cervix is at a one.
   K: Ha ha. You mean, like, it's Number One?
   M: No. Like one centimeter. As in, you still have nine more to go.
   K: (See footnote 1.)
6. Three hours later, I'm ready to go. That's right. One to ten in three hours. I'm like, in the 99th percentile for cervix opening.
7. Push. Push. Push. (See footnote 1.)
8. Done, awesome. Lovely, perfect, most handsome boy Finn Vincent slides out at 6:55 a.m., April 14, 2011.

Stats: 7 lbs. 7 oz., 19.5 inches long. It's the most awesomest thing I have ever done. And I've been to a lot of neat places, and done many awesome things. Still, it's the Most Awesome. He is, evidently NOT in the 90-something percentile for size, but he is in the 100th percentile of being mine.

Ok, so here's where it gets a little dicey.

9. An hour later a freak occurrence: I develop a hematoma (I was in the 99th percentile of hematoma size, which is two grapefruits!)

because I tore an artery in the inter-pelvic space between the birth canal and my hip. Weird, right? After receiving excellent care from my home-birthing team who is puzzled by my blood loss (hematoma burst) yet perfect blood pressure (seriously, I had lost more than a liter of blood at still had 120/70 blood pressure...let's hear it for health!), I was transported to the hospital. Had two surgeries, one blood transfusion, and I'm feeling SO much better now!

A lil' note on me: I decided when I woke up in recovery, that I would NOT view my situation as anything less than as it needed to be. While in a hazy fog I decided that I would be thankful in that moment for being flat on my back—a sensation I had come to miss while prego. Also, my husband and baby were waiting for me in my room, where the hospital let them move in and stay with me. My also super-friend-lactation expert met me five minutes after recovery, where she had the babe latched on in less than a few minutes. He's in the 99th percentile of eating. Come to think of it, so am I.

Another side note: I think breastfeeding is awesome. I'm playing with all the positions and all the methods used by hunter-gathering populations (where they nurse for three to five minutes every fifteen to twenty minutes), as well as the developed nations' methods of feeding for longer, less often. It's fun.

Here he is.

They said it would be five to seven days before I could go home.

I was healing well in two days, had refused all morphine and pain meds, so they let me go home on day three. (Let's hear it for health!)

I couldn't walk when I came home. Wow. That was a big one for me.

They said it would be twelve weeks before I would be out walking.

I took my first walk (fourteen minutes) on day three of being home.

I can now walk forty minutes (slower than normal) and do so every day.

Finn has been cuddled or carried every moment of his life so far. We are practicing with different techniques and apparatuses. (Why is the plural of apparatus apparatuses and not apparati? Language seriously bugs me sometimes.) I am enjoying wearing him and no, we don't have a stroller, which freaks people out. Which is fair, cuz there are a lot of things people do that freak me out, so we're even.

Moral of the story: I, despite making excellent beans and cornbread and knowing how to ride a horse, would have been the robust pioneer woman who didn't make it through childbirth. But thanks to my fast-acting team and some great surgeons, I'm typing a blog a week and a half later.

Another moral of the story: Striving for excellent physical health can not only help you have an excellent delivery, but can also significantly impact your recovery should you have an outcome you weren't (or even were) expecting.

Looking forward to getting back to it…with modifications. Wow, it's a lot of work! Kudos to every mom and dad out there. You are in the 99th percentile of parenting. (What the HECK is all the percentile business about, anyway? If I ever use the word "percentile" seriously, you can pour a glass of water over my head. Geesh.)

---

1. "Huge" contractions have been modified to "tiny" contractions after finishing the course.

May 12, 2011 — ABCs, BABY VERSION

A baby was born. And then no one slept for one hundred years. What is up with that??

Breastfeeding. I love everything about it except for the sore neck and shoulders. After a fatiguing birthing session, being thrown into the new-mom Olympics of hoisting and lifting and schlepping leaves the shoulder girdle needing some restoration. Try this Floor Angels stretch to open a tight chest and biceps. You can even put your babe on your chest or stomach as you do it.

NOTE: Don't just lie there and fall asleep. Move your arms around in the same way you would make a snow angel, trying to keep the backs of the hands on the floor as you slide them up toward the head.

Carrying the baby. I love it and we put a nix on strollers so there isn't any other option. Pregnancy is actually nature's strength-training program (if you know how to carry your body weight to maximize this period of time.) You know all that weight you gained during the last nine months? Some of that was muscle, in the perfect match to the weight of the baby-to-come. Why? So that when baby is born, you have extra muscle mass to be able to carry him/her/them. No, the weight isn't in your arms, but in the legs. Keeping the baby on your body allows you to keep this muscle mass and all of its metabolic wonder. Don't think of losing all that baby weight—keep the good stuff!

Diapers, cloth. They are amazing. And hold much less than disposable diapers, especially when you have a newborn with no body fat (or "chicken legs," as they are known in the Bowman family, a phenomenon that is explained by my father, who says we came from a noble line that rode horses instead of walking). Which is why we had to switch back to disposable for two weeks. Then the little man gained two pounds (he started out as a 7-n-7) in three weeks, so we were back in the cloth diaper game.

Early. That's when I get up now. And interestingly enough, it is also when I go to bed.

Feeding schedule. According to anthropological data, hunter-gathering populations today feed their babies every fifteen minutes for a few minutes at a time (called continual feeding, used by mammals that "cling" or "follow" their babies) as opposed to the longer feedings adopted by "spaced feeders." Spaced feeders are birds (who leave their babies in nests) and most of the industrialized cultures who schedule longer feedings based on the infant's napping/playing away from the mother (more than one meter). The main difference seems to be in the properties of the milk and in the effect on reproduction. Breastfeeding using the continuous method is what minimizes chances of becoming pregnant while nursing, as opposed to longer (twenty to forty minute) feedings. I'm trying the shorter, frequent method, although every fifteen minutes is a stretch. More like every thirty to forty minutes. And I just feed when the baby is looking for it. And guess what hunter-gathering moms do just after feeding?

Gymnastics! A couple times an hour, H-G babies are encouraged to move their body vigorously. This is done to tone the babies' muscles, cultivate their walking and gripping reflexes, and calm them. I, of course, started doing this right away. No, I'm not trying to get the kid to do push-ups or anything, but using the baby's natural grip, pulling the arms gently to create resistance, pressing against the soles of his feet to let him practice pushing off…that kind of stuff. And I have one of those babies who started crawling up my body in search of the party (if you know what I mean). He also started rolling to one side at two weeks and rolling to his stomach at three weeks. I believe that our culture's failure to cultivate a baby's natural walking and hand-over-hand gripping reflex (called brachiating) from infancy is what has led to our extremely poor strength-to-weight ratio. That's why, for Christmas, I want some baby brachiation bars. The ability to hold one's own body weight from one's arms has a direct impact on how much oxygen the body takes in and how much the lungs inflate with each breath. If you have a child with (or if you have) breathing issues, consider heading out to the monkey bars a few times a week, slowly building strength. It's not about having buff shoulders, it's about oxygen. And it's about time you got to that, don't you think?

My Hair. Really. What is going on with this ratted-out bun on the back of my head? It's time to chop my hair, but I don't know what I should do. What would you suggest? To help you out, here is a look at my hairstyles past. (images below) I'm particularly fond of the uneven mullet that my mom says she did not give me. I can tell she's lying. You know how I know? Because of her giant glasses in the first pic (but look at my cute haircut!).

My mom brought all of these down when she came to visit, to help us figure out who the baby looked like more. Since the boy, at this time, does not have a perm, Salon Selectives in his bangs, huge 1970s-style glasses, nor acid-washed suspenders, I'm thinking he looks more like his dad. But back to the bigger issue...WHAT TO DO WITH MY HAIR?

I can't think of any thing to put here.

Just Kidding. I can, but I'm too tired to write anything more.

Naps. Let's have one, shall we?

O. This was the shape of my belly four weeks ago. Four weeks later it is almost back to pre-baby (in appearance, not in terms of strength), which I attribute to easy-paced walking and carrying baby every day, light stretching, and breastfeeding.

Potty training. Yeah, we're doing that early too with the help of the book *Diaper Free: The Gentle Wisdom of Natural Infant Hygiene,* by Ingrid Bauer.

Q: Did you really just post a picture of baby poop? A: Why yes, yes I did.

Reflex, brachiating. See "G" for Gymnastics.

STU. As in STUpidly long alphabet. Who has time for this??

Vitamin K. In hindsight I would have taken a vitamin K supplement during my pregnancy. I am now taking it in the form of chlorophyll and it helps with blood clotting among other things.

Here is da first poop in da potty

Why do they call it vitamin "K"? For the German word *Koagulation.* This reminds me of my favorite German word, *Obstipation,* meaning constipation, which, ironically, the vitamin K also helps with...

Women's Guide to Foot Pain: The New Science of Healthy Feet is now available for pre-sale. Why do I bring it up? Because in addition to babying and aligning and walking and blogging, I'm supposed to be editing for tomorrow's deadline. Just kidding. The deadline was yesterday. Just kidding. The deadline was Tuesday. But I'm almost done, I promise.

NOTE FROM AUTHOR: Due to recent lifestyle updates, she will be modifying the alphabet to not include the letters: K, L, M, X, Y, and Z. Please accept our apologies. We have to go get our hair cut.

GREAT CHIME PUNCHER

I swore that when I had a kid, there would be no flashing, beeping, or otherwise annoying toys in my house.

Now, the great thing about these toys is that they help develop motor skills. When you touch a button, you are rewarded with a buzz or honk or something, which gives a kid a sort of a calibration. They can practice the motion until it is refined, using the sound as a sort of beacon in the dark. I don't need to mention what's no so great about the buzzing beacon, do I?

But three months have passed and it's time to give him a little more stimulation than what being carried all over town and the occasional hand puppet show provides. His muscles need to start developing motor programs, but I have this thing against plastic. Especially noisy plastic.

I also happen to have a kid that loves to punch. Seriously, I have been punched in the face more times than I'd like to recall. I needed to focus the little fists of fury.

Killing two birds with one stone, I hung up some chimes. Well, actually, my husband hung up some chimes (to an old phone line cord, to an old belt, to the ceiling). It's not the prettiest of toys, but it took about five minutes before the kid realized that the general waving of his mighty fists would make a big clang on contact.

Fast forward a week and I now hear chimes in my sleep.

Cue smile.

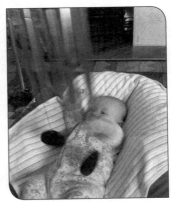

*Cue more chiming and clanging
(note the blurry fists).*

But the kid has some serious ability to connect. And what's really cool is he doesn't need to look at the chimes once he figures out where they are. He can look at other things and use his proprioception to make the movements that make a lot of noise. Isn't that awesome??

A few days later, we needed "advanced chime" exercises. This consisted of getting his feet below the chimes. This one was great too. And if the career in boxing doesn't work out, there's always soccer, or kickboxing ("sport of the future"—name that movie!). The leg-chime time is also his diaper-free time. In addition to the benefits of letting skin breathe, it is essential to have a break from things that push baby legs apart. As a culture we place babies in a lot of knees-apart situations (picture most baby apparatuses, and diapers!) and babies can't really develop deep-core musculature with a big ol' diaper on.

The constant hip-open position (called abduction) is one of the reasons many of us suffer from foot, knee, and hip issues as adults. Kicking the chimes *sans* britches helps develop the deep-core musculature along with the inner thighs. And in addition to musculoskeletal benefits, the kick-chime exercise is great exercise therapy for babies with low tone or digestion issues too.

So, chimes: toy or personal trainer?

In fact, what we really need is some adult-sized chimes. But keep your pants on, will ya?

I'll be keeping my eyes open for different chimes, as each set not only mixes it up sound-wise but will require a unique and novel motor program for the little guy. Plus, I just really like to garage sale. (As does anyone who uses "garage sale" as a verb...)

(Thank you, reader Kate Y., for sending me the Michael Olaf [Montessori] catalog! I just found out that they make a European Toy Hanger where you can suspend baby toys without having to tether them to electrical devices like ceiling fans, which I don't recommend. Smart. It does beg the question, though, is the toy hanger European, or is it a hanger for only European Toys?)

 July 11 2011 # BAREFOOT, NOT JUST FOR ADULTS!

Yesterday my niece stepped on her first nail. She survived. And I know she survived because I got this text from my sister: "Survived Rae's first nail through foot—she is doing fine."

And then I texted: "YAY! Just like mama."

Only I have an iPhone, so it autocorrected it to "Just like mana." Because everyone knows that the word *mana* is more popular than the word *mama*. (?)

And then SHE texted back "Dada" and I'm not sure why. Did her husband step on a nail too as a kid? Did one of her forty-seven kids have the phone? Had she forgotten when she stepped on her first nail?

So I texted: "No u did that too I remember the very moment."

And I did, too. She was about nine and on the phone (probably talking to a boy), when she shifted her weight forward onto a piece of wood with a nail in it. BAM. And here's the funny thing: She didn't hang up the phone. No, I remember she was trying to keep the conversation going with stuff about how she had a nail in her foot. She really liked talking on the phone.

She remembered too, because then she texted: "Yes." And then she wrote: "Barefoot in the apple hut."[1]

Only here's the thing—she wasn't barefoot. She was wearing black Keds. And I remember, because when they pulled the nail out of the foot/shoe, the shoe filled up with blood. It was cool. Probably because it wasn't my foot. So then I texted back: "No you weren't barefoot it went through your shoe and u were on the phone."

And when I read my text I can see why my little sister thinks I was bossy as a kid (only as a kid, right sis?). Plus texting always makes me sound like I'm about twelve.

And then she texted back: "Hurt like a mother."

And then I texted: "I bet."

And then we stopped texting because:

1. I have a new baby and a whole bunch of other stuff I should be doing.
2. She has sixty-seven kids and a whole bunch of other stuff she should be doing.
3. We are over thirty-five now, and why are we texting in the first place?

So, let's talk for a second about the barefoot movement. In a nutshell, feet weren't designed to be in shoes any more than hands were designed to be in mittens. Our cultural habit of wearing shoes has left us with affluent ailments (osteoarthritis, osteoporosis, knee-hip-back pain, bunions) at much higher percentages than unshod populations. And you don't have to be a genius (but you might be!) to go: Yeah, I get it.

With kids, it's even more important that they get abundant barefoot time because the proper development of natural gait patterns and deep body stabilization depends on the super-sophisticated feedback systems that come from the nerves sensing the ground and the brain.

This happens in two ways.

1. 1. There are nerves that interpret the shape of the ground by how the bones in the feet bend at thirty-three different points (joints). This creates a mental image in the brain (similar to how a dolphin uses sonar to avoid obstacles). Wearing shoes prevents any motion in these joints (except the ankle) and leaves the shoe-wearer "blind" to the environment. This is what makes stiff shoes the worst when it comes to natural development.

    And, P.S., This goes for adults too.

2. There are nerves that are sensing other things about the environment. Things like temperature and terrain. These nerves, designed to convey abundant information, are now stuck just reading the inside of your shoes. And what is that environment like? Damp and smooth? Ew. The more you expose these nerves to the same input over and over, the more their information gets put on autopilot, into a part of your nervous system called an interneuron.

More about the interneuron: You know how when you walk into a room with a freshly baked pie, it smells delicious, but after a few minutes you don't smell it anymore? Your brain moves information that is constant into a different part of your brain to free it up for other things, so the more similar your environment, the less your nerves work to perceive. And the less they work, the less healthy they are.

This is why it is important (for everyone, not just kids) to walk on different terrain, on a different path, in different weather. Carrying different weights, wearing different shoes, and eating different food. Habits are not great for biology.

Anyhow, this is why everyone—kids too—need lots of time to feel with their feet. Make the house a no-shoe zone and encourage the exploration of different areas *sans* footwear.

I brought up the nail-in-the-foot conversation because, yes, accidents happen, but in the case of my sister—and her little girl—they were wearing shoes. Wearing shoes doesn't prevent the few foot catastrophes that happen, and wearing shoes only teaches kids how to pay *less* attention to what is going on beneath their feet.

Now, that's not to say that there aren't going to be places where kids need shoes. When they do, here are a few suggestions.

1. Select super-flexible shoes with no heels. My favorite company for better kids' footwear is Soft Star Shoes. Keep in mind, shoes get expensive, and it's important to not let kids wear shoes that are too small—this is where toe-curling habits get started. So before you run out right to the store right now...

2. Don't buy "fancy shoes" in the summer. Kids will grow a lot this season and will probably not fit into their shoes at the end of the season. Summer is also the best time to be barefoot, so maximize barefoot time right now.

    The best footwear is actually (I'm not kidding) a swim shoe. They're lightweight and flexible, breathable and super-inexpensive. Get your kids a pair at the beginning of the summer to supplement their barefoot time, and donate them once they're too short. Maybe even get the next size up while they are available, for fall, if you live in a temperate climate. Swim shoes are much better than flip-flops as the biomechanical gait changes we do to keep flops on really messes with the natural gait pattern.

    An interesting note: Neurologists will often recommend wearing swim shoes year-round for kids with any sort of neurological delay, as it improves neurological function of the feet for better gait (walking pattern) development. But why only improve the gait of some kids? Let's all do it, shall we??

3. Play foot-detective games. Having your kids in their bare feet, try having them guess various objects using their feet only. The more the textures vary (wet, smooth, sandpaper, yoga mat, carpet, wood) the more the game will challenge their sensory nerves!

4. Have a little foot-massage time before going to bed. Just like book reading, add a five-minute foot rub into the bedtime tasks. Skin-on-skin touch will help stimulate the movement of each foot joint, increasing circulation of the foot's tissues.

---

1. For those of you who don't know, I grew up on an apple farm. The apples were sold in the apple hut. The Apple Hut was also the name of the business. Clever, right?

It's not often that I rant. At least outside of my head, anyway. Today, two things have inspired this blog. First is that blog list[1] I'm working off of, which says, "Write a Negative Post." I'm not sure what that means, exactly, but a rant feels right for this category. The second was a recent comment on my recent entry called "The Great Chime Puncher," from a physical therapist specializing in kids. And I want to start off by saying that I completely respect this individual's post and goodwill even though I am using her comment to make a point. The issue I have is not with the therapist, but with the set of knowledge given as curriculum for the academic programs of therapy, human development, etc.

In response to my "baby chime exercise" and diaper recommendations during playtime, this commenter wrote:

> Far more important for kids with low tone is tummy time. It's the activation of muscles in this position that serves to connect the four inner core muscles (respiratory diaphragm, pelvic floor, transversus abdominis, and multifidus), together with the other postural or outer core muscles.
>
> Also, far more than diaper/no diaper, it's the actual skeletal developmental biomechanics of the hip that create the restriction at this age. *Tummy time when they are babies, well-rounded physical activity as they grow and less sitting are what the brain and the skeleton need for a lifetime of good alignment and function.*

While there were other aspects of the comment I didn't jive with, it was really the last sentence that moved me to post today. The notion that these items—tummy time, well-rounded physical activity, and less sitting—are presented as all the human body needs for correct development is a huge oversimplification for therapists and parents. There is much more a developing human needs than these three items. And note: this is not about manipulating variables to create a physically superb child, or about being an overzealous parent. So many books and websites talk about letting a child develop in a natural way. I couldn't agree more. But the big, huge, gray, wrinkly, ELEPHANT IN THE ROOM is that we do not live in nature—and what we are calling "natural development" is actually extremely stunted and limited based on our modern lifestyles.

If I were to make a much more specific list of what babies and then kids need, it would look a lot more like this:

1. To be carried as an infant, everywhere, in the arms of another human.

   Not in a stroller. Not in a sling. Not in any contraption that prevents the infant from doing its own work. Have you ever seen any baby-transport device growing on a tree? I didn't think so.

   Why not the stroller? There are huge biological activities happening in development that require human-to-human communication that goes beyond words. This can only happen with proximity. Second, no human (at any age) can develop musculature without resistance. They need to feel their body against gravity in order to develop their brain (and proprioception) to set the muscles at the correct lengths.

   Babywearing is also a huge market with tons of products, all advertising their better positioning for better development. Well, unfortunately, human muscle doesn't develop based on position. Position is passive. If I put a cast on a part of your body into an optimal position for healing (like casting a broken leg), this ensures that the bones set evenly, but the casting actually promotes muscle atrophy (shrinking)—not muscle development.

   Now, I love babywearing more than strollers, because at least you've got the closeness and a more upright posture. But there is something much better than babywearing, and that is baby carrying.

   Why don't we do that? Probably because we are too weak (most women aren't strong enough to carry the weight of the pregnancy, hence the back pain and hip problems), or too inconvenient. How can you use your iPhone and carry your baby? Very carefully, I can tell you from experience.

2. To be carried, often.

   It's not enough to hold the baby now and then (with the baby on the ground/crib/bed most of the time). Physical anthropological data shows that women used to walk eight to nine hundred miles per year carrying their babies. These kids didn't have flopping heads. All it takes is the careful carrying of your baby for the first month or two to help the child develop correct motor skill and strength to support their hefty head.

3. No shoes.

   No human should be wearing footwear, least of all developing children. I know we all don't live in Africa, so put some socks on the kid when it's cold, but correct development of sensory nerves in the feet requires the foot skin to feel unique surfaces. And when it comes to walking, there should be nothing on the feet that would cause a baby to lose traction. The correct gait pattern has a pushing-back motion. Guess why most of us don't have it? Slippery socks on a wood or linoleum floor. Baby jumpers, and those round things that babies sit in to push themselves around? Baby walkers? What are those called? Those prevent natural gait development because they turn off the reflex for doing it "biologically best" and replace it with another computer program.

4. Cultivate the gripping reflex.

   Babies have a reflex that allows them to grip onto something and hold their body weight. As babies pass the first couple of months, they should begin to hang on to your body while walking. (Which is really cool!) This makes it less work for the baby-holder (especially when you have a three-month-old seventeen-pounder) and helps the baby develop the muscles that hold the shoulder blades down—the same motor skill that is needed in opening up the muscles between the ribs, which improves oxygen intake. (Children with respiratory issues should be working on their ability to hold their body weight with their arms!)

5. Encourage squatting.

   Westerners have less hip and knee ranges of motion than anyone else in the world. Be wary of studies that list "what humans can do with their knees and hips" based on data collection from Western populations. We're all jacked up in the body and the scary thing is, our academic texts are starting to confuse "normal" with "natural."

6. Get rid of your furniture.

   Do you have a kiddie table and chairs? Raise up the table for a standing play/work area and toss the chairs. Why would we teach our kids to sit? Really?

7. Start advocating for the removal of chairs and the use of standing tables in the classroom. Feel free to use this:

   Dear Teacher,

   Please excuse Bobby from sitting today. Research shows that sitting increases the risk of death from heart disease. I am hoping that your school does not advocate heart disease and I am providing the standing table for my kid.

Signed,

Concerned Parent.

P.S. Please don't roll your eyes at my request, talk about me in the teacher's lounge, or write off my completely science-based and logical request. I know that you know sitting isn't healthy. Who's going to be the first person to do something about it?

8. As soon as your kids can walk, keep them walking.

There is nothing that gets me more fired up (okay, so it turns out that I *do* rant a lot!) than seeing a parent strap their walking-aged child into a stroller so Parent can get their exercise. What's the message there? You sit and be still so I can get some health on? Okay, strike that. What gets me more fired up than that is the same situation, only the kid is eating a bag of Funions (a child-friendly bag of onion-flavored, onion-shaped, hydrogenated-oil-laced snack. Nice.)

I saw this one time on the beach walking path, I swear I did. W.T.H??? All humans require long-distance walking to develop the optimal amount of bone, shoulder biomechanics, respiratory function, digestion, etc. Not playing and not doing other forms of exercise, but walking. Kids can do play too, for sure, but it doesn't replace walking. Riding their bikes doesn't replace walking. Playing an exercise-video game doesn't replace walking. Biology has laws of specificity and there are physiological tasks that don't happen unless under the particular mechanical stresses and strains upright walking creates.

All right. I could go on and on. But writing out a rant has a wonderful way of removing steam (try it sometime!). And I think that there's enough here to work with, don't you?

Thanks for listening. My bigger issue is always this: We have gotten so modernized and technologically savvy, and our cultural message is so ingrained, that we fail to stop and consider the most fundamental aspects of being human.

Okay. The end. And, thanks to everyone for reading, and posting. Especially to the lady who so graciously commented to get today's post going. I am thankful you posted.

---

1. rightmixmarketing.com/right-mix-blog/blog-ideas-2-types-of-posts-to-write-be-fore-your-competitors-do/

In terms of human performance, we have developed the habit of avoiding movement (drive-thrus, cars, handbags, strollers, convenience stores) in order to get more done. Getting more done, in this case, typically means non-biological activities like working, so while it seems like we are doing more, in fact, convenience still means we are doing less.

There is a delicate balance to nature and she has a way of supplying all the right situations for human growth. When it comes to babies, the act of in-arm carrying supplies not only the correct environment for maximal baby strength development, but also the opportunity for mom to continue to peak her upper-body muscle contraction. The use of the arms not only keeps the lymph system (waste removal) process pumping about the breast area, but also helps keep the mechanics of smooth muscle lactation in prime condition.

And your arms end up looking pretty awesome, too.

What keeps many from baby holding is a lack of strength and, of course, the reality that you're going to have to get less stuff done (I'm typing this at 4:40 a.m. because I won't be able to later) while you get more of your biology on.

When arm strength isn't there, many will use all sorts of tricks to "help" hold the weight of the baby, but then suffer the consequences of tight necks and shoulders, achy hips and low backs.

Because I love you, I made this video to give your baby holding an alignment makeover.[1]

(P.S. This is me, on my new farm, where our family is trying out a slow-food project. Can two native Californians—me and my Ayurvedic practitioner DH—ditch the land of no water and grow twenty-five percent of our daily calorie intake? I think yes, because I am an optimist. And a hard worker.

One of our most recent whole-body alignment graduates, the Alignment Monkey, just posted about hanging from bars! She's got some great stuff to say. Read her post here.[2]

And, I wanted to share what is perhaps the best photo ever:

In case you couldn't tell, this is a picture of a leopard slug sliming up my leopard-print garden boots.

I can only imagine that mating conversation.

Slug: Heeeey, bay-bee. Heeey, big girl. I haven't see you around these parts before...

Shoe: _____

Slug: You playin' hard to get? How bout a big kisssssssssssssssssss.

1. youtu.be/R3H2R5Nn38g Two minutes demonstrating the incorrect and correct ways to hold a really cute naked baby.

2. alignmentmonkey.nurturance.net/2011/alignment-monkey-on-monkey-bars

 Sept 1 2011 GROWING PAINS

Kirk Cameron? Where are you? How come we are not married already?

Oh, wait. Not that *Growing Pains*. The other kind. The kind that may have resulted in leg pain when you were a kid. Or the kind that your kid may be mentioning now.

What are growing pains? No one really knows. Yet not knowing stuff has never really stopped people from saying what they think it is. One myth about growing pains that has been perpetuated (I even found it on a couple of prominent health organization sites!) is the theory that bone must be growing faster than the muscle, causing pain.

Bone: Growth spurt, woooo hoooo!

Muscle: Zzzzzzzzzzzzzzzzzz.

Bone: Hey, heeeyyy! Wake up, muscle, I'm way down heeeeeerrre. Wake uuuupppp…

Muscle: Zzzzzz…snort…*yaaaawwwwnn*…whoa, WHOA! Where'd you go, bone? Yank, yank, yank.

Little kid: Ow, OW OW OW MOMMY, my leeeeggss hurt!

In this theory, the muscle pulls tightly on the bone, causing the bone to hurt. It sounds like a good theory, only, like many theories (did you ever hear the one about Earth being flat?) it is not quite right.

Studies have shown that growing pains:

1.   Are not found in areas that are growing.

2.   Are not experienced during periods of rapid growth.

Which leads one to ask the question, "Why are they called growing pains in the first place?"

So many questions, so little time.

While no one knows what growing pains are, like many painful musculoskeletal issues, these "whatever-they-are" pains can be helped by stretching, light exercise, warmth, and massage.

If you have a kid who is whingeing about aches and pains in the legs and you've been told it is growing pains, take it to heart. Before bed, take a ten-to-fifteen-minute easy walk around the block together. Follow it up with a ten-minute soak in a warm bath, and then indulge them with five to ten minutes of a gentle kneading massage to the calves, shins, and the front and back of the thighs. Also, create a four-to-five-minute stretching routing that can be done just before getting into bed. Feel free to use these stretches pictured on the next page.

This one is good for the inner thighs.

This one is good for the quads.

This one is good for their inner thigh muscles, too.

Ahh, the hamstrings! Make sure they stick their butt up, and don't tuck their tails under…

Don't we look alike?

Here's a little story about space.

Once upon a time, there was a little bed that looked like this (image below left).

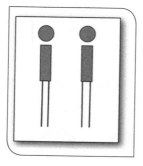

And then one day, it looked like this (image above right).

How did that happen? (Note: Don't add picture here.)

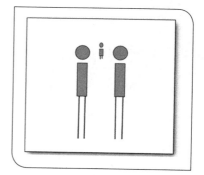

So a few days later, there was a bigger bed that looked like this (image above left).

And then, for about a million good reasons it looked like this (image above right).

And then about three months later, it looked like this (image at right).

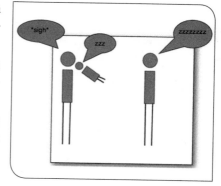

Now, I actually consider myself pretty savvy when it comes to geometry and math and stuff, but I cannot, for the life of me, figure out how every night at seven the bed looks like this (image below left):

and manages to look like this only a few hours later (image above right).

Every night.

I get a lot of emails on "best" sleep position, but I realize now (of course) that the "ideal" sleeping position has to be temporarily forgone when there are hoggers, ahem, I mean babies sleeping with you.

Why am I writing this post this morning? Would you like a little more back story?

We are a baby-carrying (as opposed to strollering or baby-wearing) family most of the time. We have about a twenty-five-pound baby. As you've seen from my baby-carrying posts, you want to make sure to not thrust your hips forward and hike your shoulders, but there are other mal-alignments I didn't cover.

You also don't want to thrust your hip out to the side like this (image at near right).

There is also one more poor baby-holding alignment that I am always guilty of. I call it the rib slide and it looks like this (image at far right).

That sneaky rib slide is pretty subtle but it adds up. This posture means I'm trying to

lessen my arms' work by holding him on my ribs. Not very spine-friendly.

I seem to carry the kid more on my left (stronger), even though I switch every few minutes. He sits more comfortably on my left—something that I have conditioned him to do, probably, so now he already has kind of a preference for twisting left to look forward. (See how simple it is to create bias in your kids' alignment even when you have all the info and the intention?)

So, back to last night. After a long week and not a lot of sleep, I woke up with a spasmy feeling in my bottom rib. I could tell, right away, that is was just the accumulation of tensing to stay on the edge of the bed, fatigued muscles, and this dopey position I had fallen asleep in while nursing.

The reason sleeping on your side isn't that great is because it causes the rib slide. It would be one thing if you could balance it out by flipping over, but you can't do that when you've got a baby. And last night, my muscles weren't going to take it any longer. Actually, it's one's spine that can't take it, which is why one's muscles spasm to protect it from one's behavior.

One meaning *me*.

I needed something to keep my ribs from rib sliding into the bed. Solution: I grabbed a cloth diaper (handy for EVERYTHING and always lying around) and folded it a good height (a few inches thick) and slid it under my ribs until I found the perfect place for the spasm to relax. Ahhhhhh.

P.S. I had a great sleep for about three minutes, until I was punched in the face. Again. My son is a bruiser, and evidence points to it not being genetic.

Some alignment tips for bed-sharing families:

1. Prop up your ribs with a folded cloth, diaper, or hand towel. Note that pillows, blankets, and long towels aren't great when you're sharing the bed with a little one. The smaller the prop, the better.

2. If you're sharing the bed with another adult (hey, how'd YOU get in here?) then switch places every night. This is not only good for the big people, but also gets the little one used to switching it up. It's all about eliminating habits! And bonus: Switching it up is better for your mattress, too!

Interested in bed sharing with your baby? My doctor recommended these safety guidelines for bed sharing (she bed-shared with her baby too!).[1]

I would totally write more but I'm so sleepy for some reason. Plus it's hard to see out of my double black eyes...

1. safebedsharing.org

Dec 14 2011 KID PAJAMAS. OFF WITH THE FEET!

This morning, I cut all the feet off of the boy's pajamas. I couldn't take it anymore. He was already shoe and sock free most of the time, but every morning I had to watch him lose traction on the linoleum or falter when walking because the grip between the floor and his feetypajamas was too much.

I'm not a big crafty lady, but give me some scissors and a math equation and I'm there:

Feetypajamas – feety = pajamas.

If you have a kid, you've probably witnessed the way each of their fingers moves in a slow-yet-definite manner when the child is trying to pick up an object big or small. Kids are working hard to wire their muscles to their brain. To do this, they need full use of their sensory and motor nerves. Footie pajamas allow foot movement, but they completely minimize the ability for the foot skin to read the environment. This data, when collected, helps establish a relationship between the foot and every other bodypart that lasts a lifetime.

The feet have the same amount of motor potential as the hands, yet we don't think much about slapping on a big fuzzy slipper when babies are learning their very first movements. This is why we all have the foot problems we do. This is also why my foot-writing is distinctively less impressive than my handwriting. And to be fair, my handwriting is not great.

In children, motor programs are being set every day, and it is very (very) hard to undo the deepest (read: earliest) motor programs we acquire. If kids take their first steps in socks on a slippery floor, they'll tense a bunch of extra muscles and that gets put into their "walking" mental file. If you want to try it, put on some fuzzy socks while on hardwood or check out how you tense various parts of your body when negotiating an icy part of your sidewalk. Bracing against slipping is a reflex. What you don't want is a child to learn *bracing* and put it into their *walking* file. This overrides their natural reflex for gait development.

Even if your kid isn't slipping about the place, putting a wrapper around one of the most sensitive parts of the human body has a nerve-deadening and muscle-atrophying effect. Would you put mittens on someone trying to learn the piano? Force a baby to wear gloves when they're trying to pick up a tiny pea off of the floor to eat? Of course not. These seem ridiculous. Yet we cover up a child's foot without thinking about it. Isn't that weird?

At our house we don't have a lot of clothing with feet, obviously, but I kept putting the boy in pajamas with feet to "keep him warmer." Boo. Off with my head. I mean my feet. His little tootsies are perfectly warm under the covers. We go barefoot in the house, as does he. And plus, it's kind of fun to cut up stuff. Aversion to cutting up "perfectly good clothing" for the sake of a child's brain development is strange, isn't it? We're a weird breed.

So, I get a lot of emails after posting stuff like this. I know you'd like to read some of the emails that make their way to my inbox. These are slightly exaggerated, but trust me, I get a lot of stuff like this:

Dear Katy,

I just read your blog post on letting babies go without socks or covered feet. It's very cold where we live. We live in an igloo, actually. There is no floor, just snow. I really want to make sure my child's feet develop properly, so do you think that he should be barefoot even though, as I said, the floor is below freezing?

I'd appreciate any help in figuring this one out!

And, I really like your blog!

From,

A concerned mom.

Dear Katy,

I love your blog! Thanks for writing such helpful information. I just read your post on letting babies go without socks or covered feet. I'm unsure what to do at our house though, and am hoping you can help me figure it out.

I am a mosaic artist and my husband is a carpenter. We live in a hardware store that has nails instead of regular flooring. And in between those nails are huge shards of glass from my pieces. I really want to make sure my child's feet develop properly, so do you think that she should still be barefoot even though our house is very dangerous for bare feet?

What would you suggest?

Thanks again. I know you're very busy, but I just can't figure this one out.

From,

A concerned mom.

See, don't you wish you were me? I suggest, concerned moms, that you use common sense. If your feet can be bare, so can your kids'.

Now, if only I could figure out what to do with all these pajama feet!

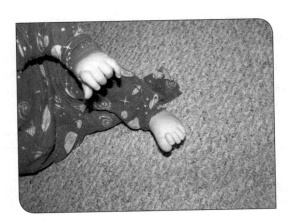

PREGNANCY AND PAIN

I'm a whole-body-alignment lovah' fo' sho', but for some reason, the pelvis and all that the pelvis does has always been a special interest of mine. And even before I had a kid—like, years and years before—I was always uber interested in the birthing pelvis.

I have also always been uber interested in those two little dots that go over the *u* in the word *uber*. How do I make those on my computer? I WANT THOSE DOTS!

Anyhow.

My fascination with pregnancy likely stems back to the fact that you, in utero, are at ground zero. The environment you are given at the time is a complex combination of what your mother is doing in terms of stressing, eating, her habitual body positions, and the amount of time she is moving. All of these things create an environment that you are responding to. You are responding in terms of your teeny-tiny baby alignment, your chemistry, etc.

There are a lot of other people way more qualified to talk about the chemical interactions of stress and poor nutrition, smoking and drugs, so I'll leave that to them. What I will talk about is the alignment factor.

Is anyone surprised at that one? Anyone? Anyone?

Pelves[1] (the plural of pelvis) are my life. And, the more you learn about the pelvis, the more you realize you can't have a "neutral pelvis" unless the rest of your body is neutral as well. But I'll write more about whole-body neutrality later, when we're discussing solutions. Right now, it's imperative that you understand the problem.

This is a two-part post. The first part is a literature review and position paper from grad school and the second is an article I'm presenting at a midwifery conference in Nashville this April.

This is a small portion of my paper titled "Postural Implications of Gestational and Postpartum Low Back and Pelvic Pain." Warning: You may find this a snoozefest unless you're into pregnancy and stuff. Chances are, if you've read this far, you're good to go. Second Warning: My copyeditor husband has suggested that I make an "I was young and didn't seem to know much grammar" disclaimer. Whatever. Everyones a critik.

Drum roll please...

Although women have been giving birth since the beginning of time, researchers have just started looking at the physiological and biomechanical changes occurring during pregnancy. A major reason for this investigation is the large occurrence of lower back and pelvic pain during and after pregnancy. Many researchers have attempted to find the mechanisms that determine hip, pelvic, and sacral pain in general populations, but few have examined this situation as it pertains specifically to the maternal female.

The cultural perception of pregnancy-related pain and resulting issues is that these conditions are a normal part of the gestation process and perhaps not research-worthy. Data shows, however, that these ailments are not natural to the state of pregnancy, only normal. Studies suggest that fifty percent of all pregnancies will begin and end with debilitating back and pelvic pain, and even more alarming, that this pain continues postpartum. The role of exercise as both a treatment and preventive measure for these conditions is the newest addition to current research.

Pre- and postnatal exercise has traditionally consisted of modified traditional aerobic and strength-training exercise. This type of exercise can be beneficial to general fitness goals but lacks the specificity in design when it comes to other requirements of the maternal female, i.e. improving structural integrity, optimizing vaginal birthing mechanics, etc. The mechanical functions of the uterus are now understood to depend on pelvic loads and pressures, yet this content has not yet made it into the academic curriculum of pertinent professional studies. Creating a movement program for this population that not only meets general health guidelines but also facilitates and improves the state of pregnancy and delivery outcomes is a valid scientific endeavor.

What have studies shown to date?

Pain is difficult to quantify for research purposes and can reduce the validity and reproductability[2] of a study. The use of the questionnaire is common protocol for data collection. For new mothers, the format is more time effective than scheduled appointments, and data is easy to collect due to simple, multiple-choice answers. Common pain assessment questionnaires are the visual analog scale (VAS), Zung, and Somatic Perception (Russell, Groves, Taub, O'Dowd, and Reynolds, 1993). Pain drawings, on transparency, are also used to quantify location (Nilsson-Wikmar, Pilo, Pahlback, and Harms-Ringdahl, 2003). The transparencies are then stacked to analyze data (2003). The most significant chronic pain patterns found in the pregnant or postnatal were listed as the posterior pelvic-sacroilliac area (PPP), lumbar spine, or a combination of the two (2003).

Experience of pain itself is, of course, an issue of discomfort and not to be discounted. The much larger issue is, however, the daily limitation of everyday activity. Within the general population, adults experiencing low back pain find bending, twisting, and lifting difficult and painful (Youdas, Garrett, Egan, and Therneau, 2000). For a new or impending mother the restriction of movement decreases the possibility of caring for and lifting a new baby, returning to movement for health purposes, or returning to work. This reality poses additional stress into an already psychologically demanding situation and could be a contributing factor to physical and psychological issues stemming from pregnancy.

[This is me, eight months prego, feigning back pain. Don't you like the expression on my face?]

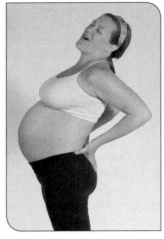

There are many circulating beliefs about where pregnancy-related pain comes from. Russell, Groves, Taun, O'Dows, and Reynolds (1993) authored the first study to look at the effects of an epidural on lower back pain. Questionnaires were sent to 1615 women who delivered their first baby at St. Thomas's Hospital in London. Following the questionnaire, the subjects with persisting pain were examined by a physiotherapist or orthopedic surgeon. It was established that LBP was equally evident in both epidural and non-epidural deliveries, and tended to be postural in nature (1993).

Pregnancy-induced posture and mechanical changes are common research topics when looking at possible causality of associated pain. What researchers have found is contrary to our beliefs about pregnancy-related ailments. The adopted posture of pregnancy is individual in nature and not created by the state of pregnancy alone. While the increase of body mass and subtle forward displacement of a subject's center of gravity is empirically valid for all pregnant women, there is no "pregnancy posture" that can be imposed on the state of pregnancy (Gilleard, Crosbie, and Smith, 2002). It is generally believed that the pregnancy gait, or waddle, is a natural change due to the gestational condition. This is not the case. When healthy pregnant subjects were assessed against healthy nulligravidae, the kinematics were found to be very similar, with only small changes in a maximum walking velocity (Wu et al., 2004). It is interesting to note that postpartum women with posterior pelvic pain showed large deviations in normal gait and large angular rotations of the hips and pelvis when compared to the healthy, pain-free pregnant population (2004). Wu et al. concluded that the pregnant female is intended to walk and move regularly, even with the extra mass of an impending baby.

These findings are very important contributions to solving pregnancy-related ailments. This data implies that it is not the physiological state of pregnancy that is responsible for pain, but the alignment and gait habits that a woman brings to her pregnancy that are the underlying causes.

Although conducting research on pregnant subjects will always be a difficult situation due to potential risk and liability, more studies are desperately needed. Studies on PPP could begin in the safer postpartum realm. More recent studies have begun to look at the benefits of strengthening exercise, specifically deep abdominal, pelvic floor, and gluteal musculature, as a possible, non-surgical method of dealing with varying pregnancy-related ailments.

More studies are needed to quantify, define, and produce postural interventions to mitigate pregnancy-related issues. Treating pregnancy-related conditions with medication and surgery is not only expensive, but also ineffective in the long term. It is obvious that studies need to be developed observing less invasive, less costly repair. One glaring hole in pregnancy research and correlating back pain are studies that deal with the postural and muscle stability loss following cesarean section. Cesareans are major abdominal surgery and are bound to result in systematic, measurable changes in strength and subsequent injuries to the lumbar spine, hips, and pelvis. With a developing trend of elective cesarean, it is imperative that as much information on recovery and long-term effects be researched.

Due to the traditional lack of women in research, medicine, and science in general, it follows that the amount of studies pertaining to women's issues be few in comparison. As this slowly changes, studies examining traditional viewpoints on the natural birthing process, especially the maternal mechanics of vaginal expulsion, be better understood by birthing professionals as well as those prescribing exercise.

Pre- and postnatal exercise is not new to kinesiology communities, yet it has always been presented as a modified aerobic and strength-training workout, allowing a participant to continue with a fitness-type exercise program during the gestational period. The numbers of those suffering through pregnancy should be taken as a signal to improve the education of birthing and kinesiology professionals. Exercise that is designed to facilitate an easy gestation period, delivery, and recovery must be studied and dispensed through the correct channels.

Due to the lack of clinical and scientific training for most kinesiologists, exercise is oversimplified when presented to the pregnant population. Physical therapists and physiotherapists typically do not see the general population until a problem has manifested, and even then are often prevented by professional guidelines to address mal-alignments elsewhere in the body, even if they are related to the presenting injury. Where are women to find this information before they realize they need it?

Most reviewed articles called for a movement in prenatal physical education. Because poor postural habits lead to low back pain in the general population (Youdas et al., 2000), it is only to be expected that participants with prior posture and lumbo-pelvic problems find that pregnancy exacerbates their problem. Deviations in alignment are indications of a failing muscular support system. This system can be improved through a restoration of muscle and joint positioning. Intervention in alignment could potentially eliminate pregnancy-related pain if good instruction and intervention is early. Potential candidates, those with LBP before pregnancy, should be given postural exercises or guidelines to offset gestational pain.

After reviewing the literature on pregnancy and related low back and pelvic pain, a common thread becomes apparent. While the physiological changes are fairly similar for various women during pregnancy, biomechanical changes seem to be of a personal nature. Of these mechanical changes, one's postural habits seem to have the greatest effect on pain. It would seem from an evolutionary standpoint that debilitating low back, hip, and pelvic pain would not be conducive to the natural birthing process.

So that's my paper. Wow. Better stand up and touch your toes. Maybe take a quick walk around the block and have a glass of water. And a nap. So now you have a bigger picture of the problem. I'll write more later. I need to stretch after reading, writing, and editing this!

---

1. Pelves is pronounced pel-veees. It does not rhyme with elves, as in a group of elfs. Or elves, I guess. Although the idea of a fairy tale about a pelvic elf has just started to form in my mind. Great. That's not going to cause me any nightmares.

2. Reproductability: Even in grad school, I was busy making up words.

Sources:

Chiarelli, P. & Cockburn, J. (2002). Promoting urinary continence in women after delivery: randomised controlled trial. British Medical Journal, 324, 1241.

Elden, H., Ladfors, L., Olsen, M., Ostgaard, H.,& Hagberg, H. (2005). Effects of acupuncture and stabilising exercises as adjunct to standard treatment in pregnant women with pelvic girdle pain: randomised single blind controlled trial. *British Medical Journal*, 330: 761.

Gilleard W., Crosbie, J., & Smith, R. (2002) Static Trunk Posture in Sitting and Standing During Pregnancy and Early Postpartum. *Archives of Physical Medical Rehabilitation*, 83, 1739–44.

Leivseth, G. & Drenup, B. (1997). Spinal shrinkage during work in a sitting posture compared to work in a standing posture. *Clinical Biomechanics*, 12, 409–418.

Nilsson-Wikmar, L., Pilo, C., Pahlback, M, & Harms-Ringdahl, K. (2003). Perceived pain and self-estimated activity limitations in women with back pain post-partum. *Physiotherapy Research International*, 8, 23–35.

MacEvilly, M., & Buggy, D. (1996). Back pain and pregnancy: a review. *Pain*, 64, 405-414.

Rodacki, C., Fowler, N., Rodacki, A, & Birch, K. (2003). Stature Loss and Recovery in Pregnant Women With and Without Low Back Pain. *Archives of Physical Medical Rehabilitation*, 84, 507–12.

Russell, R., Groves, P., Taub, N., O'Dowd, J., Reynolds, F. (1993). Assessing long term back ache after childbirth. *British Medical Journal*, 306, 1299–1304.

Sihvonen, T., Huttunen, M., Makkonen, M., Airaksinen, O. (1998). Functional Changes in Back Muscle Activity Correlate With Pain Intensity and Prediction of Low Back Pain During Pregnancy. *Archives of Physical Medical Rehabilitation*, 79, 1210–12.

Van Dongen, P., deBoer, M., Lemmend, W., & Theron, G. (1999). Hypermobility and peripartum pelvic pain syndrome in pregnant South African women. *European Journal of Obstetrics and Gynecology and Reproductive Biology*, 84, 77–82.

Wu, W., Meijer, O., Lamoth, C., Uegake, K., vanDieen, J., Wuisman, P., deVries, J., & Beek, J. (2004). Gait coordination in pregnancy: transverse pelvic and thoracic rotations and their relative phase. *Clinical Biomechanics*, 19, 480–488.

Youndas, J., Garrett, T., Egan, K., & Therneau, T. (2000). Lumbar Lordosis and Pelvic Inclination in Adults with Chronic Low Back Pain. *Physical Therapy*, 80, 3, March.

chapter 10

WALKING AND GAIT

From *De Re Militari*, by Flavius Vegetius Renatus (this book is thought to have been written between 375 and 392 AD, which is a realllly long time ago):

> The first thing the soldiers are to be taught is the military step, which can only be acquired by constant practice of marching quick and together. Nor is anything of more consequence either on the march or in the line than that they should keep their ranks with the greatest exactness. For troops who march in an irregular and disorderly manner are always in great danger of being defeated. They should march with the common military step twenty miles in five summer hours, and with the full step, which is quicker, twenty-four miles in the same number of hours. If they exceed this pace, they no longer march but run, and no certain rate can be assigned.

As a biomechanist, it seems I am cursed to enjoy my life from the perspective of a gait laboratory.

I had a great time at our town's fair parade this weekend, watching the floats, listening to the music, and analyzing the gait of the marching bands. (Doesn't everyone do that??) And it just happens that I am in book-writing mode, in the chapters of "history of gait patterns," of which MARCHING plays quite a role.

Now, most of us have been walking for years, but what you may not know is *how* your walking pattern came to be. Like all animals, we learn the bulk of "how to be" through observation. Your walking, or gait, pattern is based on how others walked around you when you were growing up. If you spent time in ballet class, cotillion, or the military, then you furthered your development by adding additional information through mimicry or direct instruction.

It is widely known to physicists that most people are not walking anymore, but are instead falling in a controlled manner. What is the difference between falling and walking? Well, the only way a body can move itself forward is by pushing off in the opposite direction. Imagine swimming, or paddling a boat. Think of the tires on your car or the thrust from the engines of the space shuttle. The motion is always in the direction opposite to the desired movement. So, actual WALKING requires you to push your leg behind you...kind of like ice skating. If you're not pushing behind you, then you're simply lifting your leg out in front of you and then leaning forward slightly to generate momentum to help you fall forward.

Calories burned = negligible

Body parts that have to dampen the force when you fall forward = knees, hips, and spine

My brothers, father, and grandfather were all in the military throughout the world. How about yours? Because gait patterns are learned primarily through visual input, our culture's rich military history has had an impact on the way we move as a nation, and the way we are ailing as a nation. Especially when it comes to ailments of the joints, bones, and musculoskeletal system. These are injuries we are creating ourselves due to the fact that we have never been taught how to move in a way that is controlled, metabolism enhancing, and joint conserving. Biologically we are designed to walk from the hips, generating a "push-off" from the large muscles of the hamstrings. Our Western gait pattern, however, has evolved to match one closer to marching, lifting the leg out in front from the hip, which makes us walk "from the knees down"—overusing the knees and underusing the hip joints. This is a similar gait pattern you get from walking on a treadmill—a big no-no if you want to walk with an efficient, non-joint-damaging gait pattern. Because the treadmill belt is moving backward, you have no choice but to lift your leg out in front of you and lean forward (it's very small, and sometimes easy to miss, but check out the torso angle on the people walking and jogging in your neighborhood or gym and see how they lean forward).

April 11 2011 SEEING THE SIGNS

Walking about town today, I was struck by the notion that someday, future populations are going to be making large assumptions about us just as we make large assumptions about populations past. These assumptions are usually not total W.A.G. (wild-a$$ guesses), but information supported by the evidence—some sort of documentation left behind on the walls of caves or etched into pottery. Feeling a little like Indiana Jones (only without the hat), I collected some evidence about North American humans c.2000:

Two things bug me about these photos (with the exception of the mannequin photos, where so many things bug me, I don't know where to start). One is the forward-pitched torso and head and the other is the bent knees. Why is the human form constantly portrayed as a forward-leaning image? What we have here is not life imitating art, but our pictographs re-

flecting the state of our condition. And what conditions would those be? Knee and hip osteoarthritis. Osteoporosis. Low-back pain. Low metabolisms. Bone spurs. Torn meniscus. Foot pain. I could go on and on.

So I will.

Walking is often said to be "controlled falling," and I would agree. I would agree that what most people are doing when they believe they are walking is accelerating their body toward the ground and quickly yank-

ing a foot in front of them to catch themselves before smacking into the pavement. You're not doing this because it's the "natural" way to walk, but because one by one, all of the reflexes you were born with have been shut off via the deadening of various motor programs or the failure to turn them on to begin with.

So, for a few quick tips on walking:

1. Don't let your torso get in front of the rest of your body. This posture uses gravity to accelerate you and your joints to buffer the landing.

2. If you want to actually do mechanical work to move forward, your hamstrings need to be at their full length. It is the leg that pushes back in a controlled fashion to move the rest of your body forward. Stretch, stretch, stretch.

3. Stop landing on a bent knee. Only beg on one. The rest of the time, the muscle strip along the outside of the hip and knee should be working to keep the leg bones fully lengthened (straight). Landing on a bent knee is an indication that the strength in your hips is less than the weight of your body.

BONUS: Use these posted signs as a reminder to straighten yourself up when you're out and about. Use them as teaching tools for your kids, as in "See that sign? Make sure your body isn't leaning forward like that when you're walking."

Quick exercise: Stand on one leg without bending either knee, without leaning forward, without thrusting your pelvis forward, and without putting your arms out to the side. You can do this with your back against a wall to help you figure out if you are leaning forward from the hips or not. Your weight should be in your heel. This exercise is a simple screening mechanism to see if your legs are strong enough to hold up the weight of your body. If you're wobbly or have to bend your knee, then you are a faller rather than a walker.

First step to becoming a walker is be able to do this on each foot for a minimum of a minute. P.S. You should be in bare feet.

And start paying attention to the signs you surround yourself with. The bulk of "learning how to be" is done via observation. Imagine what messages we are programming for "normal human condition" as opposed to "we're all broken in this boat together." Let us change our movement

A

B

patterns and then let us change our signs.

This year, I had to modify the Restorative Exercise Institute's logo from this (image A, previous page) to this (image B, previous page).

The front leg needed to be straightened to how the front leg actually *should* receive the weight of the body during a normal gait cycle, and we  gave REX a new hairstyle. Very gender neutral and frankly, much more flattering, I think. You?

In case you hadn't noticed, I now have an iPhone, which means you can expect many more photo-laden posts from me. Also, be wary of me following you around, filming the way you walk. My cat is already annoyed at my constant photo snapping.

See? He hates to be bothered while reading.

 GAIT 101

October 3 2011

October is National Walking Month, or "Walktober" for those with refined sense of humor. This is the first of at least one post on walking this month. Enjoy. ☺ I've been reading a lot of magazine articles and blog posts discussing gait patterns, and with the recent popularity of barefoot running and minimal footwear, these discussions are popping up more and more often. Of course, many of the discussions are filled with errors and inaccuracies—both mechanical and semantic—so I thought I would put out a sort of clarification piece, should anyone need to wade through the good, the bad, and the ugly with a reference guide to the accurate.

What is gait?

Originally, the term *gait* meant a manner of walking, but has since become used more generally, referencing any pattern of limb (arm and leg) movement while moving on foot. Everyone has a particular gait pattern, or way of moving. It comes from a combination of how one learned to walk as a child by mimicking those around them, a lifetime of musculoskeletal habits, and any injuries thrown in along the way.

Your gait is so specific to you that the government has worked on developing technology that recognizes a particular person from satellite video based on this pattern. Your face, you can hide. Your gait pattern, not so much.

Walking and running have been specifically defined to avoid errors in research and resulting conclusions. The clinical definition of walking requires the mover have at least one foot on the ground at all times. Human walking should be a mechanical process that moves the pelvis (your center of mass) in a smooth, forward-moving line. Your loaded knee shouldn't bend, your pelvis shouldn't twist, and your head shouldn't be bobbing along. And P.S., this smooth walking requires whole-body participation, from head to toe, and is the end result of perfect mechanical alignment.

The clinical definition of running requires the mover have a "flight phase," or period of time during each gait cycle when no foot is touching the ground. This means that the center of mass ping-pongs along, accelerating and decelerating in each cycle.

Here is a high-tech side-view image of your center of mass during the ideal form of both activities:

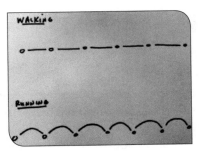

As you can see, correct walking gait patterns are entirely different from correct running patterns. The smooth translation of the pelvis while walking requires your foot to move through four points: Heel strike, foot flat, heel off, and toe off. If you don't begin with the heel while walking, it means that you can't use your posterior muscles to propel you forward, which means you don't have the natural strength you should have in your gluteal muscles. Which means weak pelvic floor muscles, too.

Running, due to the much higher G-forces and abrupt changes in whole-body direction (even though you feel like you're moving forward, the body is really traveling mostly up and down while running), requires more force absorption than walking. This is why, when you run, you should land on the front of your foot so your calf muscles can work eccentrically to slow your crash-crash-crash into the ground. What's the force difference between walking and running? I'm glad you asked.

A quick lesson in G-forces: "G" or gravitational forces are a relative unit, based on your particular mass.

Standing is a 1-G activity. If you weigh 160 pounds, then your body is dealing with 160 pounds of force.

Walking is a 1.5–1.75-G activity. If you weigh 160 pounds, then your body is dealing with 240–280 pounds of force.

Running is a 2–3-G activity. If you weigh 160 pounds, then your body is dealing with 320–480 pounds of force.

If you are in the movie *Top Gun* and your name is Maverick, you would be in a NEGATIVE G-Pushover with a Mig 28. Keeping up foreign relations. You know, giving them the finger? Yes, Goose, I know the finger.

(Just in case you were wondering what I was doing between the ages of twelve and fourteen, this was it: memorizing *Top Gun* so that when I was a naval aviator, I would know how to talk. Really.)

To quickly summarize, should you be the sort of person who flips to the back of the book:

Optimal gait pattern for walking is heel-toe.

Optimal gait pattern for running is toe-heel.

In the last three national-publication interviews I did, the journalist wanted me to explain correct gait for barefoot footwear like Vibrams or Sockwas. I asked for clarification—for walking or running? They didn't know what I meant. There wasn't room in the article for the difference. Couldn't I just quickly tell them how to use the new barefoot shoes correctly? In five sentences?

You can see why I often want to poke things into my eye.

I wrote this post to help you discern between accurate and not-so-accurate information presented in articles on running and walking gaits and to help you understand the biomechanical requirements of each. Also, don't confuse the minimal-footwear and barefoot movements with a running trend. Every human needs to be able to walk correctly before they run and shoe characteristics as we know them are interfering with the health of the entire body.

I took an awesome hike today. I hope you did too.

WALKING OR BOUNCING?

Last night I watched *Superman*, the movie. Not the new creepy one, but the old one. With Christopher Reeve, who forgot that we were supposed to get married.

I preface my post with these facts because I was trying to get my iMovie to work correctly, and it wouldn't, but it may have been due to the fact that I clicked something I wasn't supposed to while watching a hunk in blue tights.

Anyhow, it's still October. I mean, Walktober. So, more walking posts, with stuff for you to practice. A couple of weeks ago we talked about the difference between walking and running. This picture (image below left) shows the classic difference of how the center of mass travels during both.

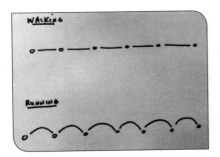 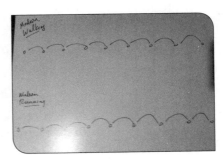

But here's the problem. People aren't really walking like this any more, as defined by clinical determinants of gait. These determinants are how biomechanists calculate forces and measure joint changes.

It turns out that most people's walk looks exactly the same thing as most people's run. (Image above right.)

When I look at people, I've always been able to see what all of their joints are doing, even when they have their clothes on. I don't know how or why, but that seems to be the thing I'm really good at. I was hoping it would be something like writing operas, or painting frescoes, but nope. That's not what I got. I have something more like this:

Katy (watching *Superman*): That guy playing Clark Kent as a kid is limping.

Husband: Shhhhh.

Katy: There's something wrong with his hip.

Husband: Shhhhh. (But reaches for iPhone.)

Katy: I love Superman.

Husband: Who doesn't?

Katy: Shhhh.

Husband (super movie nerd): I just used my IMDB app and it says that the kid tore his hamstring muscle while filming. That's why he's limping.

Ha! Katy 1, Superman 0.

Hello, sidetracked.

So ideally, in walking, your body makes smooth translation, as shown in the first drawing, which is what gives our joints longevity. This smooth gait is really determined at the knees, as these are the joints that buckle the most. You may have heard physicists calculate that people are now falling in lieu of walking or running. They are correct, and this pattern of walking is a really good way to wear out your parts really fast.

Wanna see the difference?

Here's a regular-speed clip of walking, with the leg fully extended, receiving the weight of the pelvis.[1]

Now here it is in slow motion. Get your bread and jelly, cuz this video is extra smooth.[2]

Here is a video of an unstabilized knee joint. Feel free to make "boing, boing, boing" noises as I walk by.[3]

Now here it is in slow motion.[4] I have no idea why this clip got all stretched out and weird. I spent about an hour trying to fix it, to no avail. Hopefully you can see that when the leg receives the weight of the body, it bends farther and then has to lift the body's weight back up. Each loading cycle causes the quadriceps to contract, which pulls the patella (knee cap) back into the bones behind it, creating little etchings in your cartilage. (Note: Etchings sound cute but they're not, really.)

A SMOOTH GAIT REQUIRES THESE THINGS:

1. Hamstring length. You sit a lot? Get to work on your hamstring stretches and try to stand a bit more through the day.

2. A fully extended leg and relaxed kneecaps. NOTE: Straight legs are not "locked legs"! Use this video (the second half) to help you find relaxed quads.[5]

Super-tight quad users use the boing-boing gait pattern because they have all their strength up front and none out back. Your kneecaps should always be in the down position, unless you're walking up or down hill.

3.   Lateral hip strength. To getcha some, try this Pelvic List exercise[6] from the Knees and Hips DVD.

4.   Coordination. You can't straighten your leg once your body-weight is on it. Not good for the knees, and it uses the quads to straighten the knees, not the hamstrings. The leg must straighten in the air, before landing.

Also note, the stride should be long behind you, not in front. Lifting a leg out in front of you uses that danged psoas. We do not want to use that muscle with every step. Ideally, the leg work is behind you, so it swings into the forward position, no psoas needed.

Have fun practicing! Or at least watching others to see if you can spot the boing!

---

1. youtu.be/ImyBAW_9irQ Three seconds of smooth walking.

2. youtu.be/iMkYPXXLqJs Eighteen seconds of slow-mo smooth walking.

3. youtu.be/PDBb1n5_-nk Four seconds of bouncy walking.

4. youtu.be/Js9jeGy_6NA Nineteen seconds of weird, stretched-out bouncy walking.

5. youtu.be/qcGPY4BMdIU About four and a half minutes showing how to get knees to point straight forward and relax. Lots of good sound effects, too.

6. youtu.be/wH6SNijSPgU About three and a half minutes teaching the pelvic list exercise.

chapter 11

CARDIOVASCULAR SYSTEM

I thought that it would be easy meeting my blogging schedule with a new-born, because don't they sleep a lot?

Turns out, not so much.

And it also turns out that I have too many subscribers to this blog—it can no longer email notifications of posts to you because it crashes the email system.

Go me.

Also, in other news, I found my first gray hair. Thirty-five years with nothing. Six weeks of sleep deprivation and hello Bea Arthur. I didn't pluck it. And not just because I didn't want three more to come to its funeral (is that even true?) but because my eyes no longer focus well enough to find it in the mirror.

While I'm not blogging as often as I like, I do make it to our Faceook page often to interact and answer questions. Come join us if you can. It was on FB that someone posted the infographic below (courtesy of medicalbillingandcoding.org) and asked my thoughts. Read through this and then see my thoughts, posted below.

medicalbillingandcoding.org/sitting-kills/

My thoughts (please mail penny to my home address):

I agree completely with the statement "sitting is killing you," but I would say the reasons sitting is harmful are not what they have listed and the proposed solutions (jumping jacks in place and putting your chair back into a reclining position) are not correct.

And P.S., sitting at 135 degrees is just as harmful as sitting at 90. The reason sitting is harmful HAS NOTHING TO DO WITH BODY FAT, OBESITY, etc., which is why you see the diseases associated with sitting in people of all body masses, great and small.

The link between sitting and cardiovascular disease has to do with the physics of how your blood flows through your arteries. Different types of flows promote different types of plaque accumulation.

Plaque accumulates NOT because of blood chemistry, but because of the action of the cells flowing within the blood vessels.

Normally, cells in the blood glide along down the blood vessels, like a ball rolling down a hallway. If that hallway has a turn in it, however, the ball will smack into the wall.

Blood cells do the same thing. A blood cell hitting the wall of an artery leaves a wound, which, just like a cut on your hand, needs to make a scab. Scabs in the arteries use cholesterol to seal the wound. Your cholesterol is not arbitrarily sticking to the walls of the vessels but is part of the healing mechanism your body has. That's why high cholesterol isn't the problem—the wall wounding is.

What causes the blood cells to smack into the walls of your blood vessels?

A couple of things. (There are actually more, but the physics is a bit more complex for this blog today.)

1. The type of blood flow you are creating. Blood flow can be turbulent (like turbulent air during a flight) or laminar, all of the blood smoothly flowing in the same direction.

    Turbulent flow can take a blood cell and, instead of letting it flow straight through a vessel, can slam it up or down, into the wall. BAM. Wall wound.

2. Blood vessel geometry. The good news: Your arteries, where they are located and how they branch off, are arranged in a very specific way to maximize pressure gradients and keep flow smooth. The bad news: Your artery geometry changes with your posture.

Adding twists and turns in the arteries by chronically bending joints for excessive bouts of time a day is like adding curves in your hallway. Rolling balls are guaranteed to smack into the walls in your home, just as blood cells will run into the curves of the vessels...every time. This is why research shows that exercise doesn't offset the effects of sitting. You can't undo eight hours of wounding with a run or with bigger muscles. Fitness doesn't touch the wound that has been created.

We have made the entire science of cardiovascular disease about fat and cholesterol and chemistry when it is really about geometry.

You can try to cut all the cholesterol out of your diet, but that doesn't keep the plaque from accumulating. You have to stop the WOUNDING, which means you have to fix the geometry.

Which means you have to fix your alignment.

I'm so boring saying that over and over again, right?

I go on and on about alignment, but I don't want you to think it's about aesthetics, or appearance, or even the macromechanics like muscle function and joint health. Even though it is about those things too. It's really about the micromechanics. The fact is, if you are not aligned correctly, you are creating damage on the cellular level.

And I know most of you care more about cardiovascular health in the long term, more than you care about your bunions.

And P.P.S., vigorous exercise (where your heart rate is up above sixty percent) can be just as harmful as sitting, as it creates turbulent flow.

My three-point solution for starting a healthy-heart program is this:

1. Get a standing workstation. You cannot justify sitting at work. You really can't any longer.

2. Walk (don't run!!!) four to five miles a day (not necessary to do it all at once, and don't do it on a treadmill).

3. Fix the position of your bones via stretching your muscles.

The position of the bones changes the geometry of the blood vessels—when they're not where they are supposed to be, you are sitting on a cardiovascular time bomb.

Yes, you can quote me on that.

In fact, please spread this quote around to everyone you love. And like. And the dude who delivers your pizza. Put it as your Facebook status, with a link to this post, and we'll see how long it takes to reach 25,000 people.

P.P.P.S. I think this thousand-word blog post is worth more than one penny.

And (how many times have I started a sentence with "And" in this post?) to help you out, I am having a contest where you can submit a picture of your standing workstation to get a chance to win one of my DVDs of your choice and the coveted half dome. If you don't know what a half dome is, it is the best piece of health equipment, ever, used for the best exercise for your health, ever, the Calf Stretch.

I'm not going back to edit that sentence. I'm leaving it for style purposes.

Below are a few examples of submissions.

Sorry the pics are so small. Now you know what it feels like to be me, squinting at my hair follicles.

More reading, if you are a huge nerd, have an extra $150.00, and have taken math up to differential equations (if not, it will be a waste of $): *The Physics of Cerebrovascular Diseases, Biophysical Mechanisms of Development, Diagnosis and Therapy*, by George Hademenos and Tarik Massoud.

 June 1 2011 MORE BLOOD PHYSICS

Today's word is *hemodynamics*. I actually prefer the spelling *haemodynamics*, probably because my dad is Canadian, and I like to mix it up.

Fun fact: I wanted to name my kid something original with reversed letters, two dots over the "a," or that a and e that are connected, like it was written on the old encyclopaedias I grew up with. I was vetoed.

Whatever.

I still think every person should have to use the word "diphthong" when explaining their name to the substitute teacher.

So, hemodynamics (*hemo*: relating to blood or blood vessels; *dynamics*: mechanics, "the time evolution of physical processes") is the study of, or the principles that dictate how, blood moves through the body.

Hemodynamics are similar to hydrodynamics (the principles that dictate how water moves), only different. The main difference is that water is a Newtonian fluid—meaning that all parts of water are doing the same thing at any time. Blood is a non-Newtonian fluid, meaning different parts of blood do different things. This is because blood is made up of blood cells floating in liquid. The cells have a rigid shape (well, rigid for cells), while the liquid is more fluid.

The liquid is more fluid.

Ha.

Get it?

In case you couldn't picture the difference between laminar and non-laminar (turbulent) flow in my last post, here's a picture:

We want to avoid turbulent flow because this type of current takes the blood cells and accelerates them into the walls (BAM!), creating a wounding that begins a cycle of inflammation, which leads to scabbing using cholesterol and calcium (which contribute to the hardening of the artery), which leads to more turbulent flow (and more wounding, and so on and so on).

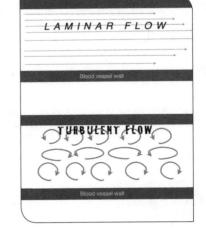

By now you are probably chomping at the bit to know what causes the turbulent flow, yes?

That's what I thought.

But before we look at the causes, you should know a little more fluid dynamics and recall a bit from your math and physics classes.

Picture a tube, like your garden hose for example.

You probably know that you can change the velocity (the speed and direction it shoots out) of the water by changing the size of the tube, i.e. putting your finger over the end.

When you are washing your car (I hope there are still people out there washing their own cars in the driveway) and want the water to squirt out in order to be a bit more abrasive getting the soap off as well as cover a greater area, you instinctively know to put your fingers in a way that reduces the area and increases the velocity. Got it? This is a basic principle of hydrodynamics. Reducing the area of flow increases or changes the velocity.

Within the arteries, non-laminar flow results in blood flowing all different directions. Turbulent flow creates a white-water wall that creates a temporary narrowing, similar to your finger over the opening of the hose. Even though water is a fluid, it can still act kind of like a solid in that it can change the direction of other fluids hitting it.

Are you with me?

To keep the blood flowing in a smooth fashion it is imperative to keep your blood laminar, man. We want groovy, easy-flowing blood. We want the Simon and Garfunkel of blood flow.

Once flow gets a bit bumpy it creates more and more turbulence, which in the long term creates plaque. But this plaque, see, is like a permanent finger over the hose so you get chronic turbulence wherever this plaque is located. Which then begets more plaque.

It's a vicious cycle.

So, back to causes of turbulent flow.

1. Smoking. Smoking causes an instantaneous reaction (because you are basically preventing your lungs from getting any oxygen in that breath, which freaks the body out) of stress chemicals that causes the blood vessels to open up instantly—creating a change in the size of the tube, which then of course causes the blood to slosh around creating...you got it, turbulent flow.

2. Stress. Stress itself causes stress hormones that do the same thing as 1.

3. Changes in blood vessel geometry. There are natural changes in BVG in the body, where arteries branch off to travel different directions and in these places, the arterial walls have programs to deal with the change. The problem occurs when you change your blood vessel geometry arbitrarily, by adapting certain postures.

   For example, the carotid arteries are oriented in a fairly vertical fashion, but if you suffer from "computer head" or "cyclist's head" or "I'm anxious to get where I'm going head,"

then you, my friend, have changed the physics of the flow in your arteries to non-optimal. And increased the wall wounding in these arteries.

Another cause of turbulent flow is anything that changes the viscosity of your blood. Viscosity is the tackiness of a fluid. If your blood wears black socks everywhere then the viscosity is high.

No, just kidding.

Tackiness as in how much a fluid sticks to itself. Water has a lower viscosity than, say, honey or motor oil. When a fluid is more viscous, it does not travel down a tube all together, but tends to clinging to itself creating...

Anyone?

Anyone?

Bueller?

Bueller?

Did you answer "turbulent flow"?

Did you say it out loud?

You are a nerd. You are reading a blog and you just said "turbulent flow" out loud, didn't you.

Heh.

Good job. You are right!

So when it comes to the viscosity of your blood, there is one thing that changes it up pretty fast and that is...

4. Blood sugar. The higher your blood sugar, the more your blood is like syrup. Like the type of syrup Starbucks puts in your vanilla latte. Think about that next time you get a frappa-mocha-pepper-mint-dulce-venti-throat massage.

So, the moral of the story?

For cardiovascular health the items you need to address before running and jumping and medicating are:

1. Stop smoking. Really. Are you still doing that?

2. Start meditating or at least do all the other stress-reducing things that you already know you should be doing.

3. Align…everything. But start by pulling your head back (what we call "ramping up" in the A&W program) until your ear is over your shoulder like this:

This will make your standing workstation EVEN BETTER.

4. Watch your blood sugar. But you already knew that too, didn't you?

5. When we talk about CV disease you will rarely hear about the mechanisms—only the symptoms (like plaque). A correct, thorough prescription for prevention (or treatment) of CV disease should START with the items listed above.

Think about it.

Okay, off to feed my regular-named child.

 5 THINGS YOU DIDN'T KNOW ABOUT EXERCISE AND HIGH BLOOD PRESSURE

Since I am on a cardiovascular roll, I thought I'd keep going.

One of the most frustrating things I experience as a scientist of human mechanics is the constant mis-prescription of exercise that is happening everywhere.

You know how, when someone points something out, like an annoying trait someone has, or like when you learn a new word, or like when you get a new car, you hear/see it everywhere? It's like that. (It's like, annoying, right?)

I cannot escape poor movement science information. It punches me in the face when reading any "health" magazine, stabs me in the ears when I listen to any radio or television segment on exercise, and, unfortunately, I often hear it out of the mouths of professionals.

A few posts back I mentioned a bit about the intensity of exercise (over sixty percent of peak) increasing turbulent flow, a precursor for arterial plaque accumulation. Big deal. Huge deal. Especially for those of you out there who think that the more you work your heart, the healthier your heart will be.

Nope. This is not correct, and it is really frustrating when I see people at the prescription counter for their blood pressure meds who are regularly doing intense to very intense exercise. Most of you out there may be aware that high blood pressure (HBP) is a major public health issue (and you who are on the meds are not "cured" of your HBP—you still have it.) In fact, the American College of Sports Medicine states that "Hypertension (HTN), one of the most common medical disorders, is associated with an increased incidence of all-cause and cardiovascular disease (CVD) mortality."[1]

Because many of you are using exercise to deal with or prevent HBP, I thought I'd clarify a few things for y'all.

1. The correct exercise intensity for those with medicated or unmedicated HBP is moderate, or forty to sixty percent, NOT seventy to eighty-five percent.

    You've heard that walking is super-hyper-awesomely beneficial to your health, right? One of the reasons it is so beneficial is that you get all the benefits of movement (a symmetrical, whole-body pattern of muscle contraction, fully weight bearing on the skeleton, increased circulation and oxygen distribution) without the plaque-increasing turbulent flow that comes with greater intensities.

    According to the American College of Sports Medicine (ACSM), "The intensity of the exercise is directly related to the hemodynamic response and myocardial VO2[oxygenation]."[2]

2. Resistance exercise lowers blood pressure and has additional benefits to "cardio." Strengthening your body is not just about being a beefcake.[3] Nor does resistance exercise need to look like lifting

weights. In fact, using your body weight as resistance gets you much more of the BP-lowering effect as well as uses way more energy (aka "burns more calories"). Pilates, yoga, calisthenics, martial arts, CrossFit, etc., are all good options, but you need to keep the intensity at forty to sixty percent to get the healthy-heart benefit.

(BUT, even if you do a lot of BP-lowering whole-body resistance training, you still have to walk. EVERY HUMAN needs to be walking at least three miles a day.)

Why does it work? The more muscle you innervate, the more the blood leaves the big tubes (the arteries) and flows into the capillaries of the working muscle. When you take a tube (artery) and remove some of the fluid (blood), it drops the pressure.

Simple. Easy. Inexpensive. Like me.

Just kidding,

I'm not that simple.

3. More exercise (above the daily forty-to-sixty-percent-intensity walk and whole-body resistance training) is not better. Medical literature has demonstrated that excessive training, or "chronic exercise" (think marathon running) can actually precipitate a heart attack in certain individuals. What causes it? Researchers aren't sure. It may be the abrupt change in heart rate, or the temporary oxygen deficiency at the heart-cell level. Or a billion other things. What increases the risk? Age. Already having coronary disease. The intensity of exercise. (See 1.)

4. If you ARE taking a medication for HBP, using a heart-rate monitor to gauge intensity doesn't work very well. Blood pressure meds can often alter the natural hemodynamics of exercise, so you can be working at a high intensity but you won't actually create a measurable increase on your measuring tool. It is much better to go with a self-gauge or perceived rate of exertion (PRE). If you feel like you are struggling but your tool shows that you are "flying safe," trust your gut and bring down the intensity until you can talk comfortably to the person next to you, but not down so low that you find yourself on the couch eating potato chips.

It's all about balance.

5. "The risk of cardiovascular complications and orthopedic injuries is higher and adherence to an exercise program is lower with

higher-intensity exercise programs."[4] Or, said another way—if you taking exercise for heart and joint health (as opposed to competition), lower intensity is more better.

I just typed "more better." For reals.

And P.S., I chose for 5 a quote from the ACSM's "Guidelines for Exercise Testing and Prescription" because some of the greatest offenders of bad exercise prescription are professional exercise prescribers.

MOVEMENT or HEALTH PROFESSIONALS, please read the abundant scientific information on exercise. Exercise is actually a very intense science. If you are training others in exercise, read the textbooks, find the literature for yourself, and be smart. Don't make reading *CARDIO* magazine or BodyBuilders.com (sorry if that's a real website—I was just making one up) your continuing education.

Now go and take a walk.

---

1. See my first suggested further reading below.

2. See it again!

3. Now that I've typed it, "beefcake" is a pretty creepy word.

4. See the ACSM's guidelines below.

American College of Sports Medicine Position Paper, Exercise and Hypertension

FRANKLIN, B. A., M. H. WHALEY, and E. T. HOWLEY (Eds.). ACSM's Guidelines for Exercise Testing and Prescription, 6th Ed. Baltimore: Lippincott Williams & Wilkins, 2000.

KELLEY, G. A., and K. S. KELLEY. Progressive resistance exercise and resting blood pressure: a meta-analysis of randomized controlled trials. Hypertens. 35:838–843, 2000.

KELLEY, G. A., K. S. KELLEY, and Z. V. TRAN. Aerobic exercise and resting blood pressure: a meta-analytic review of randomized, controlled trials. *Preventive Cardiology.* 4:73–80, 2001.

KELLEY, G. A., K. S. KELLEY, and Z. V. TRAN. Walking and resting blood pressure in adults: a meta-analysis. *Preventive Medicine* 33:120–127, 2001.

KELLEY, G. A., and K. K. SHARPE. Aerobic exercise and resting blood pressure in older adults: a meta-analytic review of randomized controlled trials. J. Gerontol. A Biol. Sci. Med. Sci. 56:M298–M303, 2001.

MACDONALD, J. R., J. M. ROSENFELD, M. A. TARNOPOLSKY, C. D. HOGBEN, C. S. BALLANTYNE, and J. D. MACDOUGALL. Post exercise hypotension is sustained during subsequent bouts of mild exercise and simulated activities of daily living. *Journal of Human Hypertension.*

KIVELOFF B., and O. HUBER. Brief maximal isometric exercise in hypertension. *Journal of the American Geriatric Society* 19:1006–1012, 1971. in humans. Hypertens. 18:575–582, 1991.

chapter 12

RETURNING TO NATURAL MOVEMENT

PUT ALL THAT "CARDIO"
NONSENSE BEHIND YOU!

Q: What is the best way to get an effective cardiovascular workout without putting a lot of demand on the quads?

A: It's going to be tough to answer this from my soapbox. Good thing I have long arms. Well, first off, I'm going to have to define "effective." You can't really measure the best or most effective way of doing something unless you know the purpose for which you are doing it. I will take the liberty of restating your question to read "What is the best way to get whole-body movement that makes my heart rate increase without using my quads?"

1. Your body should be as weight bearing as possible with minimal knee bend. No sitting on anything (that leaves the bike out...).

2. Your spine shouldn't be rounded forward, but should be upright.

3. To minimize quad use, you should be propelling yourself by pushing behind you. (Treadmills, ellipticals and other machines actually require you to move via hip flexion—and can increase osteoarthritis and decrease bone density!)

4. Your arms should be moving reciprocally with your legs for the greatest, symmetrical blood flow. (Your arms should be swinging!)

5. You should have enough loading and vibration to maximize bone generation without exceeding what your joints can tolerate without breaking down tissues. (Jogging and running are out...)

The best whole-body exercises that meet these requirements are: 1) walking, 2) ice skating, 3) cross-country skiing, and 4) roller blading.

Of course, most of these activities require advanced skill, which can further increase muscle usage, neurological integration, and balance...all things that ramp up your caloric expenditure and need for oxygen!

IF THE BILL FITS...

Have you ever exercised a bit harder because you had ice cream the night before? Ever justified making a poorer dietary decision (dessert/second helping/third cocktail) by thinking "I'll work out twice tomorrow" or

"It's okay, I ran ten miles today!"?

In recent years a new disorder has emerged. Commonly referred to as "exercise bulimia," this subset of bulimia is not really an "eating" disorder, but part of a "disordered diet" profile. It simply means exercising for the purpose of removing or "erasing" the impact of the food you have just eaten. Instead of vomiting, the exerciser will work until the perceived amount of calories taken in match those being burned off.

Regular movement for health is a noble and valid choice—the optimal function of the body depends on it. But through my lecturing, I have been coming across more and more individuals who fit the profile of exercise bulimics—many of whom work in the health and fitness industry, and most of whom suffer from real health issues such as chronic pain, foot/knee/back injuries, declining bone density, and unexplained weight gain. These issues can all be correlated to excessive or compulsive exercise behavior.

"The experience of intense guilt when exercise is missed and exercising solely or primarily for reasons of weight, shape or physical attractiveness, were the exercise behaviours that most clearly differentiated between women with eating disorders and healthy women."[1]

This condition can be a tough one to identify. Most people have made their trainer/aerobics instructor/favorite athlete their role model for good, "healthy" behavior, but their role model, their "picture of health," may be suffering from physical, chemical, and psychological strain from the quantity and intensity of exercise—yet they look "good." To help you identify your habits as "healthy" or "health-defeating," take a look at the risk factors below:

- Missing work, parties or other appointments in order to work out
- Working out with an injury or while sick
- Becoming unusually depressed if unable to exercise
- Working out for hours at a time each day
- Not taking any rest or recovery days
- Defining self-worth in terms of performance
- Justifies excessive behavior by defining self as a "special" elite athlete

My suggestion? Take a week off from intense exercise. A daily walk, even sixty to ninety minutes, is great for the body—but the intensity should not be too high, or the pace quick. Just get out and move! If you feel anxious that your movement isn't "hard enough" or won't burn off the calories you consumed the day before, then you may have highlighted a potential issue for yourself. If you are also suffering from the aforementioned health issues (chronic pain, anxiety, bone density…just to name a few), you now have more information on how your habits may be contributing to your overall health!

---

1. From the *Australia and New Zealand Journal of Psychiatry*, citation below.

For more reading:

Eat Behaviors. 2009 Jan;10(1):68–70. Men, muscles, and mood: the relationship between self-concept, dysphoria, and body image disturbances. McFarland MB, Kaminski PL.

Australian and New Zealand Journal of Psychiatry. 2009 Mar;43(3):227–34. Excessive exercise in eating disorder patients and in healthy women. Mond JM, Calogero RM.

 SEE SPOT SIT

August 19 2009

See Spot run.

See Spot sit.

How come we never see Spot squat?

Probably because Spot's hips were too tight, his knees too sore, and, of course, there was his chronic constipation and bowel problems.

Or is it the other way around??

As someone who studies optimal human function, I can tell you that there is an interconnectedness between the diseases we are suffering from and the way we move in our culture. So the question is, why don't we squat? Well, why squat if there is a perfectly good chair around…and there's ALWAYS a chair around, isn't there?

Last weekend I attended the King Tut exhibit at the De Young museum in San Francisco. And although the gold and turquoise was amazingly detailed and lovely, the real treat (for me, anyway) was the chance to see the oldest chair in existence. Believe it or not, chairs used to be rare. They were decorative and a symbol of wealth and royalty—not something for everyone, and not to be used often. Berkeley architecture professor Galen Cranz wrote a book called *The Chair: Rethinking Body, Culture and Design.* (You could *walk* to the library to pick it up!) It is a great, thorough history of the structure of the chair, the role the class system has had on chair design, and the impact sitting has had on the body.

So, how *should* I sit?

I am asked this question quite often and the simple answer is "not very much." One of the major contributing factors to our national health crisis is the fact that we don't walk or stand very much. And when we do sit, it tends to be in the exact same position and in the same chair. There is a great anthropology journal article from the 1920s in which a researcher quantified the body-resting postures of the Australian Aborigines. They had over twenty postures to choose from, which gave all joints the opportunity to bend and straighten, rotate, and rest! That all being said, I'm in a chair as I write this, and while I am sitting cross-legged right now, I will most likely change my position every few minutes. Taking my cue from the above-mentioned research, I choose multiple ways to sit so my body doesn't adapt to any one position.

The amount of time we spend sitting is determined by 1) your job/school/driving-related necessity or 2) the amount of fatigue or pain you have while standing. While most sitting feels required, evaluate things objectively. Could you create a standing workstation in your home or office? How about walking to closer errands and events? If you are sitting because you don't have the strength to stand, then standing is the exact exercise you need in order to improve! Try standing during the commercial breaks of your evening television show.

If you are sitting at work, try to adjust your pelvis so that you are sitting on the portion of the pelvis that is designed to bear your weight—and keep you off of your tailbone!

 THE BEST PIECE OF EXERCISE EQUIPMENT YOU'RE NOT USING...

Ah, I see I got your attention.

And you are wondering what I could be talking about. The treadmill in your guest bedroom? (The one with the clothes on it.) The roller blades in your closet? The semi-inflated ball in the backyard, or the weights you have lying around just in case you are inspired to do a bicep curl? Nope, none of those. Give up? This one-size-fits-all, miraculous tool of amazing design is the FLOOR!

That's right, ladies and gentlemen. The floor can do wonders for your hips, shoulders, and back tension. It can help you build bone, and decrease your blood pressure. But guess what? You have to get down on it. I know. It's an extreme thought. In fact, the hardest three feet you will ever travel is between your hips and the ground.

1. I can get down, but I can't get back up.

    True strength is the ability to manage your body mass in all situations. If you can't get off the floor, then you are lacking the strength and flexibility required to keep your body healthy. I suggest everyone gets down on the floor and up again every day—just to make sure you don't lose the ability to. Exercise teachers: Have your clients practice the functional task of climbing down to a supine (face up on your back) position, and then back up again.

2. My floor is too dirty.

    Um, vacuum it? And this reason seems more like a health-prohibiting excuse and not a valid reason. When it comes time for your annual doctor visit, how are you going to feel knowing you could have taken better care of your spine, but didn't want to spend ten minutes cleaning up?

3. There is not enough space.

    The smallest space I have ever seen was the quarters a family of five, living on a small sailboat. We were still able to roll a yoga mat out to create a usable exercise floor, just the perfect size for an adult. If you have a smaller space than that, I want proof! You many have to move things out of the way when it's time to get down, but guess what…that's exercise too. ☺

The biggest thing keeping you from coming down is your own inertia. Seriously, even athletes don't like to get down on the ground. Slows them down, as you can imagine. Your body is always trying to conserve energy. It's not that you are lazy—you are just programmed with an internal voice.

Panicky internal voice: Stop doing that…you are using too much energy…there may not be enough food to last us through the winter.

Someone really needs to figure out how to explain the, uh, recent abundance of kcals to our metabolisms.

You don't have to feel badly, you just have to have a little mind over matter. (That saying is a lot more literal than figurative in this example!) So, back to solving the problem of inertia. Step one is to get down on the floor regularly. Even if you didn't do anything else while you're down there, getting up means you just performed a whole-body exercise, lifting

___ pounds (fill your body weight in the blank). Next time you see someone showing off their chest press with fifty-pound weights, you can show them up by doing your impressive *full-body press*.

Want to sweat? Start from a standing position and then climb down to the ground, ending up face down. Climb back up to a standing position and climb back down ending with you face up. Repeat this cycle five to ten times. It's more fatiguing than you think! Like I said, those three feet are hard. Many would prefer a challenging work out that doesn't involve the floor. Just looking at it fills you with hesitance. You can comment on why YOU won't get on the floor. Unless, of course, you do!

(I just made myself do it, so, join the club!)

Nov 16 2009 MOVEMENT **IS** MEDICINE

Why is it that we follow a doctor's prescription to a "T" but then take additional advice like "NO SUGAR" and "You need to exercise" with a shrug of the shoulders? If your exercise plan was written out for you on a tiny prescription pad, would you follow it? If you had to go to the pharmacy counter to get your grocery items, would it change the way you eat?

A few weeks ago, Dr. Ranit Mishori wrote an interesting article for *Parade Magazine*[1] that discussed his personal experience with patients not following his recommendation for exercise. I, of course, agree...kind of. If your doctor tells you to exercise, chances are your doctor has read research that shows your ailment can improve with an increase in circulation, higher metabolism, lower body mass, or reduction in stress. But "You need to exercise" is simply not a sufficient prescription.

What if you went to your doctor and were told to "Take some medicine"?

Okay, sure...but can you be a little more specific?

Of course, when it comes to pharmaceuticals, your doctor IS very specific in telling you what to take, how much to take, and how often. Just as chemicals in your body have an effect on specific tissues, so do different exercise machines, movements, and modes (types) of exercise. You probably understand that some medicines make you better and some can make you worse off. This is exactly how movement works as well. And while it seems that any exercise would be beneficial, often this is simply not the case.

Do you need to move regularly? Absolutely, every day. Moving for multiple hours is optimal, although not practical for most. This is why we started to "exercise" in the first place. Because we weren't moving at all! In fact, now the term "exercise" has become synonymous with "fitness" (smaller, controlled bouts of movement at a higher intensity). Even though fitness can have a short-term benefit of weight loss, some types of exercises (mostly gym-type activities with bouts of accelerated heart rates) have a fairly significant negative impact on the long-term health of joints and the long-term function of the adrenal glands.

So where to start? Begin with the most important component to cardiovascular health, muscle strength, and immune-strengthening lymph removal—your muscle length. Your sarcomeres—those are the little units of muscle fiber tissue—are like Goldilocks. They only function optimally when they are just right—just the right length, that is! If you want to see dramatic improvements to chronic conditions such as fibromyalgia, cardiovascular disease, osteoarthritis, and Type 2 diabetes, then stretch for twenty to sixty minutes every day! Follow up your stretching with a light, easygoing fifteen-minute walk, two to three times per day.

A note to patients: The time has come to take better responsibility for your own health. Taking a regimen of pills is not taking care of your body—it's cleaning up the mess. Do better. Start today.

Some general guidelines:

1. You need to walk, every day.

2. If it hurts to walk, then you need to stretch the backs of your thighs, calves, and feet every day until you can walk comfortably. Then see number one.

3. Your strength-to-weight ratio should be 1:1, meaning you should be able to: stand comfortably on one leg for up to two minutes (without lifting your arms out to the sides or bending either knee), and do a pull-up. No way on the pull-up, right? This inability is a basic indication that you have gotten too weak for your weight or too heavy for your strength. Start with stretching your shoulders and chest—it's a good start!

A note to doctors: The time has come to take better responsibility for the advice you are giving. When you give casual suggestions for exercise, whether it be based on your own favorite activity or something you read in a magazine, understand that movement has a localized, instant impact on circulation (blood and lymph) and bone density. Advise better. Start today.

Some general guidelines:

1. Anyone with back pain or degenerative changes in the knees, hips, and spine should be instructed to immediately stop any running, jogging, or jumping! These activities will increase the rate at which the tissues degenerate and will lead to greater osteoarthritis, disk damage, etc.

2. Please explain that osteoarthritis is not the autoimmune disorder arthritis, but a self-induced condition of inflammation in the joints due to minimal joint space caused by tight musculature. No treadmills or ellipticals for anyone with OA. The backward-moving belt on a treadmill requires a hip-flexing gait pattern, increasing tension within the hip joint. Walking is good, it just needs to be on regular ground or in a mall.

3. Please explain that research shows that consistent, lower-intensity movements (walking, yoga, stretching classes) demonstrate greater long-term decreases in body fat than high-intensity and joint-damaging high-impact exercise sessions.

4. Refer responsibly. Most exercise is fad and fat-phobia driven. Find a local practitioner who offers the most research behind their program design and book a session for yourself! Introduce yourself as a local physician and say that you are looking for a place to refer your patients for a health-based program. Most will be willing to give you a complimentary session!

All right, everyone. Take two (laps) and call me in the morning!

---

1. parade.com/health/2009/11/01-when-doctors-prescribe-exercise.html

January 19 2010 SWIMMING IN THE RAIN!

I love watching Southern California weather segments. There are lots of weather warnings and STORM WATCHES. It's raining here in Ventura, in case you didn't know. We're not sure how many people will survive, but keep checking the news. So far there are thousands of worm deaths across the local soccer fields. It's just horrible. Driving seems to be very difficult as well! And it is today's weather (all two inches of it), and a few inquiries into my thoughts on swimming, that has prompted today's blog (oooh, thunder! Cool).

What do I think of swimming? Well, first off, I love it. Being in the water is one of my favorite activities and submerged playing makes me VERY happy. I was also a competitive swimmer in school. I love everything about the water. I like the buoyancy. I like the way the oil on my skin mixes with the water and makes my body slimy. I could do without the *manos del prune*. (This is Spanish for prune hands...and if anyone knows the correct Spanish term for prune hands, would you let me know?) So there you have what I think of swimming. But maybe you wanted to know what I think of swimming as an exercise program? That is a better question.

There are many reasons people exercise, so let us break those reasons down to evaluate swimming as a health program.

1. Weight loss. If you are a non-exerciser beginning a swimming program, you may notice a decrease in weight initially, but to see continuous changes in body composition, research shows that swimming is not the best choice. The temperature changes the body goes through coming in and out of the water seem to signal a complex nerve pathway that lowers metabolism. Why this is, physiologists are not sure. It is most likely due to the heat-regulating processes of the body. Being submerged in water makes it more difficult to maintain body temperature, so the body responds by using less energy. Even competitive-level swimmers— swimming multiple hours per day at high intensities—carry a greater percentage of body fat than athletes of most sports.

2. Bone density. If you are exercising to increase bone density, know that swimming is the least effective type of movement when it comes to strengthening your bones. In order to stimulate bone generation (osteogenesis), bones must be weight-bearing along the axis of gravity—which means you have to carry your body weight around while standing up! The buoyant force of water makes bone loading almost zero in the water. Better option? A thirty-minute walk to the pool is a good strong-bone choice, as is maybe alternating a swim day with an hour of walking.

3. Muscle mass. Swimming can definitely aid in increasing your muscle mass, as water is great resistance. But keep in mind that swimming as most people do it is a sport. Swimming strokes have been developed to use the water as minimally as possible to reduce any drag that may slow the athlete down. What does

this mean? It means the better a swimmer you are, the less muscle benefit you will receive from your best stroke. Water aerobics classes create exercises that use more muscle than your basic swim strokes. Another option—ditch your favorite or best stroke and start learning a new one!

4. Joint health. Healthy joints need regular use, but there's a catch. If you use your joints when your muscles are tight, movement can have a more detrimental effect than if you didn't use your joints at all! Swimming is a great way to get whole-body movement, but are your shoulders tense? If you are doing stroke after stroke and your neck and shoulder muscles are tight, the risk of developing shoulder impingement syndrome (shoulder bursitis, deltoid tendonitis, rotator-cuff tendonitis or tear, bicep tendonitis or tear) significantly increases. If you like to swim for joint health, then you need to make sure your shoulder alignment is correct. You also need to work to decrease your neck, deltoid, trapezius, and rotator cuff range of motion. My spinal alignment DVD has great corrective exercises for shoulder girdle alignment.

5. Cardiovascular health. Again, swimming is a great way to increase your circulation—always a great benefit to the heart and lungs. Something to know about your circulatory system is this: Circulation does not increase throughout the entire body, but only in the muscles you are using. To truly optimize cardiovascular function, you need to use as many muscles possible and make sure that your muscles are in their longest possible position the rest of the time. When you are done swimming, you need to stretch. And when designing your swimming workout, make sure that you rotate through different strokes. Maybe even create some of your own!

So, what do I think of swimming as an exercise? The most simple way to state it is this: Each type of exercise is like a food. How many different types of foods do you need for balanced nutrition? Many. Is swimming a good exercise? Sure. But it shouldn't be the only "food" that you eat, or you will most likely be missing the movement "nutrients" required for keeping other tissues healthy. Our understanding of exercise is oversimplified. Movement is chemistry. Every tissue in the body (bone, blood, nerves, organs, muscles) requires a unique component of exercise to keep that tissue healthy. The exercise that "nutrifies" most of your tissues is well-aligned walking. Stretching to maximize your bone placement and joint space while walking is also extremely important.

Should you swim? YES! But swim because you love it and love how it makes you feel. Know that when you get out of the pool there is still a lot of work to be done to keep your other tissues healthy. Exercise prescription is very much like planning a healthy diet! We'd all like to eat our favorite foods all the time, but in order to optimize our human machine, you have to make sure all the food groups are there.

Whoa. Sun's out. Better comb the field for survivors.

January 21 2010 · TAKE A STAND AGAINST SITTING!

The Restorative Exercise Institute is a facility dedicated to the paradigm that movement, and specific types of movement at that, are required in order to achieve optimal wellbeing. We do not subscribe to the fitness paradigm, but rather instruct that aligned, whole-body movement done throughout the day provides the required circulatory benefits, neurological stimulation, and mechanical forces to keep tissues regenerating correctly.

If the institute were a superhero, Sitting would be its archnemesis. Why? Because sitting alters the muscle lengths in the legs so greatly, the lower half of the body no longer gives muscular support to the cardiovascular system. The calves and hamstrings become so tight they cannot hold their share of blood, causing excess strain on the main arteries, increasing the risk for cardiovascular disease. The evil henchman Sitting also places increased forces on the sacrum (especially if you sit with your tailbone tucked under), which pushes this free-floating bone deeper into your pelvis, putting slack in your pelvic floor muscles.

Why doesn't one want slack pelvic floor muscles? Let me count the reasons.

1. PF muscles are the "closers" of the urinary and fecal doors to the world. *I* want proper tone in these muscles, and I hope you do too.

2. PF muscles are the only support you have for the weight of the abdominal and pelvic organs. If the PF muscles are slack and weak, the organs have no choice but to go with the gravitational flow and drop downward. Ladies, that means prolapsing organs. Menfolk, that means your organs are now increasing pressure on the prostate. Not fun for your prostate, I promise.

3. PF muscles are half of the system that stabilizes your sacrum. Slack PF muscles means that puppy is slopping around between your sacroiliac joints. Low back pain, anyone?

I could go on (you know I can), but haven't I given you enough information for one day? No? Then how about this news piece today: "Experts warn of danger of sitting." The results from a ten-year-plus study on seventeen thousand Canadians showed that risk factors and actual death occurred more in those who sat often as compared to those who didn't—regardless of whether they participated in fitness exercise or not. Exercising, even one hour per day, was not enough to reverse the metabolic damage sitting had inflicted on study participants.

Again, I will state that replacing all-day movement with intense bouts of exercise is an inappropriate prescription for health. You have to eat a certain amount of kcals and nutrients too, but you wouldn't eat an entire day's worth of energy in one hour, or only eat three hours per week, would you? Think of how that would mess up your energy regulation, fat storage, blood glucose, and pancreas activity! Fitness activities can be fun and beneficial, but in the long run, studies do not support that fitness (monitored heart rates, heart rate percentages, weight training, and aerobics) increases longevity or quality of life as quantified by medications, surgeries, or daily pain. Do you have to exercise? No. Do you need to move? Yes! All day long! You need to walk and stand, letting your leg muscles hold you up, increasing the strength of the blood vessels in the muscles, decreasing the pressure in the larger arteries, and signaling the bones to hold more minerals (bone density).

Don't know where to start? Stand up. Right now. Get out of your chair. Spend twenty dollars on plywood and make a box that holds your computer a foot off your desk so you can be part of the solution. You up? Good. Now, leave a comment, turn off the computer, and go take a ten-minute walk!

Medicine & Science in Sports & Exercise. 2009 May;41(5):998–1005. Sitting time and mortality from all causes, cardiovascular disease, and cancer. Katzmarzyk PT, Church TS, Craig CL, Bouchard C.

British Journal of Sports Medicine. 2009;43:81–83 Too much sitting: a novel and important predictor of chronic disease risk. N Owen, A Bauman, W Brown.

THE $50.00 MAKEOVER

As mentioned in previous posts, your risk of cardiovascular disease is not decreased by hours of cardiovascular exercise (sorry about that!) but by lessening total sitting time. The reason has mostly to do with the fact that sitting rearranges your muscles into a position that prevents the smaller blood vessels from holding their share of blood. The end result of sitting is more blood stays inside the larger arteries, making it harder for the heart to work. The more all-over muscle you are using throughout the day, the better it is for your cardiac health. The negative impact of sitting is so great, scientists are beginning to state that exercise as we know it (cardio and weights) is not keeping the "healthy exerciser" from dying any more than the couch potato. Muscle activity needs to be happening all day long!

In celebration of this news, I am now offering these items for fifty bucks:

- A flatter and stronger stomach
- Stronger hamstrings and glutes
- Decreased blood pressure
- Greater bone density
- Less lumbar spine compression
- A lengthened psoas
- More space in the chest for lung expansion
- Increased metabolism and caloric expenditure
- Fifty bucks sounds like a pretty good deal, right? Or, if you'd prefer: Four Easy Payments of $19.99 for the Ab Roller (or you can pick it up for $20 + shipping and handling. They're evidently not moving as fast as they were ten years ago...)
- Two sessions a week with a personal trainer at the gym to kick your butt's butt (that's a lot of butt, but, well, you know.) for $635.00
- Those beta blockers and bone-density pills that you have to take every day for the rest of your life add up, don't they? $2000 per year.
- Or, if you deal with your back pain via a glass of wine per night, then add your wine costs here: $2.00 to $150.00 per week.
- Over a lifetime, a tight psoas will cost you and your insurance about $250,000 in treatments, medication, and visits to the doctor. You'll have to take my word on this one (this is another post, or, better yet, a webinar). Lengthening your psoas should be a priority.

I could go on and on, but we're already up to hundreds of thousands of dollars and frankly, I can't find a calculator.

So, back to the fifty-dollar makeover. Is it a haircut? Or a pair of shoes? Great moisturizer? Nope. It's a standing workstation. At the institute we had simple wooden platforms built (materials were less than twenty bucks, so we opted to pay our favorite handyman to build them!) to sit on top of the desk, allowing our fabulous staff to work out on company time.

Before (above left): Short psoas, short hamstrings, low circulation, and no abdominal action!

After (above right): Muscles at the optimal length, pressure of the heart and pelvic floor, and frankly, we think she's smarter now too. (image above right)

Striving for optimal health and wellbeing is not a one-hour-a-day, four-days-a-week job. It is a do-it-as-much-as-possible kind of gig. The standing workstation is a feasible and affordable option and there is no reason you can't have one! Send me a pic once you get yours up, or call the institute if you'd like to have one built for you. We know a guy. ☺

May 6 2010 I TOLD YOU SO...

Well, I hate to say I told you so (total lie!), but if you are still sitting in front of your computer right now, you better read this: "Your Office Chair Is Killing You" from last week's Business Week.[1]

New research in the diverse fields of epidemiology, molecular biology, biomechanics, and physiology is converging toward a startling conclusion: Sitting is a public-health risk. And exercising doesn't offset it. "People need to understand that the qualitative mechanisms of sitting are completely different from walking or exercising," says University of Missouri microbiologist Marc Hamilton. "Sitting too much is not the same as exercising too little. They do completely different things to the body."

The article goes on to mention the negative impact of allowing your spine to be rounded most of the day, and the fact that even though we evolved to walking over the last million years, we are undoing years of human-body function in less that one hundred and fifty years of chair use.

The "S" Curve   The "C" Curve.

The importance of keeping the "S" curve in the spine is extremely important to the health of the nervous system (this means all the nerves and muscles. Even tiny, rhythmic actions like breathing, digestion, and vision). The "S" shape keeps forces like gravity and weight from collapsing our bodies to the ground. The "S" shape works like a spring. The structure of an "S" allows forces to be temporarily applied and then released. A "C" shape to your spine, however, doesn't have the ability to resist forces, squeezing the disks, increasing fractures in the bones, and pressing on the nerves. Ouch.

In all honesty, the chairs are not really to blame. It's the fact that we are sitting, endlessly, in them. I tell you this because people keep asking me what is the best chair. The answer is…there isn't a "best chair." The damage comes from having your legs bent. If you can find a chair that allows your legs to be straight and your spine to be vertical, then maybe I'll consider it. I'll consider it a wall.

Why shouldn't your legs be bent? The more time you spend sitting, the tighter your hamstrings become. The tighter your hamstrings, the more they pull the lowest curve out of the spine and tuck the tailbone under.

And now, for some fun math:

Tight hamstrings = "C" curve to the spine.

"C" curve to the spine = Spine degeneration, high blood pressure, and no pelvic floor strength.

To check and see if your hamstrings have pulled your spine into a "C," test yourself by bending forward to the seat of a chair, lifting your tailbone to the ceiling while keeping your legs straight (no bent knees!)

Does your spine relax in the center, with a gentle curve at the tailbone and upper back? Or does bending forward with straight legs make your spine buck up like a camel? Camel starts with the letter "C." Interesting. Test everyone in your office and let me know how it goes. Hey! Are you standing yet?

1. businessweek.com/magazine/content/10_19/b4177071221162.htm

 June 7 2010 YOUR POSITION IN LIFE

Good question in my Ask Katy today:

> "Standing workstations are clearly a good idea and I have fashioned one for myself. But what is too much standing? We all know that people whose jobs require constant standing, like restaurant servers and factory workers, are often plagued with varicose veins. Is there a balance to be struck here?"

This is a great question.

So, we've got a situation where sitting constantly is creating disease and standing constantly is creating disease. Do you see the theme? Although the research and media are going to probably miss the boat on this one, the problem isn't the sitting (or the standing, for that matter), but the constant and continuous use of a single position. Even this question smacks of someone from a North American and European perspective. As if sitting is bad and standing is bad, and the only option left must be lying down. As if there are only three choices for how you position your body. As if there aren't about a thousand different ways you could position your body.

Believe it or not, the positions you are able to get your body in were

learned via observation. Our culture's use of chairs and toilets, our beliefs in what our posture means to others (think of women who cross their legs and adjust their heads to look demure or men who jut their chests and flex their elbows to communicate authority), and even our clothing (rigid shoes, narrow skirts for women, etc.) have all resulted in self-induced joint rigidity. All the movements you have never done are movements that would have toned muscle, kept connective tissues moist and supple, and kept blood oxygen flowing evenly to all areas of the body. Instead, we have huge chunks of unused muscles, bones scraping together at the joints and increasing friction (causing osteoarthritis), and we are constantly medicating to make living possible in our physical agony. This all sounds pretty depressing, I know, but the totally awesome, super-cool and exciting thing is it can be different whenever you're ready.

Another awesome thing is, while I may seem like the only person saying these strange things, there are actually other people out there who have researched this for the last one hundred years. The big difference between then and now is that 1) there wasn't the internet back then, which must have made it very difficult to share insights collectively, and 2) there is now a wide breadth of subjects a "good education" covers. Most of the people observing the very real phenomenon of cultural postural habits and habitual uses of the body (physical anthropologists) and the people in charge of health education and prescription (the medical community) are in two completely different sciences. They don't even talk to each other, even at parties. But I am hopeful my education in the biomechanics of human movement *and* disease, coupled with my awesome typing skills and Al Gore's internet, is going to help. How is this information ever going to get to you, the people? One blog post at a time, I guess.

One extremely cool journal article from 1955, "World Distribution of Certain Postural Habits," reported the findings of physical anthropology professor Gordon W. Hewes. It is an amazing read, and if you'd like to have more than my take on it, you can get it here.[1]

As many anthropologists know, the way we move is mostly a result of our cultural inheritances, and has very little to do with genetics. Clothing, terrain, temperature, gender, class, and fear are only a few of the many factors that affect how we adjust our joints when sitting and standing. Hewes reported on about a hundred resting postures of the world, and I have posted this image (image top of previous page) so you can see, perhaps, why our Western joint health and metabolism (which is dependent on muscle length) is the poorest in the world.

So we need to think bigger. There is more than just sitting and standing. Create ten different options of each! If you have a standing workstation, stand a few different ways every hour. When you sit, sit a few different ways every hour. Open your mind and open your joints! When you get home, stay out of your chairs and try out a lot of these worldly options. (Note: If you don't have a spear, a broom may work...) Circle the ones you can't maintain for longer than five minutes and make a note to practice that posture at the beginning and the end of an exercise session. And parents, don't insist that kids sit in the same fashion as us stiffer folks. Allow them to explore other options. And join them! They can teach you something about natural movement.

Also, if you do spend a yoga or stretching class cycling through ten or so of these postures, know that while this cycle is a good thing, getting back into your usual sitting position the other six to ten hours of the day reduces your health just the same. Adjust the way you sit, as often as possible for a real, deep, and cellular change.

Hewes concludes his research:

> Physiologists, anatomists, and orthopedists, to say nothing of specialists in physical education, have dealt exhaustively with a few "ideal" postures—principally the fairly rigid attention stance beloved of the drillmaster, and students' or stenographers' habits of sitting at desks. The English postural vocabulary is mediocre—a fact which in itself inhibits our thinking about posture. Quite the opposite is true of the languages of India, where the yoga system has developed an elaborate postural terminology and rationale, perhaps the world's richest. In conclusion

I should like to stress the deficiencies in our scientific concern with postural behavior, many of which arise simply from the all too common neglect (by nonanthropologists) of cross-cultural data.

I concur.

---

1. jstor.org/pss/666393

 NANOOK OF THE NORTH

October 11
2010

Last night I watched *Nanook of the North,* on my mom's recommendation. *Nanook* is allegedly the first documentary film (filmed in 1922), and features an Inuit (what we often call Eskimo) family in the Canadian Arctic.

Okay, firstly, I don't really dig the cold. I like the *idea* of cold, mostly because it's really warm all of the time where I live and cold just sounds like vacation. You know, because you can make hot chocolate. In this film, Nanook was not ever, EVER on vacation. This family moves to wherever Nanook can hunt and get them food. The whole gang gets on a sled and just goes to where he might get a seal or fox, and then builds an icehouse there. This guy BUILDS an entire igloo (with a WINDOW) in less than an hour. These scenes alone are worth the rent.

All of you out there wondering how to hold your baby or what gadget your baby might *need,* should watch this. The Inuit wear nothing but furs and boots made out of sealskin (which keep their feet dry). If you have a baby, the baby just gets placed in the back of the fur, naked. And you're naked too. They start as teeny tiny babies and are still back there when they are three! Locomoting Inuit can't hold their babies with their hands, because 1) it can get up to −40° in the Arctic, and 2) they are always DOING something that needs to be done right now, for survival, like steer the dog sled, or help build an igloo. And yes, there is the obligatory spit bath mom gives the baby. He doesn't seem to like it any more than I did. Gross.

Of course, I am always checking out alignments when I get a chance to view older ethnographic and anthropological films, and this movie was no different. The squats were all there, the straight-leg bending from the hip was always used, and not the hybrid knee-bend, tucked-tailbone motion we have developed to "save our spine," which in fact degenerates

---

our spine and knees while allowing the hips to wither. Nanook's rugged lifestyle seems to promote better musculoskeletal health than our current model, but you watch and decide.

The best moment of this film is watching Nanook fish for salmon off of a floating ice ledge. No bait. No net. Just a string that has two tusks that he rattles in the water and then spears the fish with his other hand. That's not even the best part. The best part is when he places the twelve salmon across the top of his (handmade) kayak. Not inside the kayak. Not tethered down to the kayak. Just resting on the top. And then he gets into the kayak with nary a wobble, picks up a fisherman who needs a ride (who easily steps onto the back of the kayak, lying on top of the fish without making a wobble), and away Nanook goes. The last time I kayaked (about three months ago), I flipped my kayak in less than six inches of water. I am so far from natural living, I don't even know how I can get there, but I'm trying.

This movie (which I got from Netflix) is a great piece to watch for any of you interested in natural movement, nature, baby transport, anthropology, and survival. *Nanook of the North* was one of the first twenty-five films selected for preservation in the United States National Film Registry by the Library of Congress as being "culturally, historically, or aesthetically significant." *Nanook* is a silent film, so I think it lends itself to family viewing with discussion and gasping.

Here are some more random thoughts that came up while watching *Nanook*:

1. We have entirely WAY too much stuff.

2. We have no idea how to live with nature.

3. I would really like some fur pants.

4. There is *nothing* a knife carved out of a walrus tusk can't cut through.

5. Dogs are not that far from wolves. Okay, maybe your chihuahua is a bit far, but all non-rodent-sized dogs could Take You Out if they wanted to.

6. Cold isn't really COLD if you move constantly. I point this out to everyone in the Midwest who can't figure out how to walk without a treadmill in the winter because of the weather. This is how: Put on your polar fleece, your wool socks, gloves, and a hat. Get your favorite music going and put one foot in front of the other. You are a heat-generating machine.

7. People who kayak in the Arctic have a better center of balance than any athlete I have ever analyzed. And I have looked at the mechanics of every sport I can imagine.

8. I would really like some fur pants.

9. I wonder if fresh (and I mean fresh), raw walrus tastes like sushi.

10. Everyone is so very happy. Don't they know how COLD it is?

11. Nobody needs an electric blanket. I just watched five people take their shirts off to sleep between two furs in one igloo that has to stay below freezing.

12. I would really like some fur pants.

13. ☺

YOUR BODY— THE BIG PICTURE

THE MECHANICS OF
WELLNESS

I am going to tell you something that I think will interest you. Your body, right this minute, is attempting to do these tasks:

- Breathe
- Deliver oxygen
- Maintain your immune system
- See
- Hear
- Smell
- Generate bone
- Repair cartilage
- Digest food
- Regulate your temperature
- Make new skin
- Get rid of old skin
- Filter your blood
- Store energy
- Access stored energy
- Get rid of lactic acid
- Move an egg to the uterus
- Move an egg out of the uterus
- Move blood
- Break down food
- Move waste products out
- Repair any wound, grow a baby, grow hair, grow nails
- Regenerate all the cells on all the organs and all the tissues in the body

This is a shortlist of all the things your body is trying to do, this minute. But it's all okay because your body has the skeletal muscles to help with all of these items. Yes, your bicep group flexes your elbow, but these muscles also move blood back up from your fingers, and help prevent lymph from collecting in your armpit, where it overwhelms the lymph nodes. The muscles in the feet can wiggle your toes, but they also keep the nerves from the sacrum to the legs from dying. Foot muscles also prevent the skin around the ankles from developing sores and severe wounding. Today is Love Your Muscles Day. You won't find it on any calendar, because I just declared it just now.

Guess what? Tomorrow is Love Your Muscles Day too.

Your skeletal muscle system can help you to store long chains of energy (glycogen) so you don't have to have to store sugar in your blood, leading to high blood sugar and triglycerides, or as excess body fat. Your muscles also store the water needed for hormonal and cellular functions.

But your muscles can only do all these things when they are at a specific length. "Quoi?" you say. It's true. This information, the fact that your skeletal muscle is only fully active when it is at a specific length, could have an enormous impact on your health for the rest of your life.

As we have made huge technological leaps, our musculoskeletal function and human-body function have been withering at an alarming rate. Diseases that we North Americans perceive as age-related declines are happening to industrialized nations only and not in countries all over the world. And even within our population, health is declining more rapidly than in the generation before us. This next generation will be the first to under-perform and under-live their grandparents. Ouch.

The position (literally) we have gotten our body into is being passed down to the next generation, because we have not been instructed on how to move. We keep teaching the next generation our poor habits because we don't understand how movement and alignment are passed on (psssst... it's not genetic). The farther we have gotten from nature, the less we are moving like nature designed. Bones that regenerated while we walked a few miles a day for food didn't come with a backup plan, and now they don't stand a chance. Blood circulating into the feet, knees, and legs did so because of the many hours we spent using our leg muscles to hold ourselves up. Sitting down for eight to ten hours a day keeps the pressure in the blood vessels high, increasing the work for the heart. Research shows that fitness activities are making the situation worse, not better. The benefits of exercise are not equal across the board.

We have gotten ourselves into a predicament, but the solution is extremely simple: Get each muscle back to its perfect length, and adopt a movement program that best mimics nature's design. Managing any health issue, injury, or disease is more simple than you could have ever imagined. The Restorative Exercise Institute and the Aligned and Well program are the facility and program dedicated to bringing this Physics of Wellness information to the public.

We need more teachers, which means that first, we need more students. There are twenty-five objective markers that anyone can see with the naked eye that reveal the actual functioning position of a skeleton. When you learn how to see and evaluate these points, it becomes very clear why the human body is not functioning optimally in most people around you. There are then fifty exercises that can be learned and then mastered to significantly increase the neurological, metabolic, circulatory, and locomotion systems of the human body.

These bony alignment markers and corrective movements are the foundation of our six-month training program, available online. I strongly recommend this course from the self-healing perspective first (take this course for the benefit of your own body and those nearest to you) and for your students or patients second. Every human should have to attend a university course on their own body in their lifetime. This is that course. Get a degree in YOU. The information is not only fascinating, but hugely important for the community of people you serve.

We are interested in local students, of course, but are also in need of longer-distance students to provide this information to communities outside of our state and country!

May 7 2010 — TIGHT HAMSTRINGS SINK SHIPS

I received such a good question today, I think I'll answer it for everyone!

> Katy, I am curious about individual differences in physiology and how that affects alignment. I have, for instance, very tight hamstrings, and have had since childhood. Even when I was young and flexible and could do full back walkovers and front handsprings, I couldn't touch my toes (I'm actually closer to my toes now, at forty-six, due to a determined program of regular stretching). On the other hand, I have always had oddly flexible hips. Full pigeon is NO problem for me. At forty-six I can still almost put my foot behind my head. Obviously, I don't have to work at hip flexibility at all.

So how do you factor in these natural physiological differences? Is alignment meant to be the exact same for everyone?

Yes, everyone's alignment is meant to be the same. No body is naturally designed to have their spine in a round "C" shape, have leg bones grind at the knee, or press their sacrum into their body because they're too tight to get off of it. Muscle physiology is also about the same in everyone, with the exception of some people having a slightly higher or lower percentage of fast-twitch muscle fibers (those with more FT fibers make great sprinters and lifters!).

Although it may feel like tightness is your "natural" state, the reasons for tight muscles can go back all the way to your time in utero, which is why many people have always felt tight! When your body goes through oxygen deprivation, the brain begins shutting down different parts of its function. Motor function (communication between nerve and muscles) goes first, and vision and cognitive brain function go last. Oftentimes the birthing process goes on a bit too long and without ample oxygen, the brain lets little things like motor skill go.

The body is great, though. If you had super-tight muscles all over, you wouldn't be able to walk. As a compensation for muscle tightness, the body creates a hypermobile joint by relaxing ligaments to allow movement. If you sit on the ground in a cross-legged fashion and your legs just fall to the floor, I would suspect that your hamstrings were locked up! And for people who have pretty good hamstring mobility, they often have extra tension in the groin and hips. It's all about balance. All-over mobility isn't the goal, nor is all-over tension. We should all be trying to optimize flexibility (the range of motion that allows all aspects of the joint to articulate) and stability (strength in all ranges of motion). Many ballerinas were "broken" as children, their soft tissues ripped either over time or by a one-time force to free up the hip joints. All that mobility makes for beautiful moves on the stage. Unfortunately, when Prima ages to her fifties, her hips or knees now require replacement, because a lack of hip stability increases friction, which creates osteoarthritis. Famous Olympian Mary Lou Retton had to have one of her hips replaced too. We don't want our joints to flop around with loose ligaments, we want good ranges of motion, where we can feel the muscles reaching their end point and contract our muscles when we need to.

It seems that your "naturally" tight hamstrings are changing with regular practice. Keep up the good work and pay close attention to getting your tailbone up toward the ceiling (on a forward bend). There are also other alignment markers you can use on the back of the leg that will show you where your hamstrings are in space (this needs an entirely different blog post! Is my work never done???). Working on opening areas that have been shut for years is extremely rewarding and calming to the brain (your brain kind of panics when you have a lot of tight muscles. Your brain knows that you need long muscles to keep your body from degenerating fast!). Reallllllllly tight muscles that have "always been that way" are going to require lots of diligence, but the wonderful thing is that when they start to change, your physical and mental wellbeing improve!

To help everyone out there with issue, I'd like to add some tips:

1. For tight hamstrings, try two minutes of relaxed forward bending every hour. You can't stretch them too much, especially when you're sitting a lot or standing in the same position all them time.

2. Notice the terms like "it's my natural state," or "I was born this way," etc., above. Your motor pathways are somatic—which means you have total control over what is happening to all your muscles. When you have beliefs that reinforce where your body is, then it's going to be difficult to change. This is a new mantra for everyone: *Muscle length is completely restorable.*

3. For all you new moms, moms- and dads-to-be, or even grand-parents: Gently tickling and massaging a baby's entire body will restore the lost motor functions that may have happened during the birthing process. Gently stimulating nerve receptors on the skin and pressing into baby's muscles to increase blood flow into the smaller blood vessels in the muscle bundle will help restore any lost motor function. Special areas to pay attention to: The feet, the inner thighs, the calves and hamstrings, the waist, and between each rib.

4. Moms-to-be! You need to better prepare your body for birthing. Your tight hamstrings and calves mean a tight pelvis! Start calf stretching, hamstring lengthening, and squatting NOW to make your birthing process better for you and baby.

May 19 2010 THE IMPORTANCE OF POSTURE (IN 1965)

What if I told you (and guess what, I am!) that you don't have a degenerative disk disease (even though that's what it is called), but that your disks are wearing out where you excessively bend and compress your spine because of your posture? And that osteoarthritis is not a genetic disease (it's not, people) but an indication that your tight muscles are pulling your bones together, causing friction, which causes heat? That's what friction does, every time. And finally, what if your painful menstrual periods (ladies) or reduced ability to get an erection (gentlemen) was due to your postural habit of moving the pelvis from a vertically aligned position to angled one?

I like to go to garage sales, and am especially excited when I come across medical and health textbooks of yesteryear. A few weeks ago I found *Therapeutic Exercise* by Sidney Light, MD, from 1965. Here's what 1965 was saying:

> There are postures of lying, standing, and sitting. Although each posture is important, the most important posture, and the one that is evoked by the mere mention of the word, is the standing posture. Correct posture is an alignment of maximum physiologic and biomechanic efficiency and requires a minimum of stress and strain. Regardless of body type or size, certain factors of skeletal alignment can be correlated with variations of body contour. Except in the obviously disabled, poor posture is seldom the presenting complaint. In fact, posture seems to be the concern of only of physical educators; yet, a significant number of medical complaints may be attributed to inadequate posture, and often, the relief of symptoms can be predictable and early if the posture problem can be identified and the patient is willing to work out the physician's solution. A therapeutic program which strengthens muscles, and stretches shortened tissues, if persisted in over a sufficiently long period, is often ample reward for the tedious exercises which must be performed daily. [Did this guy just call my exercises tedious???] Unless the patient is motivated by the physician's persuasion and conviction, the exercises may be done inefficiently or abandoned. [So, if the doctor says "Try stretching a little bit and see if that makes a difference and if it doesn't improve in the next few weeks, we'll do meds or surgery," that may be underselling it a bit??]

> Posture is often referred to as good or bad. In good posture, breathing is effortless and efficient, and joint forces are kept to a minimum. Posture analysis depends primarily upon inspection and palpation, but for the beginner, the most important instrument is the plumb line.

Hmmm. I would type more but I'm strangely motivated to spend my blogging time doing some of my alignment exercises right now.

P.S. Stay tuned for what 1965 had to say about where flat feet come from! Hint: It's not genetic, but they do start developing in the first few months of life by 1) sleeping positions, 2) putting too-bulky diapers between babies' legs, and ... ☺

A friend brought me this week's *Newsweek Magazine* issue: "The SCIENCE of Healthy Living." I'm not really a fan of getting my news from *Newsweek*, but I was happy to read it, since someone suggested it. The article is titled "A Healthy Life is a Happy Life." So sweet! And to make it even more "homey," the title page looks like it was written in needlepoint. Hard-hitting science news, from somebody's grandmother.

I didn't have high hopes for this article, but I certainly wasn't prepared for the section "Healthy at Any Age: A guide to what you really need to know, and when." The word "really" targets the lead-in statement: "In the era of Google, medical advice is more confusing than ever." Yeah, I can see that. Google can give anyone cancer in about ninety seconds if you use the right search terms. "Blood in the stool" is a good search term to get to a colon cancer diagnosis. What Google should respond with is "Did you have beets yesterday?" and leave it at that.

So this article tells me what I REALLY need to know to be Healthy at Any Age, because a Healthy Life is a Happy Life. Got it. I'm going to paraphrase the easy-to-use section called "Keys to a Healthy Life" because 1) I'd like to save you the $6.95 this issue costs, should you be interested in obtaining the guide and 2) the article is mostly cartoons and pictures, because, well, it's written for people who don't read...anything, apparently.

"Keys to a Healthy Life" (BTW, the definition of "keys" is "something crucial for explaining," in this case, health. And keep in mind this is the SCIENCE article. The stuff you REALLY need to know, if you are going to be SCIENTIFIC about health.):

(USPSTF: U.S. Preventive Services Task Force. ACS: American Cancer Society)

Zero to two years

(You have to start thinking about health EARLY!)

- Developmental screening. (Talk to your doctor.)
- Blood tests. (Check hemoglobin or hematocrit.)
- Fluoride supplements. (talk to your doctor) and dental check ups at year one.
- Vaccines. Get all required (?) vaccines following the Advisory Committee on Immunization Practices' published schedule. Oh, and a yearly flu shot for children six months and older.

Two to twelve years

- Oral health. Brush, Floss, and have regular checkups. And use fluoride toothpaste starting at age two.

- Sensory screening. Screen for vision and hearing by age four and then annually or every other year after that.

- Blood pressure. Blood pressure should be checked annually starting at age three.

- Vaccines. Stay up-to-date with vaccines and boosters, including tetanus, diphtheria, and pertussis along with meningitis. Also three doses of the HPV vaccine for eleven- to twelve-year-old girls. And continue with an annual flu shot.

- Obesity. Schedule an appointment with a doctor to get your child's BMI (body mass index—research has shown that it is NOT an adequate measurement for health prescription). Schools should maintain a "sage and supportive environment for students of all body sizes" and "parents must make sure they understand what the BMI number means and how they can help their child achieve and maintain a healthy weight and lifestyle." (Parents, we're not going to tell you what that is. You have to read another article to decipher the keys in this article. Send away for a Healthy Keys Decoder Ring!)

Thirteen to eighteen years.

- Vaccines. "Get a meningitis vaccine, if you haven't already, before going to college, and women who have not yet gotten the HPV (all three doses), should get it." Oh, and did we say that everyone should have a flu shot, because we really want to stress that every person (not just babies, six-month-olds, one-year-olds, two-year-olds, three-year-olds and four-year-olds need one, but even thirteen- to eighteen-year-olds!) need to get one right now! And wait, there's more, "talk to your doctor about any other childhood vaccines you may have missed." Wow.

- Oral health. Brush, floss, and get to the dentist every six months.

- Depression. "The USPSTF recommends that adolescents get screened for depression if proper treatment is available." Huh?

- Blood pressure. Check annually. (I think I need to check mine right now...)

- Sexual health. Women, get a pap smear within three years of becoming sexually active or by twenty-one (whichever comes first). Talk to your doctor to see if you should be treated for STDs. Also, chlamydia screening for all sexually active women under twenty-five years old. (What happens at twenty-six?) Dudes: "Some docs tell male teens to check themselves for testicular cancer. The American Cancer Society recommends that doctors check for this cancer during physicals, but the USPSTF recommends against routine screening. He said, she said?

Nineteen to thirty-four years

- Oral health. Every six months, we know.

- Depression. Adults should be screened for depression if proper treatment is available.

- Heart health. Get your blood pressure checked and if you have risk factors for heart disease, have your cholesterol checked.

- Sexual health. Talk to your doctor about checking for STDs. The CDC recommends that adolescents and adults be routinely checked for HIV. Get a pap smear at least every three years until age sixty-five.

- Nutrition. (Finally!!!) If you're a woman who could become pregnant, whether or not you're planning to, take 400-800 micrograms of folic acid per day. So, nutrition is a key to health in nineteen- to thirty-four-year-olds who may get pregnant. Got it. And the only recommendation is folic acid. Check.

- Vaccines. Flu shot, forever and ever, amen. And, get a booster for tetanus and diphtheria every ten years. And "talk to your doctor about your risk for certain diseases and they might recommend additional vaccines."

- Health insurance. This is a side bar that states "it turns out the young aren't quite so invincible after all" and lists the fact that fifteen percent of this age group has one chronic disease (diabetes, cancer, asthma, heart disease, HIV, depression, or anxiety) and makes up the highest rate of ER visits (I can see that). But even if you ARE healthy and avoid accidents, there are still the hidden costs of wisdom-tooth extraction (four hundred dollars) or an appendix removal (twenty thousand dollars). "Without an insurance company to do some negotiating, those prices can climb even higher. So even if you're young and healthy, not being insured is likely to cost you more in the long run." (Especially if you are following the "keys to health" from this article.)

Thirty-five to forty-nine years

This is my category, so I'm all ears for the wisdom!

- Vaccines. Yes, Master.

- Oral health. Gum disease and flossing, flossing, flossing! (Easy and inexpensive!)

- Heart health. Blood pressure checked regularly and cholesterol test for women (ages forty-five and up) and men (ages thirty-five and up). And talk to your doctor about whether you should be taking daily aspirin.

- Diabetes. Talk to your doctor about getting checked. Everyone over forty-five should be screened.

- Nutrition. Woo-hoo! If you're a woman, your bone density peaked at thirty (WRONG), so the CDC says you should get 1000 mg of calcium per day and keep taking folate supplements, 400–800 mg per day.

- Depression. If the proper treatment is available, you should see if you need it.

- Sexual health. Regular pap smears. The USPSTF says once every three years. The ACS and ACOG say more frequently, at least until you've had three negatives in a row. You shouldn't not uncheck this issue. (Three negatives...I'm off the hook!) Talk to your doctor for help deciding what is right for you. (I strongly suggest doing something besides reading this article.)

- Mammograms. USPST says: "Women between 40 and 49 who are not at high risk of breast cancer should not necessarily get regular mammograms. They should choose, balancing the benefits of early detection against the risk of false-positives and unnecessary treatment. Every woman between 50 and 74 should be screened every two years instead of annually." There were huge backlashes from the "community" (the people whose business it is to do mammograms?) who felt annual mammograms were condemned, because, well, they were, kind of. The USPST has updated to say "the choice should be an individual one."

Fifty to sixty-four years

- Oral health.
- Mammograms.
- Heart health.
- Check blood pressure and women should talk to their doctor about taking aspirin to prevent strokes.
- Pap smears.
- Depression.
- Prostate health. Some docs say start screening, the USPSTF says evidence is inconclusive for screening younger than seventy-five years.
- Vaccines. All the same ones, and add the shingles vaccines.
- Colon-cancer screening. Start at fifty.
- Osteoporosis drugs. I can't blog about this here. It's too insane of an issue to cover, but a bit from the article: "Although bisphosphonates [osteoporosis drugs] slow down the body's resorption of older, brittle bone and reduce the risk of fractures in high-risk people by 30–40 percent, little is known about their use beyond ten years. Many experts worry that, over decades, bones could weaken as the proportion of older bone increases. The USPSTF recommends checking bone density between 60 and 64 if you're at increased risk of the disease.

Sixty-five and up

- Oral health.

- Vaccines. (Don't forget the shingles vaccine!)
- Heart health.
- Blood-pressure screening.
- Bone density. Get checked regularly.
- Mammograms. You can stop at seventy-four.
- Pap smears. You can stop at sixty-five unless you have abnormal risk.
- Colon-cancer screening. You can stop screening at seventy-five. (What are you going to do with all that free time?)
- Prostate cancer. The issue is the same as with mammograms. To check or not to check with an annual rectal probe? USPSTF says "no" and the ACS says "maybe." Men should make an informed, personal decision. And one that is not influenced, in any way, by the words "annual rectal probe."

So, that's it. These are the Keys to Health, brought to you by the recommendations of The American Academy of Pediatrics, CDC's Advisory Committee on Immunization Practices, the American Dental Association, the Centers for Disease Control and Prevention, and the U.S. Preventive Services Task Force. Based on the title of the magazine issue, there's apparently no science proving that nutrition, movement, stress management, good relationships, love, and continuous education have positive effects on health. There *is* definite correlation between the presented keys and avoiding death. Does not dying equal health? I'm not sure what definition of health the writers of this article were using, but it probably wasn't "The state of complete physical, mental and social well-being and not merely the absence of disease or infirmity."

Using this definition (guess where it came from?) I've come up with "The Other, (Non-Hundred-Billion-Dollar) Keys to Health":

The early years…Prepare well for pregnancy and birth naturally, preferably outside of a hospital. Breastfeed as long as possible. Eliminate television and sugary foods. Give adequate outdoor time, plenty of rest, and lots of time, attention, and love. Take a daily walk with your family. Minimize footwear use or opt for soft, flexible footwear.

As your kids get older, keep screen time (computer and TV) to fewer than ninety minutes per day. Keep sugar and sodas to a minimum. Be respectful. Have conversations with the entire family present. Eat plenty of local fruits and vegetables and have at least thirty percent of your diet come from good fats.

Parents of teenagers, know where your children are and BE their example. The BEST parenting moment I have ever seen was a mom and daughter attending our annual holiday party. Daughter was sixteen, and mom declined a glass of wine (just one!) because she was driving. It's very hard to explain to a sixteen-year-old why MOM could have one glass and drive but DAUGHTER was not able to do the same. Bravo, Ann G.

For your whole life, choose relationships well. Minimize drugs, sugars, and alcohol. Cope well. Meditate. Take a stress-management class. Walk every day. Stretch every day. Minimize stress. Get adequate sleep. Read. Talk. Be. Have the conversations you should. Laugh. Laugh. And, laugh some more.

P.S. The aforementioned definition of "health" comes from the World Health Organization. First created in 1946, the definition has remained unchanged since it was entered into force April 6, 1948.

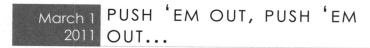

March 1 2011 PUSH 'EM OUT, PUSH 'EM OUT...

Have you ever wondered how things get out of your body? Things like food? And snot? And babies?

The human body has great equipment for expelling various items. Smooth muscle creates wave-like motions that move items through the body (like food through your digestive tract, or particles up through your sinuses). Specialty muscles like the diaphragm (generating upward forces for coughing and vomiting) and the uterus (vaginal delivery) are well-designed and can really get the job done for you.

Then there are other ways of getting stuff out of your body. These methods include YOU choosing to do something extra to help Mother Nature along. The skill most commonly utilized to "help" move things out is the Valsalva maneuver. Even if you have never heard of this world-famous maneuver (going on tour this summer with the Heimlich maneuver for a Maneuver Extravaganzaaaa!), chances are you've already used it three times today. Maybe even once while you were reading this blog, if you read blogs while on the toilet.

Do you?

If you've ever:

- cleared your ears at altitude
- strained to go to the bathroom

- held your breath or inhaled while lifting weights or doing abdominal exercises
- held your breath and PUUUUUSSSSSSHHHHEEED during vaginal delivery

then you, my friend, are quite familiar with the Valsalva (rhymes with Kalsalva) maneuver. The Valsalva is taking in a breath, sealing off any way for that breath to escape, and then pushing against that "balloon," causing it to bulge in various directions (up or down, mostly).

If you look at the physiological and biomechanical properties of the tissues that expel things, one would ask, "Why is anyone straining to get things out if there is equipment there to do it for you?"

And the answer would be...

Because we have over-developed habits that work against the natural expelling tissues, making elimination of all types more difficult.

Oh yeah, like what? (Please read in your best Tough Guy voice.)

I can tell you that you probably have one habit that you've been practicing every day, all day long, that creates a force that is messing with your Let Me Out functions.

Are you ready?

It's sucking in your stomach. No, really. That's one of the worst things you can do for your health (even though it may be one of the BEST things you can do while being photographed in your bathing suit).

Most people have replaced deep abdominal activity with "sucking their stomach in." The belief held by most is that "sucking it in" constantly uses one's abdominal muscles, but really, the sucking-in motion creates a pressure (like creating a vacuum) that pulls the abdomen's contents up (not in). It doesn't do anything for core strength (except weaken it over time) or back health (it increases the loads placed on the intervertebral disks).

Try this: Stand up in front of a mirror looking at your side-view. Now let it all go. Let it all hang out. If you pay close attention, you will see that your stomach really drops down as it moves forward. If you notice a LOT of mass there on your six-pack, then you are regularly increasing the pressure in your abdominal cavity by squeezing all your goods into a tight space. Please, do not mistake sucking in your stomach for abdominal-muscle use. It's not the same thing. Sucking in your stomach doesn't even use the abdominal muscles...it uses a vacuum (no calories were burned or strength generated by creating a vacuum!) and you can actually get LESS

TONED in the midsection as a result. So stop it.

The constant upward motion of "sucking it in":

1. Interferes with the wave-like motion of your intestines, causing a delay in digestion (excessive gas, anyone?) and the need to physically PUSH your waste into the toilet. Your waste-removal system is not designed to have you working against it, so you end up Valsalva-ing to get stuff out.

2. Reduces circulation of blood. Because of location, the constant upward tugging on your guts ends up reducing the full circulation down into the legs. This is a biggie, because the less blood that makes it into the smaller blood vessels in the trunk and lower extremities, the more blood (and the higher the blood pressure) in the main arteries.

3. Reverses your motor programming with your diaphragm. Doesn't seem like a big deal, especially if you don't know what these words mean. Let me break it down for you. The act of coughing is an expectorant—an upward movement meant to clean out items trapped in your lungs or throat. Your body uses coughing to prevent accumulated fluid or phlegm in your lung tissue.

Test your cough: Lying on your back, place your hands on your low belly. Cough. You should be creating an upward (towards your head) force. If your lower belly bulged out when you coughed (and it DID, didn't it!!?) then your constant sucking in your stomach has smashed everything up against the bottom of your diaphragm so now when you cough it isn't pushing a hundred percent up any longer, but has a downward component. (Which is why a lot of people pee their pants when they cough or laugh...)

Which brings me to the pelvis.

As last discussed in the Super-Kegel post, prolapsing organs are a huge deal. Many people are trying to fix the situation by doing various types of exercise, which is fine, but what they are NOT doing is realizing that these organs did not drop out...they were pushed out. By the owner of the aforementioned organs.

Keep in mind that pelvic issues are multi-factoral, but that there are physical situations that increase the strain on the ligaments. (No, your body weight is not one of them.)

1. A lifetime of sucking it in typically results in a regular downward strain when bathrooming. If you've dealt with chronic constipa-

tion, then at least once a day (hopefully!) you've been pushing those organs down and out when trying to get the other stuff out.

2. If you were given extremely outdated directions while birthing your baby (think being TOLD to Valsalva, "Take a deep breath, hold, and push") then you could have created a large downward force on the organs, not just the baby itself.

3. If you regularly do high-impact activities, you are straining the ligaments on each landing.

Our misguided solution to these items is to strengthen the Keep 'Em Ups to offset the damage, but, unfortunately, it doesn't work that way— much in the same way you can't do cardio exercise to offset the harmful effects of smoking.

Fixing prolapse in the long term requires that you stop the downward habits in addition to strengthening your musculature. If you are in the process of getting things back up and on (or in) the saddle, check out your downward and upward forces. Have you been messing with your mechanics???

To really stretch out your sucker-inners, do this exercise: Start on your hands and knees. Good. Now, relax your stomach alllllll the way. And, make sure you aren't tucking your tailbone under. Let the spine relax all the way too.

Check out our (brave) volunteers showing what they've actually got up front (image below left):

And then them sucking it back up...where they like it (image above right).

Looks good, buuuut, it creates disease and injury, so not such a good habit to cultivate. (And, P.S. where do you think all that stomach excess is going when you suck it in?) Better to release and learn how to really use that transverse abdominal group and really increase your metabolism, which is how you really get rid of body fat accumulation.

The moral of this story is: All expelling is not created equal. You have great (amazing!) systems in place, but we've disengaged these natural functions and are instead doing a hodgepodge of other things (maneuvers) that seem almost the same, but are quite different mechanically and will lead to tissue damage.

Pay attention to any forces you are creating that go against natural functions. Then learn to stop creating these inappropriate forces. Upward forces lead to downward pressures, which lead to ligament damage, which leads to things falling out. Of your body. Sucking-in habits in conjunction with Valsalva habits can also cause other upward strains, like in the blood vessels in your head. Or the decreased ability to inflate your lungs with each breath. It's a really big deal, man.

June 15 2011 COMPENSATORY MECHANISMS

I actually thought my book was done when I turned it in in January, but then came the first round of edits. And then the second round.

And then there was organizing the photos.

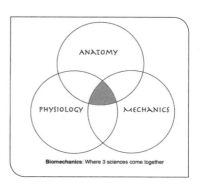

And realizing that I forgot to include a photo in the photo shoot, and then rescheduling the photo shoot.

And there were dirty diapers. And breastfeeding, and lots (and lots) of one-handed typing, which isn't as much fun as it sounds.

Today, I am supposed to finish the glossary. To prepare for doing the glossary, I thought I would do this instead: (image above right)

And this (image right):

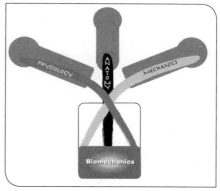

And while I was at it, I thought I'd take this picture and play a little game called: GUESS THIS BODY PART. (Image next page.)

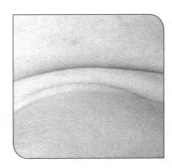

Did anyone else get *World Magazine* as a kid and love the GUESS WHAT THIS IS photo game in the back?

So anyhow, it turns out that a glossary is kinda hard. Definitions are pretty easy, right? But then I tried to define the term *compensatory mechanisms.*

I came up with: "Things we do to correct for other things that we do."

But that sounded like I was in third grade.

So I checked with Wikipedia, who had this definition listed under the engineering term *compensation*: "Compensation is planning for side effects or other unintended issues in a design."

(Why bother doing anything as long as Wikipedia is there to do it for you??)

I like this definition as it applies to the human body. There are all sorts of things we do that were unintended and not accounted for in the human genome. What kind of things? Oh, how about: Driving, jogging, bicycling, purse carrying, shoe wearing, flying, computer viewing, book reading, sitting, using a toilet, under sleeping, over caffeinating, and blogging. Just to name a few.

And all of these situations require your body to compensate, in some way, with its geometry (think alignment) and physiology. The ability to compensate, while an awesome one, is not without penalty.

When you are having any (and I mean ANY) type of health problem, create a list of all the things you are asking your body to do and ask yourself—How many of these things are natural to the human being?

Your ailments will start to be less of a mystery when you look at yourself as an engineer looks at a machine. The remedy is usually not in what to do, but what to *stop* doing.

(How about now? Can you tell what the body part is? [Image top of next page.])

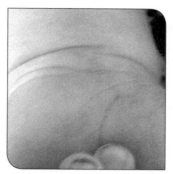

So now it's been about seventy minutes since I started working on this one definition. This is what I came up with:

"Compensatory mechanism: An action taken by the body to continue physiological function despite an alteration in natural function."

I like it. And I should be done with the rest of the glossary by the time I am ninety-four years old at this rate.

Final guesses, anyone?

Okay, the dirty fingernails give it away, right? How is it that a two-month-old is able to garden behind my back??

What habits do you have that are requiring your body to work differently than it was designed?

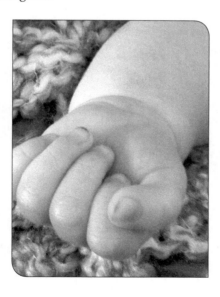

chapter 14

MIND MECHANICS

From *A User's Guide to the Brain*:

> While a new task is being learned by specific circuits in the frontal cor-
> tex, many neighboring neurons drop whatever they are doing to assist in
> the process. Once the task has been mastered and executed a number
> of times, the firing patterns become established and the behavior auto-
> matic. It no longer requires conscious attention.[1]

Perhaps this explains the plateau people experience in exercise.
When you are first learning to coordinate your body, your brain has to
work much harder. Once the brain has established a good connection to
the muscles and the pattern used for the task, you no longer have the extra
work of organizing neurological current, preparing the muscle for water
and glycogen storage, and growing capillaries. All of these tasks use kcals
as energy—so when the jobs decrease, so does the caloric expenditure.

Once this process happens, you have to go longer, or faster, increasing
the intensity to match the same calories burned as before. So it feels like
your movement program isn't working. To prevent this phenomenon,
consistently change up the activity you are performing. And never, NEV-
ER, stop learning! That is why, next month, I am taking classes on Italian,
French, yoga, and knitting. For one month, I relinquish my role as teacher
and dive head first into the much more challenging role of student.

And speaking of learning, I would like to abuse my blog-power and
extend a brief motion of appreciation to every teacher that took some time
to directly impact my path. If you are out there, I am forever obliged...

Donna Hubbard (2nd grade), Ms. Jackson (3rd grade), Mrs. Franich
(4th grade), Mr. Jim Clemmer (5th grade), Mrs. Pat Bendix (6th and 8th
grade English), Steve Wong (10th, 12th English/Drama), Dr. Tim Anderson (Biomechanics), Dr. Jacobo Morales (Ex. Phys) (B.S), Dr. William
Whiting (Biomechanics) (M.S.).

Please do not hold any of the English teachers accountable for my
writing skills. They did the best they could. ☺

If you would like to thank a teacher (or two or seven), please use this
space as an appreciation sanctuary. Go for it!

---

1. From *A User's Guide to the Brain: Perception, Attention, and the Four Theaters of
the Brain*, by John J. Ratey, M.D.

YOU ARE BOUND TO BE SUCCESSFUL

I became a Vipassana meditation student over two years ago, and I will say with a hundred percent conviction, that it was the best thing I have ever done for my long-term health. Listening to me explain what Vipassana is (and more importantly, what Vipassana *isn't*) is very similar to listening to a three-year-old explain their version of foreign policy. I don't have the words (as the experience is so much greater), nor do I really "get it" on the level that it could be possibly gotten. My explanations start with big ideas but end up with the default "It's really really really important that you do it to understand, as anything I may say about it doesn't do it justice."

Vipassana is not a religion. Vipassana is not a treatment. It is simply the instructions on how to not do anything without getting distracted. It sounds pretty easy until one realizes that in order to relax, we have to add stimulation to our system. Watch a movie, or read a book, or listen to music. We can't seem to sit without squirming around because our legs hurt...from doing nothing. At the end of the ten-day course (these courses are taught in exactly the same way, all over the world, in the exact same format) you will have learned how to do nothing—and not be freaking out about it. Why do nothing? Your physical wellbeing requires it. Got tight muscles? They can't relax. They can't stop "doing," and neither can your mind.

During the course I was pretty smug about how well I could sit (without moving anything, including swallowing!!!) for a couple hours at a time, until I realized that I was THINKING about how good I was. Thinking was still doing something. And then I was realizing I was thinking. And then I was mad that I was realizing that I was thinking. DAMN! Start again. Start again. Start again. Start again.

The mantra of the course is: *You are bound to be successful.* And you are.

I am blogging this today because I happened on the first excellent article I have ever read about Vipassana in a So. Cal publication. "The Whole Person. Vipassana Meditation, Mental Health, and Well-Being," by Paul Fleischman, M.D.[1] is additional info to help you comprehend what I am so poor at explaining.

Some course information:

All first-time courses are ten days long. Once you've done your first course, you can then take shorter courses. Ten days seems like a lot. Are you saying to yourself "There is NO WAY I could ever take that much time off away from my [pick one] kids, job, dogs, family, life"? That's what I said, too. And it doesn't change the fact that you still need to go. We believe our lives are inflexible (same as our hamstrings), but I'd bet if there were a catastrophe, or a death, you would be able to arrange someone to care for all the things you believe are "must-dos." We tend to be very savvy when it comes to reacting to a situation and unskilled at preparing for one. Trust me, you can take the time.

P.S. The course is FREE. All your food, all your housing, and the instruction. There is no charge. So the whole "I can't afford it right now" thing doesn't really work either. Damn.

And P.P.S., the food is AWESOME! You are not going into a cult or some weirdo place. The salad bar and chocolate chip cookies were so good!

P.P.P.S. There's no talking. If that's a red flag for you then consider what your need to constantly talk is preventing you from doing. Doing nothing!

P.P.P.P.S. There is no need to respond to this blog (if you have an urge to throw tomatoes at me by now, then you should REALLY be considering this course). I will ask everyone to just take a look at the schedule—you can search "Vipassana" online and find dates worldwide—and respond by posting the dates that would maybe work if the universe somehow aligned and your boss gave you the week off and your kids would survive with only one parent or family member, or someone might walk your dog for you, or if Hell froze over. Show me that you at least looked...

Be happy.

1.alignedandwell.com/components/com_wordpress/wp/wp-content/up-loads/2010/03/vipassana-meditation.pdf

 BEYOND BALANCE

August 9
2010

Can you ride a bike? Probably. And you know that riding a bike takes balance, yes? So if you ride a bike, you would probably come to the conclusion that you have pretty good balance. But what if you were just sitting on a bike, not pedaling? Just sitting there. Not falling, but not moving, either. What would that phenomenon be called? This unnamed phenomenon is what I like to call *active stillness*. This unmoving stability not only requires the greatest amount of sensory input, it also allows for the greatest sensory input. Huh? What? Tricky, right?

True, biomechanical stillness requires every one of your more than six hundred muscles to be active, precisely at the same time. This time-sensitive coordination requires muscle tissue to be at precisely the same (relative) length to maximize the current of electrical communication. Inflexible sections of the spine decrease the health of the nerves that reside there, limiting clear communication. Stiff, unyielding muscles send "fixed" information from your proprioceptors (sensory organs in the joints, tendons, and muscles). Inflexible muscles give incorrect information to the decision-making center (brain), which the processing center (not knowing that the information is based on stiffness and not external limitations to the joint) makes an overcorrection, or lurching movement, to stabilize.

When you try to "stand still," you will feel these lurching motions happen in all different directions, one right after another. They are simply your body's best guess at where you are, as the communicative pathways between the muscles and spinal cord have been allowed to die off. No bueno. And no more!

The stabilizing system of the entire body is the relationship between the proprioceptive system (information coming from the muscles, joints, and tendons) and the processing of that sensory input (what the brain tells the body to do with this info). If you send a message STAND STILL, then you should be unmoving, one hundred percent of the time you are asking your body to be still. If you aren't able to stand still, there is a problem with your nervous system, either at the sensory level (tight muscles can't determine position) or at the processing level (information isn't coming clearly through the spinal nerves, usually due to sluggish myelin regeneration…also a result of tight (spinal) muscles). Your mission, should you accept, is to stop giving yourself a nervous-tissue disease, by doing the things necessary to keep your brain-body connection open, loud and clear.

Quick Test 1: Start with your (bare!) feet pointing straight ahead. Line up the outer edges of each foot so they make the number "11" and make sure you are standing with your heels pelvis-width apart. See how stable you feel. Do you detect any moving around? A slight wobble?

Now close your eyes, and see if there is any change in your stillness.

How'd you do? Did you feel yourself move more when you closed your eyes? Here's what is going on. Your eyes are not part of the sensory input required for whole-body balance—your muscles, tendons, and bones should know where they are without looking. The more poorly they communicate with the brain, however, the more you begin to use your eyes to make corrections to your instability. This visual compensation happens so fast, you're not even aware of it. The eyes (and vestibular system) are pieces of a mechanism that balances the head, all of the time, to the horizon, no matter how jacked up the rest of the body has become. In order to repair the body (including eye-muscle fatigue, dizziness, and age-related changes in vision), you have to STOP using the eyes to do the work of the proprioceptors. This means you have to fix your body's internal sensory/communication channels.

Quick Test 2: Get yourself onto one leg, with your feet still straight. No, you can't bend either knee, no, you can't reach your arms out to the side, and no, you shouldn't be holding on to something while testing your balance. In fact, if you find yourself needing to hold onto something to help steady yourself on one leg, work on the first exercise, second level (both legs on the ground, eyes closed) and improve this skill before progressing to a single-limb stance.

Once you feel okay on one leg, close your eyes. How does that feel? Lots of lurching at the ankle?

While we've all got some whole-body mal-alignment going on, it's the failure to use our feet over our lifetimes that fundamentally messes with our stability. My point of writing all of this down is, the biological, or fight-for-survival, part of the brain feels the constant falling we are doing (even though the more conscious parts of our brain don't recognize it) with every step. Imagine the stress falling off of a hundred-story building would generate in the adrenals, which shortens the psoai, which dumps a whole lot of stress-chemicals into the blood. Now imagine you were only falling a fraction of that distance, but with every step. This small sensation (and physiological reaction) accumulates over time, leading to tissue degeneration of your "catching-yourself" parts (i.e. knee cartilage, spinal disk tissues). The reaction to falling (even a tiny reaction) leads to an overuse of the adrenaline system, a risk factor for diseases of decompensation like fibromyalgia, chronic fatigue, memory loss, cardiovascular disease, insomnia, depression, and declines in nervous-tissue health.

My problem with the word *balance* is that it has come to mean "not falling." Millions of people are taking courses and doing exercises to increase their level of balance (i.e. reducing their likelihood of falling), but because the "not falling" definition is not really physiologically good enough, most of us, thinking we are balanced, continue to fall uncontrollably through space.

Excellent neurological health means we get to pick, exactly and unfalteringly, where we would like our physical selves to be. Our body contains a complex information/coordination system, proprioception, that lets us know, without even looking, where each bone, muscle, and tendon is in space. Your brain, better than the best engineer, pilot, or computer software program, can balance you both relative to yourself, and relative to your environment...even if your environment changes moment to moment. This is the level of health that we need to train for. You have your first four exercises to practice!

May 2 2011 RATMECHANICS

Okay, before you read any further, you must watch this video.[1]

I TOLD you biomechanics was cool! So, here's my commentary.

We humans may not whisk (except for my dad, who like the rat has poor eyesight in the dark and has adapted by growing extra-long eyebrows to help him gauge the size of the furniture in a room...), but we do have a very similar system in our body. It's called the proprioceptive system.

Proprioception means "one's own perception." No, it's not like your opinion or anything like that, but it is the ability for one part of your body to know where it is relative to the other parts. Unlike the rat, who is using the deformation of its whiskers to create an image of what is external (similar to a dolphin using sonar to "see" shapes in front of it as it is swimming), we use our proprioceptive system to create an image of what is internal, or inside the skin. Propriception works in the same way as the whiskers, though. The deformation of the joints and muscles sends information to the receptors (proprioceptors) within the moving muscles and their joints. That information about a change in skeletal position then travels via our neurons to the brain to create an image.

For the rat, the more whiskers and the more mobile each whisker, the clearer the mental picture. Same goes for humans. The more muscle fibers you have firing and the more supple (not tight) the tissue, the better the proprioception. The tighter the muscle (what's going on with those hamstrings, eh?), the stiffer the joint.

Tight muscles and stiff joints give your brain a low-res image of what is happening in your body. Why is this not good? Your brain uses this image to figure out how to move in the least damaging way to the tissue. If your body is using a low-res image to make decisions, then some of the conclusions are erroneous. Which is different than erogenous, which I typed at first, then looked up, then replaced with the word erroneous.

I am good with science, not so hot with words.

Moral of today's blog: What kind of information is YOUR brain making decisions with? Hi or low res? And how can you beef up your pixels? With alignment, of course, which means your muscles are all at the right length to optimize your mental image of yourself.

EXTRA NERDY BONUS SECTION

I'm not so sure that cats or humans *can't* whisk as much as they *don't*. There are scientists in biomechanics who are looking at what human arrector pili muscles (muscles in each hair follicle) do vs. *can* do in humans.

I can't wait for more information on that. Can you?

---

1. sciencefriday.com/videos/watch/10375 Two and a half minutes showing how rats feel with their whiskers.

Sept 12 2011 SUNDAY KIND OF LOVE

I haven't had any blog-inspiration this week. Maybe the last of my book edits have tired me out. Maybe it was the trip from CA to WA (again!). Perhaps it was the fact that all of the family has a bit of a cold. Whatever it was, it finally broke today like a fever.

I go to a Catholic church. I attended mass this morning, which, as some of you may know, includes a full exercise program. If you've never been, it is exactly like this:

Stand. Sit. Stand. Sit. Sit. Sit. Check out everyone's posture. Stand. Check out everyone's posture. Sit. Stand. Stand on one leg. Stand on the other leg. Sit. Cross legs. Uncross legs. Slouch. Sit up. Stand up. Walk (so slowly you can really feel how you're really just lurching all over the place). Kneel. Stand. Sit. Eat a doughnut.

It is awesome.

It was even more awesome today because I was able to use the pew as a marker to help me stay focused on keeping my mind and body present.

Here's how I did it, and how you can do it too.

In bare feet, back your weight all the way into your heels as far as you can go while still standing comfortably. Not so far that you topple backward or have to grip hard with your quads.

Now do all of that again, but stand in front of a chair, so the seat of the chair presses lightly into the back of the legs once you're in the right (vertical with muscles on!) position (image at left).

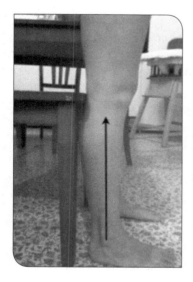

Here's the fun part.

From your vertical position, let your body relax into its usual position.

What happened? Did your pelvis drift forward a million inches (image at bottom right)?

Your relaxed (muscles off!) position tucks your pelvis, has your body weight rest on a tense psoas muscle, and places a high load on the feet. It seems like a small distance to travel, but the loading can be off by thirty to forty degrees.

Try going between each position a few times. Once you feel the distance between the two, you can get a sense of the physical difference between when you're choosing with your body and when you're relaxing into old habits.

One should be doing other things besides coming up with this while in a church or temple or ashram, right? But if you know me really well, you'll know that I see a strong correlation between body position, presence of mind, and my spirit. And for me, spirituality is all about paying attention to the quality of my thoughts and behaviors. I cannot possibly begin the extremely difficult task of constantly choosing my thoughts and actions if I cannot choose something simple as where my pelvis is in space.

Actually, the choice of keeping my pelvis stacked vertically to my ankles is not a

hard one. It's just remembering to continue to choose it that poses the challenge. I'm like, "Pelvis stacked, Katy." And then someone comes up to chat. Or a doughnut is eaten and then I check back in and wheredyougo? My pelvis has slipped into old habits!

So why do I feel good doing alignment exercises in church? In the same way a vertical leg is correct, the choice to perpetually offer love and abstain from judgment is the correct alignment for me. Readjusting my pelvis reminds me that even though I tell myself to do something (and it's even something I WANT to do!) I can fall out of form in a matter of seconds.

Like many of you, I have a constant stream of thoughts going through my head. Many of these thoughts are not in alignment with the beliefs I hold, yet they are still there. So, I love all. I have compassion for all. Except for the person who came in late (pelvis forward). Or the person who doesn't want to scoot over for me (pelvis forward). Or the person who is texting (pelvis forward). Or the person who is writing a blog in her head when she should be listening (pelvis forward). Why does my hair look so stupid today (pelvis forward)?

These thoughts are just old patterns of judgment. My pelvis being forward is just an old habit too. Eliminating habits requires constant mindfulness. It requires the ability to view your physical or mental position, check it against the alignment you desire, and then make any necessary adjustments.

I felt like I received all messages today. All it required was bringing my real (and mental) pelvis back about ten thousand times. In one hour.

I hope you had a great day.

Sept 27 2011 OPEN YOUR RIBS, OPEN YOUR MIND

We all can use better shoulder, rib, and waist mobility for optimal breathing. I'm going to teach you a stretch that helps. Why do you need to stretch these muscles? Read my post "Under Pressure (Part 2)" for more information.

Side Bend: Standing up in your own personal version of "straight," reach your arms overhead and, clasping your hands, bend your body to the side. And don't let your hips jut out to the side. Keep them anchored right above your feet.

Here's a picture to help (image at left).

Feel good? Feel any tight spots in the waist, ribs, or shoulders? Nice. Keep working on it.

Now, find a wall. And using this new tool of objectivity, get yourself lined up so that, while your feet remain a few inches from the wall, your bottom (no tucking the pelvis!) your bra strap (or bro strap, gentlemen), and the back of your head are on the wall. Now, reach your arms over your head until your hands touch the wall overhead. Make sure someone places their thumb over the camera if they take your picture, just for consistency. (Image below left.)

You might find that, keeping your body vertical, you aren't able to get your arms up as you did in the first exercise. Also, make sure you didn't tuck your pelvis when you thought you were dropping the ribs back to the wall (this is a pretty common body confusion). (Image above right.)

Now do that same side bend, using the wall to keep you in the correct alignment for the exercise.

How does being aligned during the exercise compare to when you *thought* you were aligned doing the exercise?

Objectivity is a fantastic tool that enables us to see how things really are. It is a lot easier not to look, I know, but to see something as it is, and not how you see it through the glasses of your personal experience, is the key to whole-body wellness.

Using tools like floors, walls, straight edges, and alignment markers will bring a level of awareness to your personal practice of movement. These tools keep us honest.

Now go play with the exercises and see how using the wall as tour guide helps reveal your actual muscular boundaries. See if you can travel across the distance that is separating you in your body from you in your mind. All of your yous should be in alignment with each other.

"What can we gain by sailing to the moon if we are not able to cross the abyss that separates us from ourselves? This is the most important of all voyages of discovery, and without it, all the rest are not only useless, but disastrous." —Thomas Merton

Word.

Nov 13 2011 LECTIO DIVINA

The coolest thing I ever learned via Catholicism is a meditation process called *Lectio Divina*. This is Latin (duh) for "reading divine," or "divine reading" for those who don't talk like Yoda.[1]

When you think about the term *meditation*, no doubt it brings to mind a picture of sitting cross-legged, bindhi glittering over your third eye, while Krishna Das plays in the background. This is a cool way to meditate, especially if you have Pandora. Check out the Krishna Das station—simply listening makes me feel more Zen. But the more you study meditation, the more you realize relaxation is not the goal. Being calmer is not a goal. Opening your hips is not a goal. Having goals is not the goal. Meditation is simply a chance to watch yourself and your behaviors. Some days I'm a freaky, judgmental, angry person in my mind. Some days I'm all OM. Mediation helps you get to know yourself and then see that you're no different from anyone else.

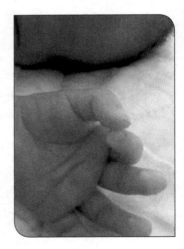

Lectio Divina is a way to meditate with the written word. This is how you can do it:

You need to do some transitional activity to prepare your mind. This can be taking a few deep breaths. It can be jumping

I love my baby.

up and down a fixed number of times. It can be whatever you want it to be. The behavior is not important, only that you are aware that your action is matching your choice of action.

Then, follow these steps:

1. Read a passage slowly, a few times. Pay attention to each word.

2. Write down any words or phrases that capture your attention.

3. Read those words and/or phrases aloud, and then ponder on them, while reading them over and over.

The theory is, the material that resonates with you is your deeper, more aware self bringing worries, ideas, or difficult topics up for you to pay attention to.

"Our bodies communicate to us clearly and specifically, if we are willing to listen." —Shakti Gawain

It's kind of like eating. You take a bite, you chew on it a bit, and then your brain takes in all of the tongue-data and prepares the body for digestion.

Of course, you can always read this way, but what makes it a meditation is that you have chosen a time in which your behaviors match your choice of behaviors exactly—hence the steps.

Your body executing your mind's wishes is mindfulness. And if you have difficulty with that (and you probably will), this is just a mirror to see how easy (not!) it is to get your body and mind on the same page.

I suggest, to tap into your body's natural wisdom, that you read a few of my blog posts using the process of Lectio Divina. It will help you discover critical information about you. Because you are who I am writing for!

---

1. If you don't know who Yoda is, you might be too young to read this blog. Check with your parents, please.

 Nov 15 2011 **HIDDEN DOUBTS**

This morning there was a great question on our Aligned and Well Facebook page:

I'm active in a pregnancy group here on Facebook, and it's so sad to hear all these women having problems with pelvic pain, leaking urine, and even prolapse. They think it's normal and just something they have to live with. I keep posting about your exercises, but can't seem to get through to them. Does the idea of healing yourself with exercise and alignment perhaps sound too good to be true?

In 1975 two researchers (Becker and Maiman) developed something called the Health Belief Model (HBM). They suggested that the likelihood of someone adopting a behavior that would prevent or correct some disease or physical issue depended on two things. And, don't worry. I'll tell you what those are two things are.

And in case you are wondering that I am some sort of information freakazoid, let me introduce you to a little garage sale find called *Advances in Exercise Adherence* by Rod K. Dishman (pictured here, with my baby, as he simultaneously reads and plays in the trash, above).

This little textbook cost me a whopping three bucks. You never know what treasures are lying around other people's houses. Which makes me sound a bit like a klepto. No judgments on my klepto readers. I like shiny things too.

So, applying the HBM to your question, a woman would need two things to be motivated enough to follow your advice on alignment and the pelvic floor. She would need to 1) believe that she had a personal vulnerability to PFD (or SI or hip pain), and 2) believe that the consequences of her condition would be severe.

The first isn't typically an issue because chances are, if these women are reading your posts (or mine), they already have an issue. There's not a lot of wondering if they're going to have PFD someday—someday is today.

The severity of the issue is more challenging. I don't think that women understand that a little PFD now means a whole lot of PFD later.

I write until I'm blue in the, uh, fingers, that if you have a little PF issue now—perhaps a bit of a sneeze-pee—then in the not-too-distant future, your organs will be falling out. Your hips will need replacing. Your knees will need replacing. Your SI joint will wobble around and push on your sciatic nerve. Your toes will go numb. Your lumbar spine will degenerate. Your organs will need to be removed. Your vaginal delivery will likely turn into a cesarean.

When I read the above paragraph, I feel like a bad person because it sounds mean or something, right? I don't say it because I'm trying to make money. I don't say it because I'm trying to make you any more tense than you already are. I say it because somebody has to. Somebody has to start saying that the way you are using your body is making you sick. And it will make you sick in the future. And it is your job—not mine—to hear it, and to take action.

I've just decided that the theme of 2012 is personal responsibility. Oh, and upper-body strength. Which reminds me that I need to go hang from my indoor monkey bars for three minutes before I spend any more time on this post.

Okay, I'm back. And buff. And definitely in that order.

Now that I've explained the Health Belief Model, I need to add some more info. Cuz you see, the HBM is not really that complete. It turns out that many people *know* that they need to get exercise to avoid dying. They *believe* that they will die if they don't do it. Yet they still don't do it. And we don't even have to go to the extreme of dying from something. Most of you, dear readers, have some ailment for which I've given you an ExRx or info to follow and you still don't really comply with a regularity that matches your desire to no longer have Issue X. Well, not *you*, who is shaking your head right now and saying "not me, I follow your advice to a 'T.'" I'm not talking about you.

You used to sit in the front row of the class, didn't you?

So why is it that we can't seem to manage to execute the behaviors that we desire to? Enter the Self-Efficacy Theory. This 1977 theory (ha, I was born just before this came out) states that "all behavioral changes are mediated by a cognitive mechanism of perceived efficacy"—which is a lot of fancy words to say that you have to believe that you can perform the recommended behavior.

I like the word "believe," because as a science-type (you know, all boring and math and lists and bad at art and no passion and left-brain and black and white and I don't see in color and I can't really taste food and my life is very straight and adventureless), I don't really know what it means. Yet I'll bet that you, like me, know what it *feels* like to be one hundred percent on board with what you believe.

Here are some indicators that might help you identify a lower level of self-efficacy:

"I'm afraid that I'm not doing it right."

"If I can't do it completely then I don't feel good doing it at all."

"I've TRIED that healthy thing before and I couldn't do it."

"My attempts at health have failed in the past."

"I'm not that coordinated."

"I've always had this _____."

Sound familiar?

Your belief system, especially in your ability to physically do something, is learned from your personal experiences as well as the good and bad experiences of those you model (parents, peer group, etc.).

The Self-Efficacy Theory has been researched and is believed to be the most successful in explaining habitual exercise behavior. Which is a lot of fancy words to say that even if your organs are falling out and your hips hurt so badly that you can no longer walk, and even if you thought that them falling out would kill you (good news: it won't) you STILL wouldn't spend a couple hours a day doing your exercises and make a standing workstation and give up your heeled shoes and value your health enough to really invest in it because you don't believe that you can change.

And therein lies the problem. It doesn't matter how many solutions you post, reader who posed the question that launched me on this post, or how many blog posts I write and emails I answer. The people with, in this case, pelvic pain, leaking urine, and POP, have to repair themselves. They have to value the repair more than you do. And that's tough, because you're excited and following my blog and posting about whole-body alignment because you've chosen to help others. They're reading your FB posts as something to fill extra time and aren't necessarily looking for answers—just interaction, which are two fundamentally different things. Interaction is often mistaken for taking action.

Then, original poster, there could be the other issue, too—that they totally believe in themselves and just don't believe what you are posting. The prevailing notion that our body injuries are a result of a single incident, and not the slow accumulation of micro-damage, makes it very difficult to even begin understanding our ailments.

In other words, if you are asked by a health practitioner to explain how your back went out, you are much more likely to say "I was moving a forty-pound bag of cement up twenty-seven flights of stairs," than you are to say "I've sat in a chair for ten hours a day for, oh, I don't know, seventeen years, and I've been under a lot of emotional stress too, and then I used my body for a huge athletic feat with no training program whatsoever."

In other other words:

We have an inexcusable lack of knowledge regarding our own function, and the inability to see that What We've Been Doing is what has led, directly, to the ailment you have now.

So, maybe these women think you are spending your valuable and precious time posting nonsense. Although I believe (with the other half of my brain that's more interesting and has a personality and passion and loves poetry and sewing and rainbows and fairies and stuff) that, if you have come here to read, or they have come to your site to read, that the material is resonating with them on a cellular level. The fact that they don't trust the info has more to do with trusting their own ability to do something with it—their efficacy.

P.S. I can't post the problem without posting a solution, can I? If you feel that you might have an issue with self-efficacy, then do a little mental investigation.

1. Do you regularly interact with people who were in your household when you were growing up?

2. Do they (not you!! Do *they*—never you—this is about *them*) regularly make excuses to avoid learning or doing something new (doesn't have to be body-related either)?

3. Even if you REALLY WANT to change something, do you have reasons that you're not doing EVERYTHING THAT IT TAKES to make that change?

4. Even if they are valid reasons, write down every reason you can't do something or didn't do something today.

5. Do you TRULY believe that you can make something better, or are you skeptical? Give your skepticism a percentage, i.e I think

that it's PROBABLY true that I can make something better, but I'm forty percent doubtful.

6. If you could only make yourself twenty-five percent better, does improving your health by twenty-five percent have value to you? Or if you can't be a hundred percent better, are you tempted to do nothing?

Once you start doing this type of research you might become aware of a more hidden efficacy issue. It's really fun.

Not really. It's pretty scary, actually, to see what's in the old upstairs closet, if you know what I mean. Do you know what I mean, "Or am I being obtuse?" (Name that movie, and you can be my new BFF.)

chapter 15

MOTIVATION/ WHEN YOU NEED A KICK IN THE PANTS

DAYLIGHT SAVINGS TIME

Daylight Savings Time starts this Sunday!

What were you going to do with that hour? Sleep? Read? Walk? The fact of the matter is that We Are Not Maximizing Our Time. Are we busy? Heck, yes! Are we productive? Maybe or maybe not.

As someone who works A LOT, I often stop and look at the way I am working. Sometimes I am moving around a lot—a flurry of paperwork, receipts, checks. Lots of phone calls, internet searching, standing in line and driving around. I think of this as kindergarten working. Busy activities that keep me constantly occupied but produce very little.

I try to work more intelligently. Of course there are some necessary tasks, but I am guilty of using the kindergarten method to keep myself from having to do the hard work. The real stuff that needs to be done that frankly, I don't want to do. Believe it or not, I make two lists. "Kindergarten tasks" and "Stuff I Need But Don't Want to Do." I put "walk one hour" on the second list.

We all have different jobs, different situations, and different priorities. But we all need to produce. Even if you are finished working for the year, or for your lifetime, you are still required to produce your health.

You are the CEO of your cell factory.

This is your first Upper Management training:

1.   Make a list.

2.   Put "Health Production" at the top.

3.   Get it done, every day.

If you don't produce, the factory shuts down.

Bad News: It's really that simple.

Good News: It's really that simple.

Okay. I have to go walk now and do my least favorite stretch (Legs On the Wall).

It had been a long week. And before that, a long month. After sixteen hours of lecturing, eight flights, two TV appearances, and one funeral, I decided that it would be a great time to move. And I thought it would be an even better idea to do it fast—lugging the couch, the easy chair, and my entire kitchen between a hip-opening class and lunchtime.

It seemed like a good idea…right up until my lumbar vertebrae shifted with an audible pop.

Now I know many of you will read this and remember yourself in a similar situation. Bent over, frozen, with a FILL IN THE BLANK (loveseat, outdoor flower pot, moving box, small child) in your arms. Feet planted. Spine rounded over in flexion. And then the brilliant, mindless decision to twist. Nice. Now I am paralyzed. Glued to the earth (or the purple easy chair, rather) for what felt like an eternity (about sixty minutes). It was bad—and I'm not going to go into it more than that—but I will say that by doing the Aligned and Well exercise protocol and many mini-sessions of Vipassana meditation to control my reaction to the situation, I was walking four easygoing miles by the next day, and comfortably upright by the day after that. With no medication.

Why am I bringing this up? Because I completely caused this situation, and I wanted to spare you the painful outcome of a mindless injury.

As soon as I heard/felt my back shift, I thought "this spasm is a reminder." That moment was the bill for every Restorative Exercise class that I swapped for twenty minutes of half-assed stretching in my living room. And every time I favored a dose of caffeine and a late-night movie over two good hours of sleep. The wail that seemed to come from my throat was really the sum total of every "YES" I had muttered when what I really meant was "NO, I don't have the time, energy, ability, desire, etc."

So here we are, at the beginning of December, with the next few weeks dragging us out of our routine. What are we to do? It always seems like participating in this time of community and joy means experiencing stress and fatigue as well. But it doesn't!

They say that the best gift is the one you give yourself:

1. Send a Christmas or holiday card to an estranged friend/family member as an act of contrition. You don't have to create anything deep—just cut and paste this: "Dear CRAZY: I am thinking of you and hoping the best for you in 2010. Sincerely, I'M ALSO

CRAZY." You can change the names if you want, but you can catch my drift.

2. Sign up for a retreat this year. Life can absolutely improve every year! I can't recommend a Vipassana meditation retreat enough. I think every human being should take this course, and it's free! Completely funded by the donations of past students, this course will give you the tools to cope with even the worst physical pain!

3. Practice un-stressful hosting. Invite a small group of friends over for a little togetherness. Let them know that you will NOT be cleaning your house, preparing foods, or giving gifts. But that you would like to be with them, if they would like to be with you. Try it! It's amazing how much more we enjoy showing off our SKILLS TO DO EVERYTHING more than we enjoy just being. Do it...I dare you. (P.S. I am no exception. I was going for the I CAN DO EVERYTHING award when I *dislocated my spine* trying. Take a hint!)

4. Don't throw your health out the window for the next month. You deserve better than that. Physical wellbeing is a pain in the butt because you can't bank it. You can't trim a little off the edges here and there and expect the same healthy-body results. Keep up your regular routine and encourage others to participate with you and learn from your good example.

I'll end with a nice email from my dad. He's actually a pretty good speller, just not the best typist": "I want to remind you, it it may be a Bowman trait to think you can move mountains, but this is constrained by the fact we have a Bowman backbone...human. You should be aware of that fact as a famous skeleton enrepenoooorrr."

 A "SHORT" CHALLENGE

January 7
2010

So it's Wednesday during the first week of the NEW YEAR and in running to my next meeting, I though I would stop in to Starbucks to pick up some tea.[1]

While waiting I decided to pick up a "guess how many empty calories/fat/sugar is in this stuff" document-that-the-law-says-they-have-to-put-out-now brochure. I was looking at all 450 things they have in the menu when I saw IT. The solution. The plan. The goal.

IT was a little word I had never seen actually printed on the big menu, presented on the counter, or uttered from the lips of any one person in any of the ten thousand times I had been at this Adult Candy Land.

SHORT.

Uh, come again?

No, that's it. I mean IT. Short. It's a size. And not just the length of a roller coaster as compared to the line you have to stand in. SHORT is a size of beverage you can get at Starbucks.

SHORT. It's right below TALL. Which is right next to GRANDE, which means big, but is really the MEDIUM at Starbucks, which is right next to the largest cup, the VENTI, which is Italian for twenty. As in ounces. Or as in pounds you can attribute to Starbucks over a lifetime or the number of monthly sleeping pills you will need after a lifetime of having one of these for lunch every day.

Interesting fact: In Japan, the SHORT *is* the small, and they don't have the option of the super-sized version, the VENTI. (I wonder if they put it in Italian so you don't have to order the GIGANTIC. Too embarrassing.)

A goal for you readers, this year. I'm not asking you to ditch the Starbucks habit, quit the coffee, or even start eating whole and fiber-rich food to fuel your body instead of your afternoon mocha. I *am* suggesting you give a SHORT a try and save yourself HALF the empty nutrition and a third of money as well. The hard part of changing is breaking a habit. Taking the SHORT challenge will minimize your discomfort by allowing you to still follow most of your animal tracks while reaping the rewards of minimizing the poor nutritional input.

Sorry, you can't get a blended-frappa-creamy-whatever in SHORT. But, a milkshake probably isn't a great breakfast, lunch, snack anyway, now is it? What would your mother say?

Also, when you take a good, up-close look at the SHORT cup, know this. The volume that fits inside that cup is almost TWICE what an appropriate glass of WINE should have in it. But that's a different post, isn't it!

---

1. I've weaned off coffee during the holiday break. Nothing that *three entire days* of sleeping couldn't handle…

It's late, I know, BUT it's also still Saturday…yay me!

A few statistics from today.

- 240: Minutes spent stretching muscles to optimal length for optimal metabolism, circulation, and joint space.
- 30: Minutes spent resting eyes and above-mentioned muscles on the couch.
- 60: Minutes spent walking and hiking at the park.
- 7: Minutes it took to make my fresh beet, carrot, apple, and parsley juice dinner.
- 16: Minutes it took to clean the juicer (that just doesn't seem right, does it?).
- 0: The amount of caloric energy expended, bone cells generated, and motor skills learned by the three-year-old girl riding an electric Big Wheel in the park.

Living optimally is extremely time consuming, I agree. With our national health status being so low, however, it is not necessary to begin with multiple hours of movement or radical dietary changes to make improvements. But for Pete's sake (who IS Pete and what does that phrase mean, anyone?), does a three-year-old child really need to be riding a motorized scooter in the park? Not on my dime.

Until everyone is doing their part, it is essential to make sure that you are doing yours. Fresh off *Avatar* the movie, it seems pretty acceptable to say that we are smaller units of one larger organ. So I ask you, my fellow liver/brain/stomach cells, let's commit to doing our daily walk, and an additional five minutes to pick up the slack. It will come back at you tenfold, I promise!

Come on, post *your* daily stats here too. I just know you did something for your health today!

 A DIFFERENT YOU

I'm giving myself ten minutes to write this blog post. After that, I'm going to take a one-hour walk because it's all about priorities.

Yesterday I went to Trapeze School, a two-hour gift given to me for my birthday earlier this year. The school's tagline is: "Forget Fear. Worry about the addiction." Um, thanks, but I think I'll hyper-focus on fear and be thankful to never come back. I can see why they've hired marketing experts. I won't recap the two-hour course and how simple and easy they made the trapeze for me and my nine classmates (including one woman in her mid-sixties!), but I will tell you how awesome it was to fly, and swing from my knees, and do a catch, and do a somersault dismount. It was not only liberating for my body (hello twenty-five percent of my shoulder muscles that never get used), it was like letting a piece of my mind out of a self-imposed prison I wasn't aware of.

People often ask me if an exercise is "good" or "bad," and I really have to say, there is no "bad" thing you can do with your body. What's detrimental is ONLY doing one thing, whether it's sitting or running or swimming or lying in bed. You've got a 360° body designed for a 360° experience. We're developing pain and disease because we are preventing full use of our muscles and joints by choosing to live within the narrow definition of Who We Are.

Top athletes and couch potatoes die at the same rate and of the same diseases because they have the same habits of choosing one motor program and using it over and over and over and over, allowing all other motor units (a "motor unit" is a nerve and the muscle fibers that nerve innervates) to wither. The mistruth we seem to all share is that some movement habits are superior to others. And while this seems to make sense (it's better to move around a lot than sit down, right?), there are many people who cannot sit. And cannot rest and recover. And cannot unplug their minds, to the detriment of their physiological and psychological health. These people would look like top athletes, yet they are physically suffering. On the flip side, there are those who cannot get up and move forward to dissipate the large accumulation of waste products in their body. Their inertia is high and usually the pain is too. These people would be called lazy, yet they are top athletes in their field, and they have a motor-program coveted by many.

In my life I have defined myself as a runner, triathelete, swimmer, cheerleader, mascot, and coach. I have been an aerobics instructor, kickboxer, burlesque dancer, and treeclimber. I've flirted with Tahitian dancing, yoga, body-building, hiking, bowling, and blogging. This month I was a squatter and every day I am a walker. And I'm proud to be a bookworm, a singer, a restorative exercise specialist, and, always, a nerd. Each of these tasks has required me to be uncomfortable, which is really...well, uncomfortable. I grew new brain with each of these definitions while I developed new motor programs. Here's the secret to aging slowly: A bit of brain is called to action with *every* new motor program, causing the nervous system's tissue to grow and firm up, to develop, just like your body.

Yesterday I was a trapeze artist, and it was the most scary thing I'd ever done, yet the most exhilarating and mind-opening. Today I am asking you to change the way you define yourself, and choose a task that is out of your comfort zone. Marathoners, be a picnicker (without running first) and try to beef up your sitting and relaxing skills. Are you confined to a chair because of stiffness and soreness in the joints? Put a blanket on the grass, in the sun, and do some easy stretching (in fact, everyone can benefit from this one). Today you are a muscle lengthener. Find a dance class or a language class that makes you feel completely silly and uncomfortable. Learn to play an instrument. If you don't read, start a book today. Anything new you do builds a unique motor program (even reading this off a computer screen requires a different set of skeletal muscles than reading off a piece of paper!)

Expand your body and mind. If you are demanding on your body, it isn't as hard to do more as it is to do less. Get it? The tough part is doing less and being okay. Getting out for a daily walk is hard when you're sedentary, but start in small increments. Wellness is not about physically hard or easy tasks, it's about not pigeonholing yourself into one mind-body pattern. Challenge yourself to become different. It's a lot harder than any Olympic event...I promise! Let me know what you've always wanted to do or what you will do today (how about sitting on the floor instead of a chair?) and put it on cyber-paper. We'll all support you! (P.S. Today I am a hobbler...wow...sore muscles ☺)

 ## June 1 2010 DEAR KATY...HALF-DOME TRIUMPH

I get a lot of katysays email with different questions, issues, etc., but my favorite Dear Katy's are those that let me know how Alignment is Changing Your Life. I got a grrrreat one today that I wanted to share. I sent a few interview questions to help flesh out her story, but she was so good, she actually wrote it out (saves me time...thanks Gail!). For those of you who feel like your age or your disease are limiting you from, well, living, and for those of you who truly believe you have "bad" knees and backs... this one's for you!

> I have been a big fan of visiting Yosemite for most of my life, but 20 years ago I decided I wanted to climb Half Dome. I was 40ish and in relatively good shape. So I started training and 2 months before the climb I injured my knee doing the "stair stepper" at the gym. After 3 years of every type of therapy (BTW, I am a chiropractor) I relented and had surgery, which had minimal results. So my exercise life was limited to short walks and short recumbent bike rides. After several years I was able to increase my exercise and started doing longer bike rides. About 3 years ago I developed acute achilles tendonitis that was so severe I was unable to even walk around my house without pain, forget any exercise! Back to trying every type of therapy there is, including apitherapy

(bee stinging—yuk!). No good. Finally a friend told me I should go to Restorative Exercise Institute and have an evaluation. Discouraged from all the failed therapy, I put it off for 3 months but desperation won out. I had an evaluation with Katy and guess what? I didn't believe her when she told me all the things wrong with my posture, in fact I tried to argue with her. She gave me the look and I said, "Right, I'll shut up now." I was internally rotated with tight psoas, hamstrings, gastrocnemius and soleus. But after 1 month of classes I was 90% better. Within 3 months I was 98% better and could walk, hike, bike and anything else I wanted. Wow!! When turned 60, I mentioned to Katy that I had always wanted to climb Half Dome and she said, "So do it!" After 5 months of training and the help of all the instructors at RExI I hiked 17+ miles up 4,800 feet and pulled myself up 500 feet of cables. Thanks Katy, you made it possible. And for all you out there with a dream, just go to RExI and never give up, you can do it.

Isn't this the best thank-you note EVER? (Image bottom left.)

And for all the brilliant program graduates of the institute, Gail has a high-altitude message for you too (image bottom right):

What an inspiration you are, Gail! And just to let everyone in a little secret—your alignment matters! Not in a little way, like your posture could *look* better, but in a huge way, that affects your cardiovascular system, your metabolism, your nerves, your mental health, and, of course, your joints! Are YOU ready to be pain free?

I encourage you to all pick something you'd like to do with your body and do it, with correct preparation, alignment, and exercise prescription. And tons of laughter. (It's so much better with laughter!)

The new year is coming up, which means you're likely thinking of what you can do for better health next year. That's good. I'm going to be writing a lot about that in the next two weeks. But before I go on, I want to share a little song I wrote. Well, I didn't really write the music or anything. The music was written by whoever wrote the song "There's a Hole in the Bucket." If you're a teacher, parent, or have ever offered anyone a solution to a problem, chances are you've met Henry. We can all be Henry, but the plan for 2012 is to keep Henry, and resistance in general, to a minimum!

Here goes:

There's a paaaain in my low back, Dear Katy, Dear Katy
There's a paaaain in my low back, Dear Katy, a pain.
So fix it, Dear Henry, Dear Henry, Dear Henry,
So fix it, Dear Henry, Dear Henry, stretch it.

Well how do I stretch it, Dear Katy, Dear Katy,
How do I stretch it, Dear Katy, with what?
Do the calf stretch, Dear Henry, Dear Henry, Dear Henry,
Do the calf stretch, Dear Henry, Dear Henry, dome stretch.
(Image A, below.)

My calves are too tight, Dear Katy, Dear Katy,
My calves are too tight, Dear Katy, too tight.
Well stop wearing heeled shoes, Dear Henry, Dear Henry,
So stop wearing heeled shoes, Dear Henry, Dear Henry, stop it.
(Image B, below.)

A   B

New shoes cost too mu-ch, Dear Katy, Dear Katy,
New shoes cost too mu-ch, Dear Katy, too much.
Go barefoot, Dear Henry, Dear Henry, Dear Henry,
Go barefoot, Dear Henry, Dear Henry, it's free.

My feet are too co-ld, Dear Katy, Dear Katy,
My feet are too co-ld, Dear Katy, too cold.
So work them, Dear Henry, Dear Henry, Dear Henry,
So work them, Dear Henry, Dear Henry, work them.
(Image C, opposite.)

My toes, they don't mo-ove, Dear Katy, Dear Katy,
My toes, they don't mo-ve, don't mo-ve, Dear Katy, so stiff.
Back your hips up, Dear Henry, Dear Henry, Dear Henry,
Get your weight back, Dear Henry, Dear Henry, weight back.
(Image D, opposite.)

This makes me feel weird now, Dear Katy, Dear Katy,
This makes me feel weird now, Dear Katy, too weird.
That's normal, Dear Henry, Dear Henry, Dear Henry,
That's normal, Dear Henry, Dear Henry, relax.

Do you think I can run now, Dear Katy, Dear Katy,
Am I ready to run now, Dear Katy, whaddya think?
Too soon my Dear Henry, Dear Henry, Dear Henry
Your spine's sick, Dear Henry, how 'bout a nice walk?
(Image E, opposite.)

I don't walk, too boring, Dear Katy, Dear Katy,
I don't walk, too easy, and boring to walk.
Well how about stretching, Dear Henry, Dear Henry?
How much do you stretch out, Dear Henry, a day?
(Image F, opposite.)

My body's too tight, Dear Katy, Dear Katy,
My body has always been too tight to stretch.
How much do you sit now, Dear Henry, Dear Henry, Dear Henry?
How much are you sitting, Dear Henry, a day?

I work in an office, Dear Katy, Dear Katy,
I sit in a desk chair, Dear Katy, all day.
So stand up, Dear Henry, Dear Henry, Dear Henry,
So stand up Dear Henry, Dear Henry, stand up.
(Image G, opposite.)

Don't think I'm allowed to, Dear Katy, Dear Katy,
Don't think I'm allowed to, Dear Katy, to stand.

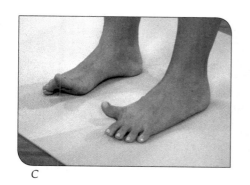

C

D

E

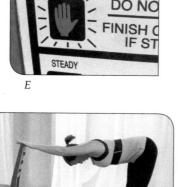

F

G

Are you serious, Dear Henry, Dear Henry, Dear Henry,
Are you serious, Dear Henry, Dear Henry, are you?

Besides that, Dear Katy, Dear Katy,
I think I need more help, Dear Katy, from you!
Just do it, Dear Henry, Dear Henry, Dear Henry,
Just do it, Dear Henry, the things I've told you.

But there's a pain in my low back, Dear Katy, Dear Katy,
A pain in my low back, Dear Katy, it hurts.

Let's keep our eyes open for our personal Henry and commit to staying open, and getting better in 2012! There is a solution. Follow it. The time is now.

Author's note: Apologies to any readers actually named Henry. You can stay, as long as you're, you know, not a Henry.

 HEALTH RECAP, 2011

Have I mentioned that there are only two more days of 2011? I wonder how many more blog posts I can jam into this year.

Did I also mention that 2012 is going to be The Year? Underline the *The* and pronounce it *thee*.

I did an interview last month for a radio show called "30 Minutes of Wisdom." I talked about alignment blah blah blah wonderful life blah blah it's that important (italicize that). You can listen to it here[1], if you're interested.

As I wrote about last month, writing stuff down, especially goals, is hugely important. I don't know why. There is just something very different about writing it—making the words concrete—as opposed to carrying it around between your ears. That's why I was über (OH YEAH!) excited to get some year-in-review questions sent to me by the hosts of the program.

I suggest you, and maybe even your entire family, go through these questions as a New Year's Eve event. Almost as good as artichoke dip, but without all the sodium. I enjoyed doing mine as it helped me to focus on the fact that this past year, I've done some amazing things with my health. I haven't been sitting around doing loads of laundry, although many days that's all I've accomplished. I bet you'll discover a little bit of your own greatness too. I'm sure you can answer these in general terms, but I am adding the word "health" to each question because that's what this blog is about.

Evaluate your health in 2011: Define health as that state of feeling happy in all aspects of yourself—in the body, in the mind, and in the soul, and being pain, disease, and medication free. I'm going to post my answers because, if writing them down causes improvement, who knows what writing them out in a blog post can do!

Completing & Remembering (Your Health in) 2011

1. What was your biggest (health) triumph in 2011?

    Delivering my baby naturally, at home was a physical feat unlike no other.

2. What was the smartest (health) decision you made in 2011?

    Moving out of urban California and into a smaller, more quiet place where I could live mo bettah.

3. What one word best sums up and describes your 2011 (health) experience?

    Maintenance. I wouldn't say I made any leaps or bounds in terms of health this year, but I spent all of my time learning to deal with being a new mom and still finding time in five-minute increments to work on stretching my hips, strengthening my shoulder girdle, etc.

4. What was the greatest lesson (about health) you learned in 2011?

    I think I just learned it last week, watching my son trying to crawl/climb up a box without putting down a wooden spoon he just found. He fell and was frustrated. I actually said aloud, "You can't do it all—you're going to have to put something down." And then I was like, "Ah $h!t. I get it. Time to put something down, Katy Ann." Not the blog, though. I picked "showering daily" instead.

5. What was the most loving service that you performed in 2011?

Probably this blog. I love my family and do everything I can for them, but every post, email answered, and comment read is how I show my love for my human counterpart. To love is to serve. I have information that can help many. I make my living out of it as well, but will always (always) spend the bulk of my time offering free service.

6. What is your biggest piece of "unfinished (health) business" in 2011?

Scar tissue—just a tiny bit (see number 10). I've been working on it, but it looks like it's going to take some of 2012 to heal.

7. What (about your health) are you most happy about completing in 2011?

I wrote a FREAKING BOOK, dude. It totally helps my brain to have it down on paper, where I can just say, You Want To Understand More? Get the book.

8. Who were the three people with the greatest impact on your (health) life in 2011?

Not counting my husband, who takes the best care of me, I'd say my midwife, my favorite bodyworkers (Anna, Cindy, Jenna, Tim), and my baby in utero, because I didn't realize that I could stop coffee, take two naps a day, work less, and feel joyful. But I could, and I did, and it was good!

That was seven people. I have a problem with limits.

9. What is the biggest (health) risk you took in 2011?

Probably driving from CA to WA. I thought my hamstrings were going to revolt and beat the crap out of me. Next time I'll walk, thanks.

10. What was your biggest (health) surprise in 2011?

That as healthy and prepared as I was, I still had a freak complication following my delivery that almost killed me. And that coming back out of it took diligent work, but that I got my body back completely, and in only about four months.

I'm not just the president, I'm also a customer!

(Me, after surgery. This is what your face looks like when you lose half your blood. I don't even have freckles most of the time!) (Image next page.)

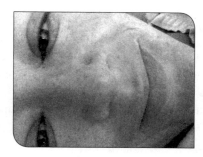

11. What important (health) relationship improved most in 2011?

    Food, food, food! Before we moved, we shopped only at farmers' markets and Trader Joe's, thinking that we were doing awesome. THEN we moved to a place that has only some giant grocers (ick), or lots of fresh produce. Now all three meals a day are made from scratch and are mostly plant-based.

12. What compliment would you have liked to receive but didn't?

    "Hey Katy, nice outfit!"

13. What compliment would you have liked to have given but didn't?

    I didn't get to tell the surgeon who saved my life, "Hey, thanks for saving my life." I know my hub wrote him a letter, but really, I should have found his address, knocked on his door, laid down without any pants on and my feet on either side of his front door and said, "Hey, do you remember me? Thank you for saving my life," because he might not recognize me otherwise.

14. What else do you need to do or say to be complete with 2011?

    Dear 2011,

    Where are all of my socks? I started this year with nine full sets. Now I'm down to four singles? What's up with that?

    Love,

    K-Bow.

Creating (Health in) 2012

1. What would you like to be your biggest (health) triumph in 2012?

    I would like to figure out my fatigue a bit more. There are still areas where I can reduce my physiological tax. I was off of caffeine, but that cup a day has crept back in.

2. What (health) advice would you like to give yourself in 2012?

Dear Katy,

You're taking on too many projects. Or, you need to not be over-whelmed by them. Do what you can do. Look at what is truly important to you and let the others go. Make sure that ANY project you take on supports your greater interest in wellbeing. If a project causes you stress, it's working against your internal desire to be stress-free. Let it go.

Oh, and take a nap at least once a week. The only one keeping you from doing that is you. I promise, nothing great is going on during that one hour anyways.

Love,

Me

3. What major effort are you planning to improve your health results in 2012? My indoor monkey bars are awesome. Now I'd really like a small climbing wall to install itself in my garage or something. Please.

4. What would you be most happy about completing in 2012?

I would like to complete a pelvic-health book, and then spend ninety days following the program to a "T."

5. What major (health) indulgence are you willing to experience in 2012?

Dear Universe,

I would like to indulge in a few weeks on a Greek island, hiking, eating, swimming, and wearing a toga. I will even teach others alignment while I'm there, if they'd like to come with me.

Thank you for your consideration,

K.A.B.

6. What would you most like to change about your health in 2012?

I would like more structured alignment time. I live alignment, but there is NOTHING like setting aside an hour a day for quiet practice. I've lost that in the last year. I will get that back! (I just typed "I'd like to get that back," which sounded a bit like a copout, so I rewrote it…)

7. What are you looking forward to learning (about health) in 2012?

I am most interested in practicing my Vipassana meditation more, and working towards non-attachment. I'm good on "stuff," not so good at allowing others to choose to be disappointed. Reading the Upanishads and a book on Jesus Christ now, hoping to learn more.

8. What do you think your biggest (health) risk will be in 2012?

We are planning a huge effort in Australia, to bring our programs there. I want it to be something I can do while not over-stressing to get it done. I'd like it to happen naturally and smoothly! Go with the flow...

9. What about are you(r health) are you most committed to changing and improving in 2012?

Upper-body strength, baby.

10. What is one as-yet-undeveloped talent you are willing to explore in 2012?

Letting others help me. I have no knack for it, yet it's getting to be quite necessary!

11. What brings you joy and health and how are you going to have more of that in 2012?

Walking, cooking, taking baths, stretching—these all make me healthier than when I'm not doing them. And, all I have to do is do them. No fancy plan needed!

12. Who or what, other than yourself, are you most committed to loving and serving in 2012?

Everyone interested in alignment science!

13. What one word would you like to have as your (health) theme in 2012?

Global.

Now it's your turn. Just go for it! Post one of your answers here to inspire others or get some cheers and praise for your general awesomeness!

(Questions compliments of Robin Blanc Mascari. Please feel free to share! She loves hearing how these are used and what happens. rbmascari@mac.com, enlightenednetworking.com.)

---

1. enlightenednetworking.com/Audio Files/18-Nov_11-36.mp3

chapter 16

MISCELLANY

MAY FLOWERS BRING WEIGHT LOSS?

As spring progresses and the days get longer we are usually renewed with more energy. Nature triggers us to shed our extra layers...including those layers under our clothes. Have you ever wondered why this is? While the sun always arcs through the sky, the height of that arc increases during the summer months, until the first day of summer, when the arc is at its highest—taking the longest time to travel. The percentage of the sun's rays making it to the earth's surface is also the greatest. More rays provide the extra energy needed for plants to grow and produce their fruit. Rays also give us additional warmth so our "extra layer" is no longer needed to keep us warm.

Cold weather requires our body to work harder to keep us warm, which means we need to eat more in the winter. Because our eating patterns become so habitual as opposed to being regulated by natural cycles (as they should be) we often miss the fact that we don't need as much calorie-dense food at this time. Over-controlling the temperature of our environments (A/C in the car and at work) is another way we reduce the natural sweat/waste removal process of the season. Paying attention to your actual hunger levels and spending a little time each day letting your body get used to the actual warmth will only serve your lymph-removal system and your waistline!

HOW MUCH SLEEP DO **YOU** NEED?

The correct answer is "as much as it takes." The quantity of sleep needed is equal to the amount of cells you used throughout the day. Ask yourself:

- How many times did you eat? Talk? Blink?
- How much time do you spend looking at fine print? In non-natural light?
- How much did you exercise or exert your physical body?
- How many thoughts or problems did you process or mull over today?
- How much time did you spend in "stressful" situations, secreting chemicals that need to be processed?
- How many heartbeats, breaths, and steps did you take today?

All of this "living" work expends energy, uses up cells, and leaves waste behind. Sleep is the body's time for repairing and restocking all your cellular tissues.

You can use a fairly simple equation.

Sleep hours needed = # hours required to undo damage.

Say you had a particularly tough day and didn't get enough rest. The next morning you wake up with damage left over from the day before. Now you are facing the next day with only ninety-five percent of your cells regenerated. This damage accumulates! One good sleep still isn't enough to repair hours of missed sleep. When you don't get enough sleep at night you encroach on your body's regeneration time. You have to be asleep for your body to repair. Didn't get a full night's sleep? Then you didn't regenerate your cells entirely, either. Over time this damage accumulates. If you've slept an hour less than you need to for say, oh, the last seven years, then your body is behind 2555 hours of sleep. And this sleep debt doesn't ever go away, and slowly begins to stress the body. Following decades of missed hours, you are left with a body that can be running on half of its full potential—leaving organs and tissues working at below optimal levels and the task of regeneration to the daytime (which leaves you feeling sluggish and fatigued). Many clinical studies have demonstrated the effect of a lack of "enough" sleep on happiness, health, blood pressure, injury, accidents, and STRESS! If you skimp on the sleep long enough, you may find your body shutting down in the middle of the day—trying to get the rest it requires in order to continue to live.

Q. What do we do with our midday fatigue?

a. Head to Starbucks for a Frapp-Cost-a-lot double shot and a cookie.

b. Throw a temper tantrum in the office, accidently send a personal email to the boss, and stub your toe on your desk.

c. Take note, skip the TV that night, and be in bed by ten.

Uh, the answer is "c." Meeting your requirement means tuning in to what your body is trying to tell you. If you have a hard time waking up in the morning, then you are probably staying up too late! We are not nocturnal animals, so we should wake with the sun and rest upon nightfall— that's why our eyes have developed the way they have, and are not meant to be open after the sun goes down. If you are staying up into the wee hours of the night, then turn off the computer and television. It may take a few nights to restore your natural sleep rhythm, but most "insomniacs" make the problem worse by stimulating their exhausted selves.

I think many of us skip sleep because as kids we understood that

"There's a lot of fun stuff going on after bedtime that I'm not allowed." You know, like TV shows, parties, and pony rides in the living room.

Suggestion: Check out the activities you are prioritizing over RE-GENERATING YOUR CELLS.

There isn't anything on YouTube that's worth it. Trust me.

21 IS THE NEW 81

They say a picture is worth a thousand words. I say this picture is worth a thousand pills!

That's right, ladies. Get your fashion on...right after your adult diaper. Can't breathe because your jeans are too tight, or because your lungs are not able to inflate all the way due to the fact that your diaphragm is up to your armpits? And what goes best with this outfit? Cute, dangling earrings OR a shiny new walker. That's right—gold is in, so you better start saving!

(YES, this picture was taken in the local mall this week. Yes, this is a real mannequin, and YES, Houston, we have a problem...)

YOUR MISSION, SHOULD YOU CHOOSE TO ACCEPT

As a scientific-minded person, I am usually pondering the deep, unanswered questions about how we move, why certain a disease occurs, and how culture affects a population's health. And about a month ago, I came up with a REALLY good idea for a scientific experiment.

What would happen to one's body if they were to get a massage *EVERY DAY* for a week?

I know, I know. I am dedicated to my work, it's true.

While many people have had a massage at some point, I am always surprised at those who have not. Massage is an extremely old practice, with Egyptian hieroglyphics depicting this ancient therapy. *The Yellow Emperor's Classic of Internal Medicine* (2,700 B.C.) recommends the massage of skin and flesh as a treatment for various ailments, and Hippocrates in the 5th century B.C. "The Physician Must Be Experienced In Many Things," wrote "but assuredly in rubbing...for rubbing can bind a joint that is too loose, and loosen a joint that is too rigid."

In 1874 medical doctor Andrew Taylor Still began a practice that we now call osteopathy. This practice had strong roots in physical manipulation for the purpose of reducing any anatomical deviations. The common practice still operates today under the premise that mal-alignment of tissues creates disease.

A quick lesson on circulation: A healthy body is constantly circulating. And while we know the cardiovascular system has the heart as a pump, it is really the action of muscle tissue that helps the blood evenly distribute throughout the body. If you have muscles in your body that don't contract (and most of us are only using a small percentage of the total muscles available) then those areas don't have fresh, circulating blood. These unused areas tend to have an accumulation of waste product, and can be pressure-sensitive.

The process of massage is a "faking it" of sorts. By pressing on a muscle, or pulling on the skin or fascia (complex tissue underneath the skin and throughout the body), a massage therapist gets the muscle to compress the vessels located inside the area being manipulated. When compressed, the smallest blood vessels respond by opening up, allowing blood to come in and waste to leave. Your massage therapist is basically doing the work for your muscles, which is why bodywork is best done in conjunction with improving your muscle-using habits—either using them more, or using them more correctly.

So, going back to massage history, how did this health-care system fall into the category of "luxury"? And don't get me wrong. I love the heated massage table, padded face cradle, aromatherapy candles, and the Zen soundtrack. But again I'll ask, when did this simple, healing treatment get placed under the same category as a spa manicure?

With all of this in mind, I started out to see what a daily dose of ancient medicine felt like.

Day one: My health issues tend to happen between the ears. Headaches, jaw tension, and constant throat fatigue are my ailments of choice. After discussing these with Cindy R., she began what I can only describe as the best head and neck massage I have ever felt. The muscles in our heads don't do much—even though there are many movements they are designed to do. We tend to use the same facial expressions and speak the same language (learning French is KILLER on the muscles around the chin!) so there tends to be a lot of tension here. Using a myofascial release technique developed by John Barnes, Cindy R. did more than an hour of mostly head, neck, and shoulder work. She found an extremely tense place in my mid-back and we discovered I had an old, ingrained habit of lifting my own shoulders…all the time! At the end of the session, Cindy told me not to try to "fix" the problem with exercise, or even really think about it, but just to allow the time to feel myself in relation to my shoulders. Got it.

About ninety minutes later, while hanging from the monkey bars to fix and stretch my tight shoulders, I got a clue. Oooohhhhh, this is what she was talking about. So I dropped off the bars and walked home right then. I'm thickheaded, it is true. Maybe that's why I'm always tight there. But less tight that day, for sure.

Day two: I met Michael K. in his office for a Swedish massage that really demonstrated his expertise in anatomical science. When I got on the table I noticed the head cradle was fairly low. How clever, I thought. His table allows your head to hang forward to relax your neck during the massage. About fifteen minutes into the session, he said "Wow, that head cradle is really low." I asked him if he arranged it that way. "No," he said, "I borrowed the table from [a friend]," he said. And then this was the best part…

"It's really old." He paused. "From the seventies."

I, too, am from the seventies.

It probably goes without saying that Michael K. is not.

Anyhow, the second part of the massage consisted of excellent work on the sternal area. The sternum (where your necklace usually hits) is a part that doesn't get worked very often, especially deep and in between the ribs that attach there. By the end of my session, and for the entire next day, I could feel this area expanding with every breath…just like it is supposed to. Evidently my tight neck and shoulders were keeping my ribs from moving to their full distance! But not for the rest of the week, I noticed.

Day three: Ronelle W. is also trained in John Barnes' myofascial technique. I was feeling a bit "monthly," if you know what I mean, and I was having a bit of abdominal and pelvic cramping at the time. I let her know what was up, and she started by working on my abdomen. Now I know that many massage therapists don't touch the belly much—especially if it is your first session or two—but I have to tell you, most of us have so much tension in the psoas (the deep abdominal and hip flexor located here) that abdominal work is a necessary thing! Ask your LMT (Licensed Massage Therapist) if they are trained to and will work in this area—it makes a wonderful treatment.

After an hour, Ronelle had released my psoas, the fronts of my thighs, and my external pelvic muscles. I walked into her office feeling like I might need a Tylenol within the next hour, and I left feeling great. It turns out I didn't need to put anything into my body to feel better. I only needed something taken out...tightness! Thank you, Ronelle!

Day four: Something strange is happening today. Relaxation is making me whine a lot. I was actually too *tired* to get a massage this day. It was *too much trouble* to get to my appointment, and I *needed to get my work done*, and *why did I have to stop working* to go *all the way* to Tim's office (which, did I mention, is upstairs inside my office building?) to get my body worked on?

This says a lot about me, and it may say something about you too. I am so used to working frantically and constantly, the notion of health (enter the word exercise, yoga class, cooking a good meal, getting a massage) seems too overwhelming. One more thing to do. But just like all of the things listed above, once you do it, you feel much better than you ever expected!

So I whined and whinged up the stairs (all twelve of them) to more than an hour of some of the deepest leg, calf, and foot massage I have ever had. And because it was massage number four, very little of it hurt. I could tell he was working, and working deep. Tim is an anatomy teacher, so he was tracing in between every muscle. This normally would have caused me to yelp, but because my external muscle armor was relaxed from the previous three sessions, I had very little resistance.

At the end of the session, not only was my superficial tension gone, the deeper tension (which I hadn't even been aware of the week before) was beginning to fade. Not whining anymore, am I?

Day five: I went to visit Anna. "Oh Anna, what are you going to do to me today?" I thought. But Anna is a pro, and knew that after four hours of massage, it would be time for an easygoing, all-over session that supported and helped integrate any muscle-fiber changes. In the middle of this excellent session, on this, the final day of my massage experiment,

I felt like she wasn't even touching *my* body. If you've ever had a massage (or even a shoulder rub), you know which areas of your body will be sore. But all of my "spots" were GONE! As she worked my entire body, everywhere I expected to hurt, did not. And near the end of the session, she had really honed in on some new spots—areas of old, OLD injury that had gone unnoticed for I don't know how long. But now I am aware, so let the healing begin!

A couple of comments:

1. By the end of the week I observed that I was extremely relaxed in my mind and in my attitude. Things that would have normally brought on a headache or jaw and neck tension didn't even phase me. And my breathing pattern was very deep and slow.

2. About the cost. I have taken spa weekends and paid three to four times more for much less than I received during my week of massage. A massage can range from sixty-five to ninety dollars an hour, and I felt extremely positive keeping the energy of my money in my local community. I also appreciate the hours of training and overhead these businesspeople have to pay for. A week-long commitment to my muscle and neurological health and physical self-analysis was well worth every penny!

Flashing forward to today (a week later), I'm still feeling super-aware and super-toned. I use my muscles A LOT, but there were some puppies in there that hadn't been doing a thing. With the five hours of "faking it" via massage, my muscles are much more toned and defined (thanks guys!) and my sternum is still moving when I inhale. Ahhhhhhhh. I've decided to do a week of massages once a quarter, to continue to address my own level of health. I think of it as my contribution to our national health-care costs.

Next week, take a look at how my second brainchild goes...A Week of Milkshakes.

 **LES VOYAGES**

Oct 22 2009

I'm back from my vacation, and what an adventure it was.

And in addition to being great, I found it completely amazing how physically exhausting a vacation can be. Perhaps we should use different terms—*vacations* (not working, no blackberry, no iPhone, no email, and no appointments!) and *trips* (not working, but tour bus pick up at 6:30 a.m., plane rides, bus rides, and scary car episodes with a French taxi driver who insists on taking all the Corsican island curves—which, did I mention are only one lane—at one hundred kilometers an hour while looking BACK at you so that you can confirm that yes, the island is beautiful and yes, I am completely relaxed…just not right at this moment).

Sooo, I guess I'm back from my trip.

Some trip details:

- Countries visited: two (Italy, France)
- Servings of gelato: ten (pistachio, strawberry, nutella, lemon, hazelnut, pistachio, hazelnut, hazelnut, tiramisu, hazelnut)
- Best, oh-so-Italian graffiti (it was written in English): "I like myself." I'm not kidding. That's what it said, written on the cement tunnel wall, leaving the Rome airport train station.
- Times I've said "You know in Rome, they (do it, think it, say it, drink it, etc.) like this…" in the last four days: 1,352

So we all know that a vacation or a trip is great. But what keeps many folks from doing it often (besides the job, the family, and the money) is how physically taxing the whole thing can be on your body!

I've put together a few tips on traveling to lighten the proverbial load of travel.

1. Save the Neck! If you tend to be stiff in the neck and shoulders, racing through the airport, carrying excessive weight, and sleeping in an uncomfortable plane seat can make it worse—and who wants to start their vacation or trip on an anti-inflammatory?

   Watch your hand placement on your roller bag. Many people pull their luggage with their shoulder joint internally rotated—not good for the rotator cuff or the trapezius muscles. How can you make sure your joint's externally rotated? Reach back to grab the handle keeping your thumb pointing away from you, as opposed to holding your roller with the thumb pointing toward the thigh (images on next page).

   Non-symmetrical loading can tax your shoulder girdle. Even if you prefer a purse when you arrive at your destination, consider packing a backpack for the plane, and place a smaller, single-shoulder bag in your bigger luggage. Evenly distributing the weight can help you use your larger core muscles to carry the weight, and prevent the smaller muscles on one side of the neck from tensing and pulling your neck vertebrae out of alignment.

2. Take care of yourself on the cellular level. Flying is very dehydrating and the constant vibration can tax the muscular and nervous system.

   Avoid alcohol and caffeinated beverages. Skip the glass of wine and instead opt for herbal tea. I always pack my own peppermint tea bags, as all planes have hot water.

   Did you know the temperature is VERY low at thirty thousand feet? Help regulate your body temperature by packing warm wool socks. After I'm airborne, I like to kick off my shoes and slip on something a little more cozy. It also helps me sleep on those long hauls.

   Moisturize well before departure—that means inside too! Right before I fly I super-dose my skin with a good moisturizer (I use sesame or coconut oil) all over. I also eat an oily breakfast, as cellular hydration requires a healthy and solid lipid layer! Avoid dry pretzels and crackers and instead opt for healthy-oil-rich nuts, avocado, or a bag of olives.

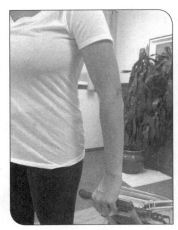

Holding your bag this way can damage your shoulder.

This is much better—your neck and shoulder muscles will thank you!

Good luck on your next travel adventure.

You know, *in Rome* people take a minimum of four to six weeks of vacation per year.

I'm just sayin'.

I just finished moving and I am shocked to say that I own over a hundred cups. Not just cups, but glasses. Mugs. Tumblers. Christmas sets. Garage sale treasures from the seventies—wacky gold-rimmed and peacock-feathered etched glass—and my coveted Tom and Jerry's eggnog set from the fifties. Now, I'm not a family of ten, or five, or even two. I am just me. And while I like to stay hydrated for sure, there is no way I drink out of more than my three favorites: my pink mug I bought for five bucks after a great, memorable meal with my dad, a plastic outdoor tumbler set my mom bought off the bargain rack at Target (you can drop them and they don't break!), and an old wine glass when I want to make my Trader Joe's cherry juice feel "fancy."

I started thinking about this "cup" phenomenon, looking for additional meaning. Why am I drawn to them? Thinking about this while unpacking (I had a lot of time to think, putting them all away!), I remembered something else. When I was six or seven, my friend Susie Carvalho had me over to her house after school. Her mom made us dinner and poured us all a tall glass of Kool-Aid (OH YEAAAAAH!). I kid you not, I drank the entire glass before she finished serving up our food. And then she refilled my glass. I drank the entire thing in less than two minutes. I remembered stopping to look, noticing that I was the only one who had had more than a few sips. I wondered if there was something wrong with me. Why was I so thirsty?

When I was in college, I drank an entire gallon of water a day, easily. I was also fairly athletic and training daily, so I chalked it up to my required water intake. If you take me out to breakfast or dinner, you can watch me unconsciously consume four to five glasses in thirty minutes (I do like to be taken out to breakfast or dinner, so please call if you'd like to check out this amazing phenomenon). And I can tell you that I am not thirsty. But if you put water in front of me, I will inhale it.

I'm not actually thirsty (I don't go looking for fluid, but if it's there, I will take all of it in), yet am strangely drawn to fluid. I spent years studying the physics of flow—electricity, blood, and lymph. I was a competitive swimmer. If there is a pool around, I have to be in it for hours at a time. I need to live by a large body of water, or I feel trapped. Oh, and I'm a Pisces.

So, back to the cups. I never knew I had so many cups—I am not a conscious collector. Why am I drawn to them? I can tell you with one hundred percent confidence that it is not due to the fact that I like to do dishes! There is no conventional explanation for WHY, but there may be a very simple "Occam's Razor"-like explanation. Anyone? Or, if you become aware of a similar phenomenon with yourself—being drawn to a certain substance or item for seemingly inexplicable reasons, let me know. Think of it as data collection. And data collection makes one thirsty... right? Or is that just me?

Dec 30 2009 — ALTHOUGH OXYGEN IS FLOWING, THE BAG MAY NOT INFLATE!

I love traveling, but I don't really enjoy flying all that much. It's the sitting and the confinement and...the turbulence. But I have to do it quite a bit, so I figured I would try to get over my dislike. I tried ignoring it at first, but I think the Universe was trying to teach me a lesson on awareness, so it decided to shake things up, just to make sure I was paying attention. Then I tried hypnotherapy, which wasn't at all like I thought it would be. It was a simply an hour set aside to help me understand what I didn't like about being in an aluminum object shuddering at 37,000 feet. while moving at 350 miles per hour.

I once had someone tell me that his *favorite* part about flying was the turbulence.

I wanted to punch him in the ribs.

What my session with a hypnotherapist allowed me to experience was how tense my muscles became when the bumpiness occurred. In fact, the mere thought of choppy air during my session was enough to get my muscles all bunched up. My muscle tension was actually making the movements seem greater than they actually were because I was resisting the motion of the plane instead of going with it. THAT made a lot of sense to me and I am happy to say that now when a bumpy sky is on my flight plan, I can relax into it and barely feel the movement. I suggest you try it next time you fly, because it really works! While flying, I now play with my muscle tension and pay attention to my shoulders (keeping them down from my ears) and my legs (not gripping them like I'm going to keep the airplane in the air with my tight quads).

So, I just learned something new just now (I'm flying right now, if you hadn't realized...) and wanted to share it with you. *Turbulence is not a continuous activity.* If you really pay attention, you'll see that within a bumpy bout is a pause of gentle floating. The only thing you are feeling during turbulence is the shift from one quiet position to another. Thinking that turbulence is a minute-long or an hour-long activity means you aren't paying attention to all the smooth times in between!

Hmmmm…where else could we apply this concept?

Buckle up, it's going to be bumpy life!

April 22 2010 ARE YOU **FIT** FOR AN EMERGENCY?

I haven't had a TV for a while now. It's a pretty good feeling most of the time, except when you realize you know nothing about the volcano in Iceland. My friends thought I was joking, but I'm pleading guilty on this one. Evidently my hours on the internet are so super-directed to the task at hand, that I'm not even getting to Yahoo News. I've been fooled there before. But back to the volcano. And to the earthquake in China this week. And the earthquake in Chile before that. And the earthquake in Haiti before that. Stuff is going down on planet Earth, don't you think?

Pressure is a really cool thing. It's not so cool when pressure is regulating itself underneath the ocean near your coastline or under the foundation of your house. We understand regulating self pressure, as demonstrated in comments like "I was just blowing off steam" or "I need a moment to decompress." We know what the planet is doing. We all know teenagers, or at least were one once.

I grew up about eight miles from the San Andreas Fault. I have had my share of minor earthquakes and regular aftershocks, and I was there for the big one that knocked my hometown over. Yet I don't have a plan. When I see pictures of Haiti, though, it's clear we need to have a plan. Today would be a good time to heed all warnings.

Plan 1: Pack a backpack with an old pair of walking shoes, a change of underwear and socks, matches, a few cans of food (and can opener), toilet paper, duct tape, water (which you should change regularly), candles (or flashlight and batteries), old gardening gloves, thermal wrap, spare glasses or contacts, hard candy and fruit leather, a water purifier kit, medication backups, and photocopies of important documents.

Plan 2: Set a meeting place. If something happens within your town or state, where should members of your family meet up? Imagine not having phones or email. What would you want everyone in your family to know about where to head, either by car or foot? Who is an out-of-state person you can all try to leave word with? Choose and share this information with everyone in the family.

Plan 3: Your physical health is your greatest asset in an emergency. Are you strong enough to haul yourself or a family member up and out of a structure? Can you lift a reasonable amount of weight? If you need to walk five or six miles to a "clear zone," are your joints pain-free and muscles strong enough handle it? Try a five-mile walk with your family once a week to develop muscular endurance of your legs and when you come across a park or school with monkey bars, give yourself a strength assessment.

Are YOU fit for an emergency?

May 13 2010 TODAY'S WORD IS: MINIMALISM

Did you know that we can walk to many of the places we go? Even in beautiful-weather California, people are driving distances less that two or three miles, seven or eight times a day. We know walking is better for our health, for our environmental conditions, and for prosperity, but, come on man. It's my car. Lesson: Walking improves the life of the planet, but we're pretty dependent on our cars. (To clarify: Walking for exercise doesn't do anything for the environment. You have to walk *instead* of driving.)

June 1 is National Go Barefoot Day. Some people are saying this day is to bring awareness to the fact that there are many suffering without shoes. Some say this day is dedicated to the notion that your feet can adequately to support your body. No extra equipment needed. And minimalistic footwear is moving beyond a trend and becoming part of a whole-body health regime, like taking a vitamin. Have you taken a barefoot walk today? Even around the house? Have you stretched and flexed all the muscles in your feet? Did you know that twenty-five percent of your body's muscles are from the ankle down? Lesson: Barefoot walking improves the nerve and muscle health of the entire body, but we're pretty dependent on our shoes at this point.

Discovery Channel's *Last Man Standing* competitor Corey Rennell started a company following his graduation from the nutritional department at Harvard. The company makes whole-food bars based on the food we have required for two hundred thousand years: large quantities of plants, grasses, and grains, and very little meat. I suggest that everyone read this synopsis of his research.[1] He's brilliant, and his all-plant, whole-food Defender bar (raisin!) is the best meal replacement I have ever had (order some, you won't regret it).

Corey's driving force is the knowledge that many of the affluent diseases are a result of the overprocessed, over-sugared, and over-caffeinated calories we consume (she types as she finishes her frozen yogurt). Lesson: Whole, unprocessed food gives us optimal health, but we're too busy and too dependent on fast and less-nutritious foods.

Last night a group of pals and I went to see *Babies*, the new movie documenting a year in the life of four babies across the planet: San Francisco, Mongolia, Tokyo, and Namibia. As a human-movement scientist and aspiring medical anthropologist, everything about this movie fascinated me, from the nursing positions (this African mother nursed her baby in the vertical position) to crawling styles to the learning of gait patterns. By the end of the movie, I wanted to move to Mongolia and have my babies there (where interestingly enough, getting cellphone technology is much less of a task than developing a sewer system). After what can only be referred to as the best all-baby episode of *Siblings' Funniest Home Videos*, it seemed pretty clear that we are overdoing it in terms of parenting, with the classes and the noise makers and the _____. You fill it in. My favorite part was the mom was reading *Becoming the Parent You Want to Be* while her baby sat at the end of the bed. Classic.

Lesson: People have been giving birth, all over the planet, to happy and healthy babies who do not require special equipment, joggers, changing tables, or Johnson & Johnson soap, apparently. One mother squirted breastmilk on the face of her baby to wash it. Good idea, as breastmilk contains heavy oils that remove dirt from the surface while keeping baby's own skin oil from being stripped away (which is good, cuz that's your first level of immunity!). I hated when my mom licked her thumb and rubbed my face clean, so you'll have to pardon my gut reaction of GROSS! But from a scientific perspective, it makes sense.

At a quick glance, it seems that in our seeking of optimal wellness, we are not, in fact, discovering new things, but discovering that where we once were was where we need to be. I love aspects of the modern world (I have a very sick and intimate relationship with my MacBook), and who doesn't love antibiotics? Not when you have a little sniffle, of course, but if you have ever been stabbed with a rusty knife, or impaled on rusty farm equipment, trust me, you're going to happily hobble to the local hospital and be grateful for the trained medical professionals. But even with all these "perks" of advancement, I have a healthy-and-perhaps-genetic longing for things not readily found here, like space, a quest for food, and a barefoot walk over dusty rocks with a baby strapped to my back. Freedom from anxiety.

---

1. corefoods.com/?q=research

HAPPY ANNIVERSARY, 21

I posted this picture about a year ago, but after shopping in a department store yesterday, I was motivated to post it again. I took this photo at a Forever 21 store in the mall. Evidently, 21 is the new 81.

I can't say what it is exactly that disturbs me most about this photo. Can you? While this mannequin, to many, simply shows a glum lady in a dress, "she" is actually speaking volumes about the state of health of our future generation. The curve to her spine screams "disks quickly degenerating here!" Inside that tilted pelvis lives a tilted uterus. And with all the torso's weight out in front of the hips, you can say bye-bye to the bone density in those vertebrae and hips. Geez, remember when being 21 was fun?

According to Robert Hoskins of the Fashion Institute of Technology, "Getting people to identify with mannequins has always been paramount. They must convey idealized images of ourselves, what we aspire to rather than what we actually are."

Um, I'm sorry, but I don't *aspire* to appear this defeated. Sometimes it just happens. ☺

And FYI, if I could be Forever Anything, it would probably be 34.

This weekend I have met the enemy, and its name is the 1908 Sears and Roebuck's Catalog.

While traveling this weekend I found this old catalog and started to read through it. If you like history and anthropology like I do, you will find that nothing captures a period of time like the stuff for sale. This catalog is phenomenal. You can find just about anything in it, from farm equipment to mandolins to toiletries. This book is like the internet. If you can't find it in here, then it probably didn't exist in 1908.

As a people, we all believe certain things because we think these ideas have always been, but when studying things like this catalog, an anthropologist can collect data showing the dates and global locations such notions developed. This book is an anthropological gold mine, so I thought I would share a few of the most mindblowing.

First off, there is an entire section (pages and pages) of trusses. What's a truss? Yeah, I didn't know either, but from the picture it looked like some kind of belt. Turns out, a truss *is* a belt. A belt with adjustable pads to change where the wearer would like pressure applied to the abdomen, for the purposes of containing a rupture, like an abdominal hernia, diastasis, etc. It turns out people didn't know how to use their transverse abdominals in 1908, either. You can get one of these bad boys for infants, those who walked a lot, or my favorite, the appendicitis truss. This guy "is newly invented and now meets a long felt want. It affords just the right pressure and protection needed for all persons subject to attacks of this dangerous disease." Ow, my appendix is inflaming! Don't worry, I'll just put on this here belt and that should take care of this dangerous disease. Nice.

All you redheads out there will also be happy with C.H. Berry's Freckle Ointment. And I quote:

> Nothing makes or mars a woman's face more than the quality of her complexion and it is absolutely impossible to have a really pretty face, a complexion the envy of her friends, if the face is covered by freckles.

Wow. That's good to know. Thanks a lot, S & R spirit-breakers. I love freckles. Doesn't everyone?

Okay. It's gets better. How about the Genuine Old Comfort Body Brace. Hold on to your hats...

A woman's general health, strength, grace, erectness and beauty of form, are regained and retained by wearing a properly adjusted Genuine Old Comfort Body Brace. It will meet and remove the cause of weaknesses and organic displacements by applying its strengthening influence and natural support to parts of the body where it is most needed [i.e., where your muscles are]. The Genuine Old Comfort Body Brace is highly recommended by physicians for all women suffering from general weaknesses to persons whose shoulders droop and whose posture is neither natural nor correct. It is a proper and comfortable brace for fat people. A large abdomen is often reduced a few inches per month as a result of relief afforded by the brace to the stretched and overloaded muscles.

Men, don't fret. There's stuff in here for you too! No, nature did not design you well, or else you would have better natural support to all of your, um, suspending items. Yes, there are over twenty different suspensories to choose from. Silk, cotton, special models for slim fellas. But the best is the Genuine O.P.C. Suspensory.

The best suspensory made. This is the one to buy. This Genuine O.P.C. Suspensory should be worn by every healthy normal man. The vital organs need a suspensory to sustain the nervous vitality, energy and force and prevent strain. It accurately fits the parts [because nothing says non-vital like constant fidgeting with one's suspensory], supports without strain or pressure. It is a kingdom of comfort in itself, a source of satisfaction at all times.

A kingdom of comfort? I am totally in. I'll take two. Oh, and P.S., the Genuine O.P.C. advertises that there are no buckles on sack. That's nice.

There are over twenty-eight styles of corsets to choose from, depending on the type of figure you'd like to have. (Image opposite.) Abdominal reducing (recommended for stout figures), easy fitting (for those who have a tendency to split them down the sides), perfect form (good news, the bust will not cave in and this one is approved and endorsed by physicians and health reformers!), and my favorite, The Martha Washington Misses' Corset Waist For Girls. This corset "combines all the good qualities of a waist, is trimmed at the top with pretty edging, and creates a waist that is helpful and will help the girl grow as she should grow."

Need I say more? And whatever happened to Roebuck?

Sept 21 2010   NIGHT OWL

People are always asking me things.

"Katy, what type of shoes should I wear?"

"Katy, what does the psoas actually do?"

"Katy…can you please *share* the ice cream?"

Out of the many questions I get, there seems to be a lot of interest in how we should be sleeping. Well, you should probably be sleeping more than you are right now, says the most current research on maintaining physiological homeostasis. Because we are a "night-lighting" culture, we stay up way past our biological bedtime (sundown), and have for most of our lives. And, P.S., sleep deprivation accumulates. If you partied hard in college, have a hundred kids, or "heart" Nick at Nite, then you may be in debt. Pay up in little chunks, and P.P.S., naps don't count. It has to be the REM sleep (rapid eye movement—not sleeping while listening to "Night Swimming") in order to repair damage done to your tissues while you were awake. Why REM? Because your body does not move during this sleep (except for your eyes, as the name would imply, which are moving rapidly…) This relaxed state of skeletal muscle is essential for long-term biological function (which is the fancy term for health).

I was obviously NOT getting my REM last night, as I woke up first at 2:00 a.m., with the top bedsheet wound completely around my arm. Must have been dreaming about being a pitcher. The baseball kind, not the Kool-Aid kind. OH YEAH!!! (Sorry if you didn't get that joke, as it was really good. For me.) I got up, went to the bathroom, got a drink of water, rearranged the sheet, and went back to sleep.

I believe I made a mistake last evening by "winding down" with an hour checking up on friends and family on Facebook. Sounds like a good idea, right? Well, after I fell asleep, I had the following dream:

I was driving to the mall. But it wasn't a regular mall, it was a mall inside a parking garage, and it was full of post-fair equipment (thanks, whoever posted their FB pics showing rides from the fair). I wanted to buy a lemonade and a cinnamon roll. You know, because that's what people do at the mall. I had to walk around a lot because there weren't any actual stores in the parking-garage mall, only closed fair booths, so it was a good thing that I had my portable DVD player playing the movie *Oklahoma*. It was the part where Will comes back from Kansas City. Lots of good singing and roping. It was such a good part, a little girl (who was kind of like that creepy kid from *Close Encounters*, which I watched the last twenty minutes of before I went to bed because *somebody* saw a UFO that very evening and when I started singing the creepy music part from *C.E.*, he didn't know the reference, and needed to be educated) asked if she could watch the movie. I said yes and left her on a giant tire with the DVD player, which was now a stuffed unicorn, while searching for refreshments.

I then found myself hanging out with Jill Miller, of *Yoga Tune Up* fame (hey Jill, I saw your post on FB—happy anniversary! Is it weird to see your life in someone else's dream?) talking about skeletons, and then on to a party in a cabin, which turned out to be Gil Hedley's house. Please note, I've never met either of these people, nor been invited to their houses (and don't expect to now). At Gil's house, however, I needed to use the bathroom, but they were all out in the open, around the edges of giant lecture rooms. And all of his toilets were fuzzy or hairy or something. Ew. I'm pretty sure this came from a pre-bedtime view of Gil's video (watch some cool cadaver and fascial science here if you're not too squeamish[1]), and the hairy thing? From his beard, of course. Duh. After looking all around the lame open-toilet party cabin and not finding any food…

Woke up. Have to eat something now. Something with peanut butter. And a lot of water.

Try to go back to fall asleep but now I can't. Everything is too hot and too achy to sleep on my right. Or on my left. Or on my back. Plus, I'm trying to remember the dream so I can blog about it later. Now my mind is going. Oh, and did I mention I'm pregnant? Wait, do I have to pee… again?

You've all been in this situation of an un-restful night's sleep. If you get here, don't panic. Panic just makes it worse. You're going to do a LOT better if you don't layer stressful thoughts about the fact you're not sleeping onto the fact that you're still awake. Try arranging the covers to minimize asymmetrical pressure. You might have to get out of bed to do this, but calmly even out blankets, sheets, and pillows, smooth PJs, and climb back into bed. Stretch yourself out in savasana (corpse pose in yoga), and with your eyes closed, let your brain take a look at each part of your body. Some parts are easier to "see" than others. It's a great way to improve the brain-body connection, and if you really like this game, sign up for a Vipassana meditation course, where you get to play with this connection all day, every day, for ten days. It's lifechanging, really.

Sleeping the correct amount (or at least longer than you typically do) is a good place to start when trying to get to the root of *any* health issue. As for body postures, it's best (meaning, does the least to shorten muscles and stiffen joints) to sleep flat on your back, no pillow, on a firm mattress. Sleeping in this way (or just getting into this position on the floor when you're awake) reveals a lot about your chronic joint position. If you need something under your knees to be comfortable, your psoas is too short for your height. If you need a pillow under your head to keep your chin from elevating, the cervical (neck) extensors are too short for the length of your spine. If you take yoga, executing the supine savasana posture takes a good chunk of time to work up to. You've been practicing chair-asana (the art of sitting long hours with your hips and knees at ninety degrees, head forward to the spine, and chin elevated) ten hours a day, it's no wonder you feel stiff getting into bed. Do five minutes of light stretching followed by a "floor assessment" of your tension patterns. Set the timer for ten minutes and relax your parts to the floor, breathing quietly and thoroughly, before hopping into bed. This should make getting into REM state that much easier.

Nighty-nite!

---

1. youtube.com/watch?v=_FtSP-tkSug

IMMUNITY BOOST

It's getting to be that time on Mother Nature's watch. Indian summer winds down and the winds kick up. Just realized right now that wind (like a clock) and wind (like blowing in a storm) are spelled the same way. Which is tripping me up and making me lose my train of thought.

I flew home through a thunder and lightning storm last week. Thunder and lightning in California. In L.A., if you can imagine. It was awesome, once I was no longer flying through it. I haven't noticed a large population of sick people yet, but as winter weather starts to develop, colds tends to develop too. That's why I thought it would be fun to talk about your immune system today.

What is this "immune system," anyway? If I had about thirty hours and the ability to type ten thousand words I could tell you, but frankly, I have to do laundry. So, to simplify, your immune system is all of the organs and actions that go into protecting you from the super-tiny, almost invisible things in your environment. There are a lot of things that can weaken these systems, including:

- poor sleeping habits
- alcohol consumption
- stress
- high-intensity exercise, and
- poor nutrition.

There is also one major mechanical component to a well-functioning immunity. No, it's not wearing that necklace from *Survivor*. The superhero of health this season is your lymphatic system. The fluid in this system, lymph, is the worker bee of your immune response. When this fluid fails to circulate, the body has a decreased ability to fight off various ailments.

"Circulate" is the key word here. The word implies movement, and a lot of it.

Here's the deal with lymph. It doesn't move very well on its own. It doesn't have a great pump like your heart, so the less you move your muscles,

the more your lymph resembles scuzzy pond water.

How are we feeling?

Another deal with circulation is you have to move all of your muscles in their ranges of motion to get the lymph to move. When your muscles are tight, limiting the motion of a joint, the lymph tends to back up in that area. Some not-so-great news: Your largest lymph node clusters are at located at the areas we tend to be tightest—the neck, the armpits and chest, the groin, and in the ribs.

Sorry.

The end.

No, just kidding. The good news is this month's Martha Stewart's *Whole Living* magazine has a feature with eight (EIGHT!) of my exercises, designed to improve the mechanics of your lymphatic system.

The article is fantastic, thanks to Kate Hanley, health writer for the Gods. Or at least for all the health magazines. The photos are great and easy to follow. Go get your copy, they're flying off the shelves. I know this because I bought two copies and I dropped one, hence the flying.

My favorite creation for this piece was my Active Breathing exercise: Using an old pair of nylons or elastic exercise tubing, tie firmly (but not so tight you cause your torso to fall off) around your rib cage approximately where a bra strap or heartrate monitor would go. Standing, breathe deeply, feeling your rib cage expand into the resistance. Take ten to fifteen breaths, and repeat three to four times. You WILL take in more oxygen than normal, which is a good thing, but oxygen monitors in your brain will take a while to adjust. Start with one set and slowly increase as you feel more comfortable.

This exercise increases the strength of your intercostal muscles (muscles that live between the ribs), which actually help you take in more oxygen and expel out more waste, increasing the effectiveness of a cough. Perfect for those with allergies, a cold, or chronic respiratory issues like asthma.

Oh, and P.S., since you're out buying magazines already (aren't you?), check out the November issue of *Fitness Magazine*. It has a great chart I developed for dealing with common irritations while walking or running, like swelling fingers, achy knees, tingly feet, and low back pain. Of course, if you read this blog, you'll already know about most of those…

And thanks all of you for being so great! If you happened to get a thousand katysays emails over the weekend, I apologize. Evidently this blog has become too popular for WordPress—it cannot handle the amount of subscribers we have (1743 billion). It is now fixed and I am no longer being considered a spam machine.

Which makes me feel better about myself.

February 2 2011 SHE'S BAAAAAACK!

HEY! Where have you been? Oh, wait. Sorry. I was talking to myself.

I can't BELIEVE it has been two months since my last post, BUT, here are some things I have done in my free time:

1. Writing (and turning in) a book.

2. Building a new, GIANT community and educational website.

3. Moving my entire institute ONLINE so everyone can have access from anywhere.

4. Bought Christmas presents.

5. Wrapped Christmas presents. Well, really, I just put them into bags and smashed some tissue paper on top, but I meant well.

6. I didn't skimp on any walking or sleeping.

7. I grew about a thousand more inches around my midsection.

8. I went to Hawaii. For. Two. Weeks.

And let me tell you something. I am very cold now that I am home. Seriously. I am wearing socks and sweats and I've been home less than twenty-four hours.

I want to send out a huge THANK YOU for everyone who has sent me emails checking on my (and my family's) health. We are all fine. And I really missed posting, but something had to go. And, it couldn't be my walking or sleeping, so…that's where I have been.

I am back now and let me say, we have A LOT of items to discuss. Here are just a few things I need to cover in the next couple of weeks:

1. Dr. Oz's Super-Kegel. People, you should know I came THIS CLOSE to poking something into my eyeball when he said that the best way to build pelvic floor strength is crossing your legs (like kids do when they have to pee), and squeezing your butt and inner thighs as hard as you can.

   I wonder if he know how many thousands of hours physical therapists have spent trying to get women to STOP using their adductors instead of their pelvic floor. And how many thousands of hours on a biofeedback machine have been used to get women to stop clenching and start firing correctly. I'm glad I didn't poke my eyes out because I will need them to write this all down for you.

   Look for another top-to-bottom pelvic floor blog because I know you still don't get it entirely. Which is okay.

2. I need to post the winners of my holiday gift blog. Obviously, when I said "holiday," I meant Valentine's Day. You knew that, right?

3. We need to discuss the *American Journal of Medicine*'s recent article "Achieving hunter-gatherer fitness in the 21(st) century: back to the future" (read the abstract here[1]). Looks like getting back to long-distance walking, outdoors, squatting, natural surfaces, minimal footwear, maximizing strength-to-weight ratios, and keeping intensities low are the key. But you readers already knew that, now didn't you!

4. We need to talk about why you can't spot-fix your body. You have to fix the entire thing in order to optimize your health. It's time to get serious.

5. I need to start on my next book.

Oh yeah, and I need to give birth here pretty soon too.

But I'm on it.

---

1. ncbi.nlm.nih.gov/pubmed/20843503

I love coconut water, don't you? For my dosha (pitta) and my lifestyle (active), the large quantities of minerals, anti-fungal properties, and hydration in the fluid of young coconuts suits me to a "T." (Does anyone know what "Suits me to a 'T'" means?)

I get my water out of fresh, handily peeled coconuts found in the grocery store or out of a can. My favorite is Amy and Brian's coconut juice. I just like the flavor best, I guess. The ease of getting coconut water is very delightful, especially since no coconuts actually grow where I live. On vacation, however, I found myself on a deserted Moloka'i beach. Imagine huge, crashing Polynesian waves. Ragged cliffs. Turquoise water. Palm trees. Hot sand. Imagine extreme boredom, if you will. Turns out there is nothing to DO in paradise when you are on vacation. So we decided that we would try to open a coconut. Without a knife. Without a machete. Without any idea about which coconuts were good to drink from. It was awesome.

If you've ever watched a one-year-old try to figure out how to put a star-shaped block through a star-shaped hole and thought smugly to yourself how easy it is, I am telling you it is because you have forgotten what it's like to figure something out for the first time.

When you think of a coconut, you probably imagine the part that, when cut in half, makes a nice top for a hula dancer, right? That brown, fuzzy part is actually INSIDE a thick, fibrous husk made of titanium. I can't tell you exactly what we must have looked like, but I can only imagine it was a little reminiscent of the scene from the movie *2001* where the primates frustratedly thumped rocks against the ground.

There was banging the coconut onto the sharp edge of a rock.

There was lots of squatting and bending and twisting.

Then there was placing the coconut on a rock and bashing it with a second rock from above.

Then there was hurtling the coconut against the rock…over and over again.

Then there were all of these things again, only with curse words.

Then there was sitting (and I'll deny it later, but there was also some whining), in a sweaty, unrefreshed pile…watching the husband do it.

Finally, an hour later, the husk was peelable, as long as you had one person to step on it and one person to yank on the skin. Then, with sore shoulders and heaving chests, we tipped it over to find...nothing. Some sticky residue inside and some moldy coconut meat. I was tired, BUT I picked myself up and walked back over to the coconuts. After all, I am the gatherer, right? This time, instead of just grabbing a fruit, I used my senses to tell me which would be the best. I shook and smelled and compared the fruits, much like selecting a cantaloupe in a store. I didn't KNOW which one would be good, but I just went with my intuition. The second attempt went a lot faster (only twenty-five minutes this time!). We had learned from the first pass and knew which techniques helped (bashing between two rocks and using a small rock like a knife) and which didn't (whining about how hoooooooot it was). Selecting the coconut based on things I knew about other fruit (hearing a lot of fluid slosh around and smelling the ends for fragrance) gave us a full cup of coconut juice when it finally opened, and after further cracking, gave us each a slab of fresh, nutritious coconut meat. I was stuffed from eating and drinking half of a one-coconut bounty. We were tired, but also very "in the moment."

As more people are moving towards the natural movement movement (tee-hee), there seems to be the desire to drag the old paradigm with us. "What kinds of exercise did people use to do? Should I do more flexibility? More cardio? More strength?" The answer is, these categories of movement don't really exist. Biology does not develop in a vacuum. If anthropologists see a heap of coconut shells, they don't jump to the conclusion that opening them had similar benefits to a movement program. It is very hard to consider "eating fruit" a conditioning program when you have never done more than gone and picked yours off of a shelf.

My recent experience in being physically exhausted by executing one survival task also shows me how much *more* my body is capable of—and I already have lots of traditionally defined strength, endurance, and mobility. There is a huge physiological difference, though, between being able to run ten miles and do ten pullups and thirty pushups and being able to open a coconut. An entirely different pattern of muscle use that comes with a unique and novel skill acquisition. I can't wait to watch this baby-to-be learn stuff every day. It's going to be SO AWESOME!

Active. As in, vs. passive. The thing that screws up your body the most is the insane quantity of time we spend with almost no electricity flowing through our muscles. Passive tissues are doing nothing for your structural integrity or your metabolism. Sitting in a chair = passive. Holding yourself in a chair position sans chair = active. Baby sitting in a sling = passive. Baby participating in its transportation by holding itself up and holding on = active. Active tissue is required for health, and large volumes of passive time cannot be made up with an hour or two of active time.

Basics. Way before you need to worry about the deeper levels of tissues, how about checking the big axes of the body? Are your feet pointing straight like the wheels of a car? Do you spend six to ten hours with your knees in flexion (bent, like while sitting in a chair)? Fix those right away.

Calf muscles. The backs of your lower legs are tight. Very tight. More tight than you realize, and doing considerable damage to your ability to stabilize the rest of your body. The calf stretch you are probably doing (think runner's calf stretch) has nothing to do with the way the foot and lower leg function while being upright and walking. Try the first exercise in this blog post[1], and to make it more challenging, DOUBLE your yoga mat or towel thickness.

Did you know I started writing this at four this morning?

Ears should be over the shoulders, but not by tipping your head back. Try sliding your chin back like making a double. Chin, not latte. Sliding your head back also stretches the back of your neck bones and muscles out to the appropriate length. If you do a lot of computer work or driving, chin jutting is a common "passive" neck-tensing habit. Drop your chin to your chest every twenty to thirty minutes for about a minute to lengthen the soft tissues, and then adjust head posture after stretching. Note: This will NOT improve your health if you do it while you are driving. Duh.

Four. As in a.m.

Glasses are difficult to find at 4:00 a.m., especially when you can't see. Anyone know what I'm talking about?

Humerus = name of your upper arm bone. Everything that stabilizes the shoulder girdle, changes the pressure gradient for your cardiovascular functions, and provides the integrity of the connection between your upper, core, and lower body requires the humerus to be swinging in full range of motion while walking (miles a day). And you probably just hang your purse on it.

Ischial tuberosities are the name of the bones you should be sitting on. Please stop calling them SITZ bones. The less we think of ourselves as five-year-olds with hoo-hoos and dinkles and SITZ bones, the more serious we will be taken when it comes to health care, don't you think?

Just enough time. Whether you are eight months or eighty-nine years old, you have just enough time to make a serious reversal of cellular death of various tissues by changing the flow of electricity, blood, and lymph in your body. Anyone can do it at any age.

Katy is my name, biomechanics is my game. No, KATY is not a nickname, it is what is printed on my birth certificate. I was named for my great grandmother, Kathe. Only she had two little dots over the "a." I wish my parents had given me two dots too. How would I type those, I wonder?

Ligaments connect bone to bone. Tendons connect muscle to bone. If you injure a ligament it is called a SPRAIN. If you injure a tendon, it is called a STRAIN. These words are commonly misused (and often by people who should know better) to mean the degree of injury. For example: It's just a strain, I think I can walk it off. Or, I'm pretty sure I sprained it, so I'm going to need the day off of work and a banana milkshake. Wrong. The term simply implies what *type* of tissue was damaged. Note: It is never wrong to have a banana milkshake.

Metabolism is another poorly understood word. Actually quite complex, your metabolism is the sum total of every chemical action it takes just to be you. When we talk about our metabolism, however, we are usually thinking about our basal metabolic rate (BMR), or the average amount of calories we expend just being ourselves. There is great variance in metabolism across the board. What most people don't realize is, your metabolism is mostly affected by what your muscles are doing at rest (and has much less to do with how much you exercise). It is your muscle's resting metabolism that makes up the greatest portion of your BMR. So, how do you increase your muscle's resting metabolism? Actively lengthen all of your muscles! Hypertonic muscles (*hyper*: too much; *tonicity*: tension) are not as metabolically active as long, full-force-generating muscles. This recommendation goes beyond stretching. It also means learning to use your muscles in their full ranges of motion—not just the same short, linear motions over and over and over again.

Newton (as in Isaac) contributed greatly to the anatomical knowledge of the time by applying mechanical principles to understand bodily functions like fluid dynamics, pressure, etc. I'm a huge fan.

OMG, it is early.

Pee. When you laugh or jump or sneeze, while perhaps NORMAL to our current population, wetting thyself is not a NATURAL state of things. It's a sign of what's to come...like, organs moving out of your body. Take the hint seriously and fix it.

Quadratus lumborum. Get to know it. If you are a rib thruster, then this bad boy gets passively tense (there's that passive again!) and increases lumbar disk compression and pelvic mal-alignment, and weakens the core musculature. Stretching it will help, but the longterm antidote is allowing the rib cage to stay where it goes. No more lifting your ribs to flatten your stomach!

Rotator-cuff muscles are the infraspinatus, supraspinatus, teres minor, and the subscapularis. But, you don't use them because most of your time is spent with your arms internally rotated (passively) using the keyboard, or driving, etc., and the rotator-cuff muscles are stiff and underused.

Shank is what we often call the lower leg in biomechanics. Did you know that your shank can rotate relative to your thigh bone? I think that's my next video blog, cuz if your lower leg doesn't rotate (and most don't anymore), then the knee and ankle joint as well as the nerves to the feet are all compromised. I'll put this topic on the to-do list.

Thermography. Look into it, and if you like to invest in stuff, I'd bet that thermography becomes "the predictor" (i.e. early detection) of the future. It's a ways off in terms of developing cohesive variables to evaluate, but the raw material is pretty good.

U. Yeah, U, reading this blog right now. You're awesome...I just wanted U to know. XOXO

Varicose veins are created when the amount of downward force is greater than upward force in the blood vessels, breaking the valves within the vein itself. In the optimal situation, the muscles around each vein should be contracting and relaxing rhythmically to get the blood back up to the heart and lungs. When muscles are passive (passive, passive, passive... how many times will I type that word this morning?), the vein has to carry the blood by "stair stepping" it up the body, against gravity, using little valve flaps. If you use the flaps (teeny-tiny moving parts) instead of your muscles (giant, force-generating parts) these little parts give out and break, allowing the blood to bulge and mis-shape the vein from the inside. Relief from the discomfort is commonly found by elevating the legs to help the blood to flow without having to climb. But *fixing* the problem requires the real issue (your muscles are freaky-deaky tight, people!) to be addressed. Most commonly found in the lower leg (see letter "C" for calves), bulging veins come with high pressures caused by the muscle tension, whether the muscle tension comes from standing long hours, being pregnant, etc. The solution is always the same—get the muscles long and flowing with electricity, take the burden of the broken valves, get the blood back up and out of the bulging vein.

Water. It's good for you. Fun fact: Water has a lower viscosity (fluid thickness and resistance to flow) than, say, honey does. Did you know we want our blood to be more like water and less like honey? Why? Because as the viscosity of blood goes up, the ability for blood to circulate goes down. Being dehydrated is one way to reduce your blood-plasma-to-blood-cell ratio, giving a thicker blood stream. Also, high blood sugar makes your blood stickier (less like blood and more like a melted popsicle) and increases the viscosity as well. Just thought you'd like to know these things.

X. Yeah, I got nothing.

Yay! I'm almost done! If only I weren't so sleeeeeeppppy...

Z. Zzzzzzzzzzzzzzzzzzzzzzzz

---

1. katysays.com/2010/06/02/you-dont-know-squat/

 April 4 2011 # FIFTEEN BLOG POSTS IN SIXTY MINUTES

In my head-cold haze last week, I developed some sort of writer's block. In an attempt to get the juices flowing, I asked on the Aligned and Well Facebook page for blog topics of burning interest. My goal, should I choose to accept, is to answer all of the requests (fifteen!) in under four minutes each. Writing this part doesn't count.
Ready, Set,...

> 1. How about ergonomics and living room furniture?? We buy pretty stuff that makes Nate Berkus proud... but what is it doing to our skeletons and alignment? I'm pretty convinced that my couches are to blame for my hip discomfort.

Okay, I don't have enough time to Google Nate Berkus. Who is that? And, why would such a man (?) be proud that your couches are creating your hip discomfort?

Your hunch is correct, kind of. It is our excessive use of external structures to support our bodies that has allowed our muscles to atrophy beyond the point of holding us up correctly. I know you're all strong enough to stand, kind of, but not without resting on a bunch of your hip ligaments or thrusting your pelvis forward using the connective tissues of your psoas, etc.

I say "kind of" because the couch is not really to be blamed (being an inanimate object and all) for your weakness, but rather *you* are, for your habit of sitting in it. Stop looking for furniture that is "ergonomically best" and start looking to the internal set of furniture you carry with you all of the time—the fully collapsible and expandable YOU. Your bones, joints, and muscles make you a virtual IKEA catalog. Actually better, because with regular use of your full catalog (getting all of your muscles to the right length to allow all of your joints to move), you'll be able to hold up more than the twenty-seven pounds that IKEA's finest plastic-covered plywood can handle.

2. Best sleeping position, pillow, etc. for head and neck?

Answered this last year. [See the Night Owl post earlier this chapter.]

I just saved one minute, woohoo!

3. Rocker shoes... Popular, but effective or damaging?

I'm assuming you are talking about the new unstable-bottomed toning shoes with which Carl's Jr. is competing to have the most pornography-inspired commercials of all time.

Yes, they are super popular. So popular, in fact, that I can spend a fraction of every day walking behind someone expecting to catch them when their collapsing ankles give out at a moment's notice.

I personally don't care for the toning shoes that have developed from shoe companies creating "wobble" or "rocker surfaces" to force the body to work more. While muscle tone and development are great, there is actually a correct amount of strength you need to have in various muscles for the sake of the health of other tissues like joints, bones, and nerves.

For example, doing one thousand shoulder shrugs a day will tone your trapezius muscle (between your neck and shoulders), but over time, this muscle mass will move your shoulders closer to your ears, compressing your neck vertabrae. Toned shoulders are good, right up until they degenerate your spine.

Toning shoes can increase lower leg tone, but not necessarily in a "healthy way." The over-squish of a shoe is not good, as it alters the natural vibrations that come from walking. Bone-building vibrations. A rolled or contoured bottom interferes with the natural gait pattern needed to keep ankles, knees, and hips generating correctly. Blasting through ankle ligaments (start watching the feet and ankles of someone wearing these if you are lucky enough to be walking behind someone as their ankles wobble inward/outward) is also an issue.

"Arbitrary strength and movements are not good for the structural integrity of the skeleton, muscles, and connective tissue in the long term." –Katy Bowman.

I just quoted myself. Feel free to forward it to your friends.

So, toning shoes. May get you tone in some places, but lifelong damage in others. I'm not paying for that, no matter how creepy your commercials are.

> 4. How to help get your kiddos into doing things to encourage proper alignment? My toddler is 2 and I really want to work with her, but I don't really even know where to start! (I know you've covered some lifestyle things, like going barefoot a lot, but it would be nice to see some ideas all in one place.) ☺

Before you can encourage your kid to proper alignment, it helps if you, the teacher, know what that is. So, step one: Find out what proper alignment is. You can check out my academic website, restorativeexercise.com, and learn quite a bit.

Here's the clincher. In order to help your toddler, you really have to help yourself. She will mimic everything you do with your body, because that's how biology works. Your gait pattern will become hers. Your affinity for walking (or lack of) will become hers.

Start taking walks with your daughter, either opting for socks or some of the new shoeless shoes they are making for kids (which will be another post here soon!). While the barefoot part is important, it's really the WALKING part that is so crucial to her development. Walk—no carrying or strolling. Just let her strengthen her own body to transport her long distances (like a half of a mile).

And, try sitting on the floor for picnic-style dining and create a standing play station instead of teaching her that chairs and "sitting quietly" are the optimal choices. Cuz they aren't.

> 5. Exercises, stretching, and alignment for better marital relations.

Wait. What?

Oh.

Ooooh.

Ooooooooooooh. I get it.

Well, the good news is, about one third of this blog seems to be about pelvic floor health. Even though you may not have a PF "medical" issue, to optimize other (ahem) functions of the pelvic girdle you need to do the same things. The mechanics of having an orgasm require all of the same blood and neurological flow as everyday functions. In the end, it often comes down to how much tension you carry in this area. Which is also a result of how much tension you carry in your mind, which is how much tension you carry in your life. There are some good resources:

1. *Headache in the Pelvis* by David Wise and Rodney Anderson, a kind of "clinical" book that talks about some of the psychosomatic properties of PF function

2. Mayan abdominal massage

3. The Aligned and Well DVDs, Down There for Women and Below the Belt for Men. (These are mine—the exercises are the same on both the women and men's disks BUT we don't recommend you buy your man an exercise DVD where he has to listen to the word "vagina" a lot. It's better to get him the "for men" copy.)

Good on the covert "marital relations"? Unless you were talking about something else, in which case 1) I'm kind of embarrassed and 2) what ARE you talking about?

> 6. I'd like to hear about proper sleep positions for pregnancy. I know you've said before that ideal sleeping position is flat on your back, but that is not recommended in pregnancy, and besides that, every time I wake up on my back (I'm 34 weeks) I also have horrible leg/foot cramps, presumably due to the lack of circulation.

No, you do NOT have to sleep on your back while you're pregnant. I'm currently sleeping on my side. And then the other side. And then the other side, and then the...

For achy hips, try putting a pillow between your knees that keeps them the same width as your pelvis. Also, you may need a slightly higher pillow under your head and maybe even one to hug while your shoulder girdle has a bit more tension than normal (it gets hard to breathe once the little intruder gets thirty-four weeks big...)

And keep up your calf stretches, doing them throughout the day to avoid lower-leg discomfort throughout your pregnancy.

> 7. I don't remember any purse/backpack/briefcase blogs.

Is this a request, or just a statement of your current mental recollection?

As you can all imagine, loading one side of your body on a regular basis has detrimental side effects. To reduce the impact of asymmetrical loading, you can either continue to switch, from one side to the other throughout the day, try to lighten the weight of your bag significantly, or find a bag that spreads its load more evenly over the body.

I like to use backpacks or a bag that crosses over one shoulder and rests on the opposite hip. I think it's called a messenger bag, but I can't remember. I have long since given up the bulky purse, especially when I travel with my computer, and opt for a backpack that allows me to swing both arms while walking and interferes with my gait pattern as little as possible.

You also have to develop a lack of caring about how your bag matches your outfit. And that's really what keeps us from making a healthy choice so often, I have found.

> 8. Transverse stomach muscles. You have taked alot about PFD, but my dysfunction is related more to just the transverse.

Pelvic floor disorder is never about one thing. There is a dynamic relationship between every muscle in the moving body, and each is affected by the other. Because most people have no gluteal function during locomotion (even if you're busting out a bunch of gym exercises) the pelvic floor is missing half of its regular stimulation, i.e. the pull the glutes would be doing were you walking in the correct way. Or at all.

The transverse abdominals can be involved due to their lack of stabilizing the pelvis, as well as the adductors (inner thighs), tension in the hamstrings and calves that tuck the pelvis, excessive downward forces created during incorrect pushing during birth, chronic constipation, high-impact running habits, or just the habit of sucking in your stomach.

> 9. Have you written about long-lasting plantar fasciitis?

Why yes, yes I have.

Have you given up positive heeled shoes, done the Fix Your Feet DVD, stopped thrusting your pelvis forward, learned how to align your feet while walking, created a standing workstation, and learned how to rotate your thighs correctly?

10. Results from the Christmas giveaway? ☺

I am so busted. This week...I promise!

11. Stretching for prevention.

Yes. Do it. Do it every day. Need to know which stretches to do? Go to www.alignedandwell.com.

12. What can be neurologically/psychologically/emotionally preventing a person from objectively looking at their alignment while doing movements?

To discover that we are doing something physically different than what we imagined we were doing is probably a huge disconnect for the mind. When we ask our mind to look at something objectively—say, at ourselves in a mirror—the resistance to calibrating "what is" to "what we thought was" must create a certain amount of internal conflict. Maintaining status quo is convenient. But so are cars and fast food and technology. No one ever said you couldn't convenience yourself to death.

13. More for pregnant ladies! sleep positions & exercises for sore hips and/or sciatica and/or SI joint pain, perhaps? fetal positioning? believe it or not I am NOT preggers ☺ are you???

I've got some more on the way on these topics. And yes, I am very, very preggo, at thirty-seven weeks yesterday, see?

The good news is, you can be this pregnant and still hike over four miles (which I had just finished doing in this pic) and not hurt in the feet and back, pelvis or hips. But you do have to train for correct alignment.

14. Body alignment in relation to brain health, mental health. Body alignment and effect on endorphins?

In the end, what alignment affects the most is circulation. Circulation is about oxygen saturation. The more correct your alignment, the better oxygenated your tissues are. People currently need cardiovascular exercise to help circulate oxygen-rich blood because their total movement is very low, which gives them very low oxygenation when compared to human potential. The endorphins that kick in from an intense bout of exercise are there for you all the time when you have your correctly lengthened muscles literally pulling the blood throughout your body instead of using your heart to do all of the work. Higher levels of oxygen keep your brain happy, as chronic low O2 creates a very stressful state. Can you imagine your brain watching you sit there basically choking to death, slowly?

> 15. How to furnish your house to be beneficial to your alignment? I have some ideas on what we will do when we get into our own place, like no chairs or sofas, just some throw pillows on the floor and maybe a blow-up chair for family that comes over and wants to sit.

Do YOU know who Nate Berkus is? Maybe you should call him.

You don't have to get rid of all of your furniture, just the notion of how you should position yourself in it. Use the floor as well. Pillows, beanbag chairs, etc., are all great too. I like a bare, open family room floor as it beckons me to stretch and do various exercises when I'm "relaxing." Also, having a bare wall for stretching against (like Legs on the Wall) or doing various balance challenges is a good idea too. You'd be surprised how many people tell me they have no floor or wall space to exercise with.

Priorities, people!

BONUS!

> I am a massage therapist, and I often send my clients articles from your blogs that would help them. I have read your bio, and I would be interested in more details about your path to your profession and how you would suggest a person become a biomechanist.

Becoming a biomechanist is similar to becoming a chemist—you should have a degree in the science. Most universities will have a degree (I have a BS and and MS) in biomechanics—or find an engineering or kinesiology program with the option to specialize in biomechanics.

I was originally going to be a mathematician, then switched to physics, then ultimately biomechanics, because I also loved my anatomy and physiology...I just love looking at the biological sciences from a mechanical perspective.

To become a biomechanist, then, would require a bunch more schooling. I do teach an extensive course (six to seven months) on biomechanics for the purpose of teaching people how to apply mechanics to their own allied and medical health professions, which makes people feel like they've received an entire education beyond the original scope of their practice!

Not to toot my own horn or anything, but I DID manage to finish the new Restorative Exercise Institute ONLINE webpage before the baby got here, which was my dream. If you wouldn't mind checking it out, www.restorativeexercise.com, there are a lot of free lectures and downloads available as well as the FIRST TWO HOURS OF OUR WHOLE BODY ALIGNMENT PROGRAM all for free. I hope you enjoy it as much as I enjoy seeing anyone make a simple change with significant results.

I'm not even going to proof this…

PUBLISH!

June 30 2011   5 THINGS YOU (PROBABLY) DIDN'T KNOW ABOUT OSTEOPOROSIS

1. Johann Friedrich Georg Christian Martin Lobstein. While you may not recognize the name, you surely know the name this French pathologist gave to bone he found with large, porous holes in it: osteoporosis (*osteo* = bone, *poros* = Greek word for passage).

    So the next time you spot the word, make sure you give a shout out to:

    Jean-Freed-rik-Gay-org-Martin-Lobs. That's my name tooooooooo.

    LA LA LA LA LA LA LA!!!!!!

    Sorry. I couldn't help myself.

2. You can actually spot-treat osteoporosis. Osteoporosis is not a systemic or whole-body disease, but an indication of where your bone is not loaded correctly. Osteoporosis doesn't mean that you have a bad-bone condition—your bone loss is not happening over your entire skeleton but in a few key places. The areas that most people experience bone loss are:
    * ribs
    * wrists
    * vertebrae, and
    * head of the femur (top of the thigh bone, commonly mis-referred to as the hip bone. There is no "hip bone." The hip is a joint made from the pelvis and the thigh bone, called the femur.)

In order to get bone to generate, you have to know your sites. If you are given a diagnosis of osteoporosis or osteopenia (a little bone loss), ask WHERE the density is low.

And, P.S., the only test really valid to give you a diagnosis is a DEXA. In-office bone screenings are not valid enough to give you a diagnosis. The margin of error is too high. They are simply supposed to be a screening technique for referral to the better DEXA.

3. Exercise trumps nutrition. Many people think osteoporosis is a result of poor nutrition. However, poor nutrition doesn't explain the fact that most bone tissue, even in the bones with osteoporosis, is doing just fine. Just here and there is there a problem. The failure for bone to generate at its correct rate is really a mechanical one—which is why I am blogging about it, of course.

    While proper nutrition is absolutely a requirement for healthy bone, the signal for bone to grow is mechanical in nature. The "GROW BONE" signal starts with a cell being squished within the bone. Without that squish of these mechanoreceptors (sensors sensitive to physical deformation), the nutrients that support bone growth can't do their job. Your body cannot use them without the signal. Taking supplements is half of the correct prescription for osteogenesis (bone growing). The other half is exercise, and let's have some words about exercise and bone development...

4. Weight-bearing exercise doesn't mean "using weights." For optimal bone regeneration, you need as much squish in the bone-growth-signaling cells as possible. In order to get the greatest amount of squish, you need to keep your bones holding the proper amount of weight—not too much and not too little. The research done on bone and exercise shows that moving around while weight bearing gives the greatest response to bone development in the right places.

    Why do I say "in the right places"?

    Because any resistance exercise will help build bone. The act of a muscle working, pulling on the bone, is enough to stimulate bone growth. This is the rationale for doing weights or other types of non-functional resistance exercises in gyms, classes, etc.

    So what's the problem?

    Bone loss at muscle attachment sites *is not the problem* in osteoporosis. Bone loss in the hips (and the eventual break) is the biggest problem in osteoporosis, with the second-largest issue being bone loss (and then fracture) in the spine. "Weights" don't hit these areas in a way that reduces the risk of breaks or fractures—even though you may be adding bone tissue in other places.

Best bone-building weight-bearing exercise = walking. Walking is better than running because bone building favors lower-impact loading as opposed to high-impact and frequent loading cycles.

Non-weight-bearing exercise = swimming, cycling, blogging. That's why I am only doing one more fact—my bones are not impressed with me right now.

5. The "hump" is not caused by osteoporosis. The "hump" is actually a *cause* of osteoporosis. It was believed for a long time that the excessive curve in the upper back (called the dowager's hump) was the result of weak, porous bones unable to hold up the weight of the spine. This weakness in the bones allowed the spine to collapse forward. It makes sense, in a way. It makes so much sense that no one actually researched it—it's just a belief that has been perpetuated and is now on the title page of almost every bone website under "Bone Facts." Which is a bummer, because the actual research shows that this curve of the spine, called hyperkyphosis, is a risk factor for osteoporosis and not the other way around. Which makes sense, because osteoporosis is caused by a lack of loading. The farther something curves forward, the less weight it places on what sits below it.

6. Your alignment is actually a huge deal when it comes to bone growth. You can make your daily walk more weight-bearing by stacking your body. See those walkers with their torso out in front of them? Their walk is less weight bearing than it could be. Have too much curve in your upper spine? Wear your head out in front of the rest of the body?

These are all things that decrease bone regeneration, and fixing them is as simple as stretching your muscles to their correct length and using your body as it was designed.

Off to walk…

You want to know a little bit more about me? This is what I am reading right now.

1. *The BIG Necessity: The Unmentionable World of Human Waste and Why it Matters,* by Rose George. Chances are, the last time you read about excrement was back in the days of *Everybody Poops.* Yes, this book is about poop, and where it goes once you are done with it. It is a fascinating look at the underworld of the sewer systems and how disease would quickly spread were it not for these systems. And guess where you can read this little gem!

   Fun fact: 850,000 cell phones are flushed down British toilets a year.

   I believe it. I almost drop my phone in the toilet a few times *per day.*

2. *Nourishing Traditions,* by Sally Fallon with Mary G. Enig. If you love chemistry or food science, you'll love this book. This book is pretty in your face about the politicization of food and is all about da fat in your diet. I'm a high-fat diet kind of girl. Sixty to seventy percent of my food comes from fat—and no, not sugary or processed fats, but nuts, avocados, and cold-filtered oils. I'm mostly plant-based, but this is a great book that not only has tons of recipes, but a lot of fun info—especially for those interested in the history of food culture.

   Favorite recipe: How to make your own coconut milk. This lactose-intolerant lady loves plant-based milk. How cool to make it in your kitchen!

3. *The Barefoot Book: 50 Great Reasons to Kick Off Your Shoes,* by Daniel Howell. Reason number one is not "because Katy says so." Can you believe it? There is lots of interesting stuff on bare feet (the author is a barefoot runner). Some stuff I agree with and some I don't—and I know he feels the same way about my book (he read an early edition of mine and was kind enough to do a blurb for me). His basic premise is excellent—human feet were not designed to wear shoes. Right on. My bigger issue is that what starts out as science-based material turns into opinion quickly. Pet peeve: He feels that flip-flops are a better alternative to bare feet than fully attached shoes. I'm guessing this is because he doesn't

thoroughly understand the biomechanical changes that come with toe gripping (really bad for wanting to keep healthy nerve conduction to feet) and what "natural gait" really is. It's okay. He's a biochemist, not a biomechanist. I don't hold it against him and I'm hoping that when my book comes out later this year, we can talk about it via blogging for y'all to benefit from. If you want to go totally barefoot, this is a great resource.

Cool stuff: As a biochemist, he has a lot of info on the "real facts" behind what diseases you can catch walking around barefoot. Hint: Nothing.

4. *Guns, Germs, and Steel,* by Jared Diamond. This is my favorite book of all time. Don't know what else to say, but go and get it. In a nutshell, this book looks at what happened to people once we stopped hunter-gathering and started farming. Hint: Guns, Germs, Steel. But you saw that coming, right?

    It's a little textbook-y, so if you don't like to read your facts straight from the professor's mouth, then read this book instead:

5. *Ishmael,* by Daniel Quinn. This book is essentially *Guns, Germs, and Steel* as a nice bedtime story. In fact, start with *Ishmael* and get to *G,G,&S* next year.

    Best quote: "Any species that exempts itself from the rules of competition ends up destroying the community in order to support its own expansion." Heavy. But the story features a talking gorilla, which softens the info a bit. ☺

6. *Euclid's Elements* (all thirteen books in one volume!). In case you didn't know, the geometry class you took in HS that seemed boring was really the life's work of Greek mathematician Euclid. If you like math and love philosophy, you'll love reading how Euclid wrote out his books more like poetry and less like "if A, then B."

    "If a unit measure any number, and another number measure any other number the same number of times, alternately also, the unit will measure the third number the same number of times that the second measures the fourth."

    I know, right?

7. *Baby in a Car,* by Monica Wellington. I can't tell you how captivating this book is. See, there's this baby in a car. What does baby see? Taxi cabs. Garbage trucks. Flags. Fire trucks. This is obvi-

ously a baby in a car in Manhattan. And there's a plot twist, too. I won't give it away, though I will say it has something to do with an umbrella and a dog.

8. *The Pleasure of Finding Things Out*, by Richard P. Feynman. KB+RF. There, I said it. I love RPF. He is smart, funny, and cool. I love science. Real science—not biased results from data collection by people who have forgotten to stay openminded. My favorite Feynman quote: "It doesn't matter what things are called. There is nothing important about what things are called. Science is not memorizing what things are called—that is classification. Science is understanding how nature works. Knowing the word photosynthesis doesn't mean a thing if you don't understand that this word means that plants are able to grow by getting their mass from the air." (Yes, they do!)

Brilliant.

So, what are YOU reading this summer? I'm noticing a trend in my books after writing this. Can anyone recommend a novel or something? (I like those too!)

August 22 2011 — CORRECTING ANATOMY MISCONCEPTIONS

1. There is no hip bone. The hip is a joint, or the place where two bones interface. In the case of the hip joints, the bones in question are the pelvis and the femur (thigh bone). When people fall and break their "hip," they typically break the femur. The neck of the femur (the space just below the "ball" or "head" of this leg bone), or just below, is where fractures or breaks happen most frequently.

2. There is no shoulder bone. The term "shoulders" refers to a general area about the top of the arm. The actual anatomical term for this point is the glenohumeral joint. This joint is made up of the arm bone and the shoulder blade (scapula). Also part of the shoulder girdle are the clavicles, which connect the sternum to

the scapulae. Because there are so many bones in the upper body, this area is way more complicated than the lower, I think. When alignment here is off, it affects breathing, neck and shoulder-girdle pain, upper-body strength…

But when it's working, it feels awesome!

3. You don't have "carpal tunnel," you have carpal tunnel syndrome. You've got carpal tunnel? So do I! EVERYONE has a carpal tunnel. It's the space or tunnel among the wrist bones (carpals), muscle, fascia and other tissues through which the hands and brain communicate, via the medial nerve.

   If you have carpal tunnel syndrome, then your space has narrowed, placing excessive pressure on the median nerve running through it. The tunnel space is reduced by wrist position, swelling of the tunnel contents, or by tension in the forearm and hand muscles.

   This is why your wrists should not look "broken" when using your computer (image below left):

   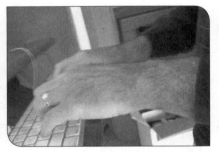

   but should stay in alignment with the forearms (image above right).

   Remind me to wax my hands later.

   Tense hands and arms? Try the My Hands Hurt DVD for five stretches to help open up the tunnel again.

4. You don't have TMJ, you have a TMJ Disorder. You've got TMJ? So do I! TMJ stands for temporal (the bony plate on the side of the face that runs under the temple to about the level of your cheeks) mandibular (the mandible is the jaw bone—*mandere* in Latin, "to chew") joint (where these two bones interact).

   Everyone has a temporomandibular joint. It comes in handy if you talk and eat a lot like I do. ☺

As with carpal tunnel syndrome, space in the joint narrows, causing excessive friction, inflammation, and pain. Tight scalp, face, neck, and shoulder muscles all make temporomandibular joint disorder worse, as does stress, which can lead to jaw clenching.

Solution: Spring for a weekly upper-body massage session. Many times you can find a fifteen-dollars-for-fifteen-minutes chair-massage deal. The effects of massage work are cumulative, so maybe try three times a week for a couple weeks to see if you notice a difference. If you do, then it's just muscle tension, baby. Fix it.

5. There aren't any arches in your feet. You should have the *shape* of an arch in your feet, yes, but if you cut your feet open, it's not like there's a fixed anatomical part that is shaped like an arch.

    The shape of the arch is created by healthy tone in the many muscles of the foot, pulling bones here and there until the arch is intact. So, no arches? Just work on strengthening the small *intrinsic* muscles of the feet until you start seeing the glimmer of a shape.

6. Men have pelvic floor muscles too. Really, they do. And when the pelvis, hip, and sacrum are not in alignment, they get pelvic floor disorders just like the ladies. So men, do a search on past blog posts for finding neutral pelvis when you are sitting and standing.

7. The Adam's apple isn't really an apple. Just in case you thought it was. It isn't.

Sept 15 2011 QUICK QUESTIONS

One of the benefits of being an Aligned and Well Facebook friend, besides the obvious prestige, is participating in Q&A sessions. Last week I requested friends to ask a quick question. You can see what I got below. Some are easy to answer, and others are more like, Hey Katy, I know that instead of doing anything else, you'd like to write a small novel.

Here we go:

> 1. Quick assessment tool or physical practice to use while stressed out and waiting in line?

Um, please state all questions in the form of a question. I'm assuming you meant to ask, "What is a quick assessment tool or physical practice I can use while stressed out and waiting in line?" Right?

Step one, accept your reality. You have chosen to stand in line. If you didn't choose it, you could leave. So being stressed is your choice. Opt to not be stressed. Then, once you're like "COOL, I've been wanting a little free time to work on my alignment and now I can cuz I'm awesomely standing in this line,"

- straighten your feet
- make sure your feet are pelvis width apart
- externally rotate your thighs[1]
- back your hips up to put your weight back in your heels
- drop your ribs
- ramp up your head.[2]

> 2. What is a good method to treat and prevent plantar fasciitis?

Stretch your calves. Stretch your hamstrings. Get out of heeled shoes. Get your pelvis back where it belongs. Do your intrinsic foot exercises. Buy my book, *Every Woman's Guide To Foot Pain Relief*. If you're a man, buy the book "for your friend," take it home, and read it in secret. Info applies to the dudes too.

> 3. Please discuss the patterning that would cause a person's elbow to be perpetually flexed. (Hands hanging slightly forward of torso.) Anything other than tight biceps?

The slightly bent, slightly orangutan-like appearance of the arms (shoulder joints internally rotated, elbows bent) often starts as a choice of postural habit. Posture is often used to communicate, and this choice of arm position is often used by the male animal population to convey increased size. Many creatures in nature will use this technique to secure alpha status when in combat.

In addition to preferred stance, unbalanced exercise programs limited to biceps, chest, and upward trapezius contraction, coupled with sitting, driving, and computering habits will change the length of the tissues making relaxation of this posture difficult.

> 4. How should it feel when you stretch your hamstrings? I experience 2 things, burning/tingling/pain, and then a deeper long pull which doesn't hurt, both sensations located in the calf, behind the knee, and up the back of the thigh.

It's going to feel how it feels. That all sounds about right though. Let me know if you feel it in your left ear. Then we should talk.

> 5. Working squats into your day?

Question. I said QUESTION.

You can squat to go to the bathroom.

You can squat to play with your kids.

You can squat to fold laundry.

You can squat to play craps illegally with your neighbors.

You can squat in your office chair, but only if your office has blinds.

6. What makes people's big toes come off the ground when they squat?

Incorrect muscle recruitment. The same thing that causes your face to contort when trying to do something difficult.

7. I have just like one fingerwidth of diastasis recti going on and I was hoping you could make some suggestions on how to make that go away (without surgery...).

It can go away without surgery, but you need to fix your alignment and find your core musculature. Stop thrusting your pelvis forward. Get out of positive-heeled shoes. Strengthen your posterior muscles. Stop lifting the ribs. Stop sucking in your stomach. Find your TVA. Take my $15 ball class for the core-exercise and rib-alignment part.

8. How about that post about varicose veins?

How about it, indeed! Prompt noted.

9. Can bunions be corrected?

Yes. But you need to learn how to stop wearing traditional shoes with teeny toe boxes and heels. Straighten your feet. Learn about torsion of the lower leg. Stretch your calves. Wear My-Happy Feet socks whenever you're kicking it around the house and at night. Fix your whole-body alignment. Did I mention my book, *Every Woman's Guide to Foot Pain Relief*?

10. Any tips for turning a posterior baby?!

Well, first of all, they need to turn themselves. It is your job, though, to provide the maximum amount of space by having relaxed psoas muscles, keeping your ribs down (instead of lifting them up!), and backing your pelvis up until it stacks over your heels – not the fronts of the feet. And, go to spinningbabies.com immediately! They have tons of great information!

11. Yes, diastasis help, please. I'm doing physio, which is helping a great deal, but am finding differing opinions on whether the connective tissue can heal at this stage (10 mo. pp). Good exercises? Bad movements to avoid?

See answer above.

12. With winter coming up, what shoes would you recommend for a toddler? She's been barefoot so far, but we'll need something to keep her feet warm and dry (and still developing properly!)

I really like Soft Star Shoes. I think they have some teeny tiny toddler winterwear.

13. How to get your heels on the ground while squatting.

Lots and lots and lots (and lots) of soleus and gastrocnemius stretch. Do the regular calf stretch (image below left):

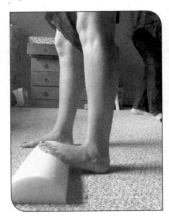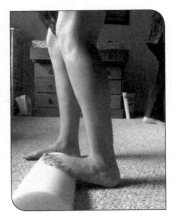

and then bend your knee while standing on your mat/towel/half dome to move the stretch to the soleus (the lower of the calf muscles) (image above right).

Also, back your hips up when you're not squatting. Stop wearing positive-heeled shoes. Increase your stretch frequency by three times.

14. Standing with feet pointed forward and rotating the legs outward to avoid knees knocking works great—how do you apply this to walking? I can point the foot better, but turning the legs seems hard.

I like this "question" because it makes me feel like 1) you have actually read other blog posts looking for answers yourself and 2) you are ready to take the next step. The reason the thighs keep internally rotating when you're walking is that the front of the foot is too stiff in our shoe-wearing population to articulate by itself. Ideally the ball of the foot should drop down to the floor without bringing the ankle with it, but our feet are pretty lumped together. The front of the foot acts like an anchor, dragging the thigh with it. Keeping the legs in "neutral" while walking really requires a whole lot of foot-joint mobility and intrinsic strength. Doing a lot of intrinsic foot exercises and stretches will help. Walking in minimal or "barefoot" shoes will make this more and more easy for you. And if you are really interested, check out the whole-body course. At least watch the first couple of hours...it's free!

15. List of yoga poses and exercises to avoid if you have POP, please.

You're probably not going to like this answer. You shouldn't be doing anything that increases the downward pressure on the organs. If you don't know how to breathe, walk, exercise, or load your body in a way that doesn't cause you to bear down, then every physical activity increases prolapse severity. It's not about an exercise making it worse, it's about the way you use your body making it worse. For example, a spinal twist can be done in a thousand different ways, depending on your tension patterns. One person can twist, causing their abdomen to bulge, pressing down on the organs, and someone can do the twist in a way that that improves PFD. So, I can't tell you what exercises to do or not do, but only that you need to learn about pressures, how they are created by certain habits (like sucking in the stomach, walking with hip flexion, straining during exercise, etc.), and how to move in a way that improves your health. And that, my friend, is much bigger than a blog post.

Peace out.

---

1. youtu.be/qcGPY4BMdIU. About four and a half minute describing how to get feet, knees, and hips into proper rotation.

2. youtu.be/TtQ-hW0_3Qg Just over three minutes showing how to Ramp Up.

Oct 5 2011 DON'T BE SO HARD ON YOUR JOINTS!

> There is an exact relation between the joint and the muscles which move the joint. Whatever the joint is capable of performing, the muscle is capable of moving.
>
> —George Bridgman

Joints get a really bad rap. Every day I hear it around me, and I'll bet you do too. "I have bad knees," "My back is killing me," "My tennis elbow is acting up," "I would love to squat if someone would just get me a new pair of hips," "I just got a new pair of hips!" (and I still can't squat).

I'll say it as clearly as I can: Your joints are not the problem.

Your problem, which is entirely fixable, is the length of your muscles. Your shortened muscles are exerting constant tension—a pulling force—at their bony attachment points, at least one of which is near a joint. In response to this tension, joints buckle, requiring other muscles in the area to work harder to keep the body upright. This means a lot of force is being exerted, accomplishing no actual movement or "work," and the joint absorbs the misplaced effort. This results in friction, and the inflammation that friction creates.

Our everyday habits have led to a shortening of our muscles. Your joints are not genetically flawed, they are just showing the effects of your habits—like excessive sitting, typing, driving—and the limited number of ways you use them.

Try finding some pictures of rock climbers, as they demonstrate joint mobility in addition to amazing strength. Functional, whole-body strength requires mobility. For each picture, imagine attempting the same movement. If your joints wouldn't open up as much as rock climbers, then zero in on which of your muscles are inhibiting your full range of movement. Try copying rock-climbing positions on the floor, where no real strength is needed, and see—is it strength, or is it mobility that's the issue?

Then make your joints healthy by stretching out those muscles in the correct direction the joint requires.

October 27 2011 THINGS I FIND VERY SCARY

1. Mannequins. Not because they are painfully more fashionable than I, but because they are good at reflecting the status quo. And right now, the status quo looks like this:

2. The fact that one of my book editors wanted to exchange the word "vertical" for the word "straight." To make it easier to understand.

Um. This is straight:

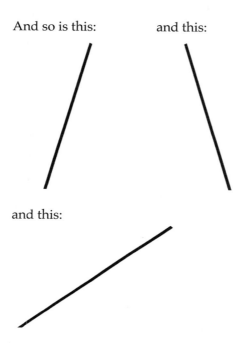

And so is this: and this:

and this:

But none of these are vertical. Which looks like this:

Hint: "Vertical" is as simple as it gets. Please stop making things wrong in the name of simplicity. Health is not simple. It requires math. At least enough math to know the difference between vertical and straight.

3. Rocker-bottom shoes. The bottom of your rocking chair is curved for one reason: momentum. Minimal effort, maximal movement. Rocker-bottom shoes create an ease of motion for very little muscular output. Meaning, these shoes use less muscle than a regular shoe. That's not what it says on the box, is it?

4. The standing workstation full-body brace. Come on. Really? How about just using your muscles for a change. Geez.[1]

5. My spelling. It's atroshuss.

6. The fact that ninety-nine percent of women's health professionals don't know that the sacrum moves independently from the pelvis

and that having a neutral pelvis is only half of the pelvic-floor equation. There is a "neutral sacrum" that needs to be addressed for optimal pelvic floor health.

7. Bikini t-shirts.

   Which is weird because bikinis are fine and so are t-shirts, but when you put them together it's a little freaky. Am I alone on this one?

8. The fact that this story got any press at all: "Wearing high heels is like having an orgasm: Christian Louboutin on why his shoes are so popular with women."[2]

9. Lateral hip weakness. The long musculature that runs down the length of your thigh is essentially the tissue responsible for up-right human walking. In most people, this tissue is so weak they use their spines to keep them upright, which not only creates overuse injuries to the discs in the spine but also underuse injury to the sacroiliac joint, pelvic floor, and hips. There is a solution, of course...[3]

What do you find scary?

---

1. See a picture of this monstrosity at admagic.com/chair/high.jpg. Seriously, go look. Then go for a walk.

2. dailymail.co.uk/femail/article-2050182/Christian-Louboutin-shoes-popular-women.html?ito=feeds-newsxml

3. youtu.be/7d9eCnBnfqU Four and a half minutes of the scaaaariest way to strengthen your lateral hip muscles.

 A HEALTH QUIZ

Nov 2 2011

Just for fun, see how you do! Try it WITHOUT Googling the answers, yo! Geez.

1. How many muscles flex the elbow (think bicep curl)?

   a. 1
   b. 2
   c. 3
   d. 17

2. How many muscles make up the calf group (lower leg)?

    a. 1
    b. 2
    c. 3
    d. 17

3. Which is a greater weight-bearing exercise, walking or running?

    a. Walking
    b. Running
    c. Walk-Run combo
    d. Jogging

4. The correct exercise intensity for those with high blood pressure or on beta blockers for high blood pressure is:

    a. fifty to seventy percent
    b. forty to sixty percent
    c. sixty-five to seventy-five percent.
    d. it doesn't matter, the beta blocker is taking care of everything, where's the spin bike?

5. What is the difference between a sprain and a strain?

    a. A strain is damage to a lesser degree than a sprain
    b. A sprain requires surgery.
    c. There is no real physiological difference.
    d. A strain implies damage to a tendon, a sprain implies damage to a ligament

6. The signal for bone to develop is:

    a. triggered by calcium
    b. triggered by calcium and vitamin D
    c. triggered by pulling on bone
    d. triggered by pushing on bone

7. Select the false statement:

    Pregnant women:
    a. can't be left alone in the house with ice cream
    b. have different alignment markers than non-pregnancy peoples

  c. should be given a shoulder and/or foot massage daily, without prompt

  d. should be stretching the leg muscles a few times a day

8. Back pain can be created by tight tissues in the:

  a. back
  b. shoulders
  c. calves
  d. office

9. Bone growth stops at a certain age (about age twenty).

  a. true
  b. false
  c. This is a trick question, isn't it?

10. Katy is

  a. smart
  b. a smart a$$
  Hmmm. Tricky.

  Answers are posted upside down, which isn't as easy to do as you'd imagine, as well as posted in reverse order in case you're inclined to cheat. Which is weird, because you'd only be fooling yourself, which is also weird and something you might want to reflect on.

1. Answer is "c": brachialis, biceps brachii, brachioradialis. **2.** Answer is "c": gastrocnemius, soleus, plantaris, **running. 3.** Answer "b". The slight knee bending to "soften" the excessive loads of a flight phase **5.** Answer "d". **4.** Answer is "a". also dampen the force of weight. Thick, cushiony shoes also reduce weight bearing quality of both walking and respond to a mechanical (not chemical) growth signal which is generated by either a compressive or a tensile force. **6.** Answers "c" and "d" are correct. Bone of muscle weakness and not a result of the physiological state of pregnancy. If it does, then this is an indication **7.** Answer "b" is false. Body alignment should not change during pregnancy. above. **8.** All of the reach a fixed length at about age 20, but growth in the width direction continues throughout a lifetime. Bones will **9.** False. Growth in length and width are two difference processes. **10.** I'm more inclined to "b", of course.

Nov 6 2011 GARAGE SALE PROPS

I love garage sales.

  One of the benefits of moving to a retirement community is that no one really works around here, which means there are always garage sales during the week. On Thursday (Thursday!) we went to a big church rummage sale. I like the big ones because a ton of people bring their stuff so it makes it easier than having to drive all over.

I don't really need anything. We just moved and purged all non-essentials and really like having less stuff in the house. I do have certain weaknesses though, like science textbooks and old books on health like my doctor's handbook from the 1950s. Hilarious and informative.

In addition to books, I am always on the lookout for non-traditional jungle-gym stuff for our giant play-living room. The kid is trying to climb and traditional furniture lacks little grooves for tiny fingers. We are always modifying stuff with thin dowels so the little man can start hanging using his upper body. Now that he's climbing, the totally flat surface (can you say booor-ing) isn't offering him any challenge besides rug burn.

I found a wooden platform that offers a height and texture variance (for both sensory and motor input) for you-know-who and some treasures for the entire family, meaning me.

I'm not a fan of furniture, having found that, once I actually bit the bullet and got rid of the living-room staples, there wasn't a need to do extra hip-opening exercises all the time. Just living on the floor opens up all sorts of opportunities for stretching in a relaxed, being-healthier-every-minute kind of way. Healthy Time and Regular Life are looking more and more like the same category. Woot woot!

What was I talking about? Oh, yeah, so while furniture is out, I am a huge fan of things that offer various heights for various body-prop interactions.

Behold, the two-dollar cubes.

These lime green and hot pink monoliths stand about eighteen inches high and were two bucks a pop. I washed their canvas covers and have since found every excuse to do this (image previous page right):

and this: and this:

about seventy-five minutes a day.

I'm obviously stretching instead of cleaning my house.

Hey. In case you're nosy, and trying to strain your eyes to peek at my table of photos, here they are up close: (below right)

P.S. If you looked hard at this picture, you might be a stalker. See if you can spot the bell (gift from favorite yoga teacher, Zan) and the empty picture frame that's been sitting there for six weeks.

Here is my favorite picture of my dad (image above right):

or as I like to call him, Bagpiping Bad-A$$.

Not really. I don't cuss in front of my dad, still.

As a total non sequitur, and only because I cannot stop taking pictures with my iPhone—here is how I am typing this blog post right now (at right):

Just in case you were worried about me spending too much time on the computer—I don't let it interfere with my body to-do list. Except that I realized that I always prefer my left leg up, SO that means it's time to switch it up.

Anyhow, garage sales and thrift stores are great places to find stuff for a healthy home. Wellbeing doesn't need to be expensive, and the hunt can be kind of fun too.

Nov 9 2011 VARUS AND VALGUS

One of my favorite college courses was Medical Latin. I loved it because speaking Latin made me think of gods and goddesses, togas and baklava, and riding a moped after swimming in the Mediterranean. Which, now that I think of it, are all Greek things, actually.

It turns out I love all Greek things and can't wait until I move to my tiny Greek island/olive-tree farm/alignment center. You're all invited.

Most anatomy/physiology students take Medical Latin at some point in their college career. We take it because we have to memorize hundreds of names of bones and muscles and tissues and functions. And most names are in Latin. Knowing even a little bit of Latin helps because then you don't have to memorize everything.

You want to know an interesting story? A teaching doctor became perplexed during orthopaedic rounds. Holding up an X-ray of a bowlegged patient and asking medical students for the diagnosis, he was disappointed to see that a third of the students got the answer right, one third got the answer wrong, and the remaining one third shook their heads. They didn't know. They were confused.

You might have heard of these terms—*knock-kneed* and *bowlegged*. But if you were way smarter and better edumacated, then you would use the terms *valgus* and *varus*. Or *varus* and *valgus*, which is my point today.

The doctor scolded his students (I just made that part up), told them the correct answer, and then drove home to look up the Latin to find out what the problem was. And here was the problem. Somewhere down the line, the orthopaedic terms had become reversed from the Latin definitions.

Valgus is Latin for "having the knees angled outward." *Varus* is Latin for "bent or grown inward." Shocked that the definitions were the opposite of his knowledge, he checked twenty-four current orthopaedic textbooks and found that twenty-three used the reversed terms, and only one made note of the historical vs. current use.

"This particular pair of terms has caused more confusion than any other pair, partly because the original Latin terms had the opposite meaning to that which is now universally accepted."—The doctor guy.

And if you were worried that this guy did a lot of hard work for naught, don't be. He wrote a paper on it and got it published in *The New England Journal of Medicine.* Nice!

Now I'm going to tell you *how* the terms got reversed. I can do this because I am a huge nerd and have my very own copy of *The Lexicon of Orthopaedic Etymology.* I know you're jealous.

In the original definition, the terms *valgus* and *varus* referred to the position of the joint. So, using the Latin definition of valgus, the bent outward knee would look like this (below left):

And Latin definition of varus would look like this (below right):

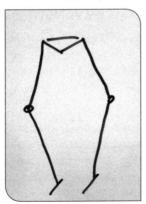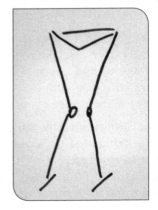

Another good example of Latin is Mr. Antonio Banderas. And Mr. Javier Bardem.

I miss Rome.

Anyhow, at some point, someone translated these terms for a modern orthopaedic text. Instead of the definition referring to the joint, they applied it instead to what the lower segment of the leg (shin bone) was doing.

Replacing the word "knee" with "lower leg," the new definition for valgus read "having the lower leg angled outward," which is the opposite of the original configuration.

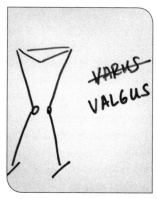

By accidentally changing one word in the definition, the meaning was reversed. Crazy, right?

A Katy factoid: I love the science of semantics. I feel that a huge breakdown in health and wellness is that students have not learned anatomy in the most correct way, but have memorized the smudgy modern version instead.

Anyhow, his paper calls for the elimination of the varus and valgus terms altogether, as they are used incorrectly. The doctor's paper also suggested using the uber-clear terms *bowlegged* and *knock-kneed* to keep everyone on the same (and correct) page.

This also makes me feel better as a patient. Say I had one varus and one valgus leg and needed the varus one amputated.

WHICH ONE WOULD THEY TAKE?

Scary. I don't like to think about stuff like that.

I like to think about stuff like Antonio Banderas instead.

Just kidding.

Maybe.

"Occasional Notes: Varus and Valgus – No Wonder They are Confused" by C.S. Houston & L.E. Swischuk, 1980, *The New England Journal of Medicine*, 302(8) (p. 471–472).

Author's note: My dad wanted me to add that this blog post may put you to sleep or leave you with a sense of "so what?" I almost decided not to post it, but then realized that there are people out there who do not prefer Yahoo News-style stories, or "The 5 Easiest Ways to Water Down Health Science to Make it Readable and Unfortunately Incorrect." There are people who want the bigger, more academic piece. So I decided to post.

Once upon a time I was a graduate student. And I had to write a lot of papers and stuff. Even back then, I was fascinated with the how and why of stuff. All of my papers read like investigative journalism. I think I missed my calling.

This paper, "Cultural Factors That May Affect Current and Future Prevalence of Osteoarthritis in Various Geographical Regions," by Katy "I'm a Huge Nerd" Bowman, was a huge catalyst in both my squat work and why I eat so much olive oil. And I mean a lot. I get about sixty percent of my daily calories from fat. Eaten, mostly, while sitting on the floor.

Anyhow, the original paper is very academic and not written well. I've become a much better writer since I became a writer. Turns out there's something to the concept of practice! I've tidied it up a bit, but left all of the smarty-pants references in, so you could cut and paste and copy it and turn it in to your biomechanics teacher and get an F for plagiarizing. Shame on you. Go write your own paper, punk.

Here's the abstract. This is what people typically read instead of actually reading the whole thing. I never recommend doing that when it comes to scientific content because you need to read the meat of the article to really get it. But, if you're busy I totally understand.

Abstract

Osteoarthritis (OA) is widespread in the United States, and is currently increasing worldwide. The reason for variance of OA throughout the world has not yet been determined. Economical and sociological factors such as literacy rate and gross national product would seem to have some correlation to the level of incidence, but this is not evident in the current literature. There is, however, evidence that cultural factors such as movement and dietary habits can possibly contribute to the likelihood of a particular region developing more OA than another. Forecasts show that without strong intervention, this global burden of injury will have huge financial implications. Increase of OA is likely due to a combination of sociological, ecological, economical, and cultural factors. Improvement in preventative treatments such as education, exercise, and nutritional programs could possibly be hindered by the sociological and psychological perception of injury, as well as the fact that current medical treatment for OA generates billions of dollars of revenue each year.

That is a lot of words to say that American risk factors for osteoarthritis like gender and age don't hold up in other countries, which means the physiological state of a human—being a male or a female or of a certain age—isn't really the problem. So, what are the risk factors, really, that people should be aware of? And is this information about real risk factors being hidden from us so that Big Corporations can continue to make a gajillion dollars?

As you can see, I used to be a lot more "conspiracy theory" than I am now. I don't think that anyone is deliberately keeping known information from the people (in this case, anyway) but I do think that the heads of health information have become so ingrained in their own culturally-based info that they can't hear a different story, even when scientifically valid.

So, here's the paper.

In the year 2000, the World Health Organization (WHO) released the beginning of a series of documents quantifying the scope of worldwide injury and the corresponding economic burden. This data showed musculoskeletal conditions (osteoporosis, rheumatoid arthritis, osteoarthritis, and low-back pain) to be the most common source of physical disability and chronic pain (13). The total cost of these diseases in the year 2000 has been estimated at $254 billion dollars in the United States alone. In developing countries receiving foreign aid, this cost was $100 billion dollars for injury care, an amount far exceeding the amount of total foreign aid awarded (14). High costs being a catalyst, the WHO, in collaboration with the Bone and Joint Decade Monitor Project (a group of international scientific and medical experts), began a project of global assessment. Health-care costs and prevalence rates for all countries were assessed, as well as potential treatments, provisions of care, and possible prevention.

For the first time, efforts were made to calculate current and projected prevalence of various musculoskeletal conditions from every region of the globe. Examining osteoarthritis (OA) closely, data showed developing countries of the Americas and Europe ranked at the top, almost even with the rate of industrialized nations—North America, Europe, Australia, New Zealand and Japan (13). Regions with the lowest current and projected prevalence of OA were countries of the Eastern Mediterranean, North Africa, and South East Asia (13).

Osteoarthritis, a loss of articular cartilage most common to the knee, hip, and hands, is most likely caused by a combination of mechanical, chemical, or physiological reasons. Physical disability and pain generated by this condition affects 4–5% of the Western adult population and is increasing as the number of older adults increase. With this musculoskeletal disorder growing at such varying rates throughout the world, the economical, sociological, and cultural differences between these regions could offer insight into the mechanism of OA, guide potential treatments, and aid in developing preventive programs.

Movement habits vary around the world. Postures required for daily living differ due to cultural factors, level of industrialization, and personal preference. Sitting preferences of cultures with less OA prevalence, such as those found in South East Asia and the Eastern Mediterranean, are very different from those of North America and Europe; studies on joint kinetics created during daily knee movements in populations with high prevalence of OA are few and far between. Ranges of motion and kinematics for the knee in cross-legged floor sitting, have been calculated for knee replacements in Eastern regions of the world. These regions currently reject arthroplasty as a treatment due to the current joint angle limitations of replacement hardware.

Traditional squatting, cross-legged sitting, and kneeling postures require an average maximum flexion of 111–165° (4, 5, and 7). Hip ranges of motion also have large hip flexion requirements of 90–110° (4, 7). Cultures that sit in chairs excessively are found to have both a higher incidence of OA and joint angles much less than their floor-sitting human counterparts. Other unique movement factors of these low-OA-prevalent regions are the religious practices found there. Many countries of North Africa, Eastern Mediterranean, and South East Asia are home to the largest populations of Muslims, a religion with mandatory prayer rituals that require a dynamic flow of squatting, kneeling, and bending (10). This bout of activity occurs five times daily, for five to ten minutes, from childhood to the end of lifespan. This cultural habit creates a region-wide "movement program" that could be responsible for joint health among this population.

Another factor, less mechanical in nature, is the dietary intake of the Eastern Mediterranean (EM) region. Over 90% of the world's olive oil is produced in this area of the world, and not surprisingly, much of it is also consumed in this area. Average percentage of dietary fat for this region is 40% of total energy intake, with almost all of it in the form of olive oil (8). Due to the population's low occurrence of heart disease, especially the decrease in heart valve inflammation, this type of diet has often been researched. During one of these research studies, an isolation of the compound oleocanthal, naturally derived from extra virgin olive oil, was identified (11). This compound has proven to be a non-steroidal anti-inflammatory, performing much like the types of non-steroid anti-inflammatory drugs (NSAIDs) given to treat symptoms of OA. Data has never been collected on the possible dietary link between oleocanthal intake and its effect on OA, but some connection could possibly exist. The same dietary intake of olive oil can also be found in Northern Africa, a low osteoarthritic region, due to the proximity of the Middle East and cultural influences of large former EM populations found within this region.

As compared to the developed nations of the world and the developing countries of the Americas, North Africa (NA), the Eastern Mediterranean (EM), and South East Asia (SEA) all show a significantly less osteoarthritic population. Why this is true is undetermined at this time. With almost no OA of the hip, and half the rate of knee OA than the countries of highest prevalence, these regions share similar cultural factors. Joint ranges of motion are consistently greater than those of the countries with the highest prevalence of OA. Bouts of structured deep hip and knee flexion in the form of religious activities are consistent throughout the entire life expectancy, as opposed to the sedentary lifestyles of the countries (or urban regions of countries) with the highest prevalence. Whether or not daily intake of anti-inflammatory compounds has any effect on a condition with inflammatory responses is still undetermined, but worth examining. Osteoarthritis is increasing all over the world and at best, the populations where it is less prevalent should serve as a window into potential preventative treatments with little to no cost or effort.

Gender, age, certain occupations, and obesity have all been credited as OA risk factors. Worldwide, women have almost twice the rate of OA (18%) than men (9.6%) (13, 14). This could be linked to the obesity rate among females, which is also almost twice that of males (2, 6, 12), or perhaps the use of higher-heeled shoes, as research has begun to show correlations between heel heights and knee and hip degeneration. Certain occupations, such as farming, also show a greater incidence of hip OA (13). Of course age, providing a greater period of loading, is also a risk factor for all joint locations of OA.

For unknown reasons, OA, as well as rheumatoid arthritis, have always flourished in industrialized countries as well as urban areas of less-developed countries (13). As developing countries begin to mimic Western culture, socioeconomic changes are impacting daily movement habits as well as dietary ones.

Prevalence of OA has not been shown to correlate directly with a region's economic status, level of industrialization, or literacy rate. If OA prevalence was based on those things, then the United States, one of the wealthiest, most schooled, and technologically savvy countries, should have the lowest rates of OA. Unfortunately the opposite is true. Preventative programs involving diet and exercise intervention are minimally researched and implemented even less. This lack of proactive treatments has been identified by Australia, and efforts are being made there to demonstrate the difference in economic spending between preventative and medical treatments to the government as well as its peoples (9).

Some research has been conducted on exercise intervention as a treatment for OA. A decrease in leg strength is often found in those with OA, especially in females (1, 3). Quadriceps-strengthening exercise programs for knee OA have been researched, but have shown little to no improvements. This could be due, however, to the intensity (high) of the exercise programs tested. A review of current exercise treatments shows that supervised exercise sessions are more effective when compared to home programs, and the combined effect of weight loss and exercise is superior to exercise alone (1).

Kinesthetic and balancing-type exercises showed significantly greater improvement in measured functional status and quality of life for those group participants than those of the strength-exercise-only group (3). Muscle stretching movements and balance/motor-skill enhancement meet the WHO's call for novel exercise treatments to be developed for the prevention and treatment of OA (1). There are many theories regarding gait patterns and placement of the patella on the knee as potentially increasing the risk of OA. There is no definitive information on how the force vector of the patellar tendon can change the mechanics of the knee, or possibly increase the chance of developing OA. If the chance of developing OA is increased with mal-alignment of particular force vectors or problematic gait, then treatments that address center of pressure, joint flexibility, and motor skill acquisition would make a positive contribution to this widespread problem. Using regions with little occurrence of OA as a model, cultural movement habits should be examined further to find possible mechanical, structural, and physiological benefits to regular deep bending. Exercise programs to balance mal-alignments and physiological changes stemming from one's culture (especially a long-term restoration to the natural lengths of human muscle found in floor-sitting cultures) has not been researched. The effect of joint mechanics, especially in regards to articular cartilage's proteoglycans, hyaluronic acid, and lubricin content and synthesis, is a large void in modern medicine.

Restorative- and somatic-type exercises currently fall outside the curriculum presented in general rehabilitation and current therapy training, and teachers of this type of movement lack the training found in anatomical and mechanical science. It is no surprise that skilled practitioners in culturally free therapeutic exercise prescription are missing from current treatment options.

As kinesiologists, we can no longer allow the path of human movement science to steer primarily towards athletic performance and rehabilitation from athletic injury. Exercise treatments are needed worldwide. Preventative exercise program development, implementation, and prescription all fall under the wide breadth of the science of kinesiology. Other sciences, such as those that focus on post-injury rehabilitation exercise, are not responsible for this task. Responsibility for this issue cannot be continually placed on the shoulders of the medical community. This is a costly problem of lifestyle-based illness and injury. Glimpses into the possibly beneficial movement and dietary habits of other places and cultures can inspire possible research that can ease financial burden globally, or potentially generate or improve treatments for a population of epic (and growing!) proportions.

WAKE UP!!!

Me again. Are you reading with one eye closed? I guess interestingness is in the mind of the beholder. I find this stuff a-mazing. Because it means that the sense of fatalism we have about the state of our health is a conditioned, culture-based notion. It is also a huge wake-up call to stop doing things just because we always have always done them. There are huge changes that need to happen in the bigger sense, but also inside our tiny yet limitless minds. Start by getting rid of your furniture. And adding a third of a cup of olive oil to your day. It doesn't get much easier than that!

1. Bennell, K., Hinman, R. (2005). Exercise as a treatment for osteoarthritis. *Current Opinion in Rheumatology*, 17:634–40.

2. Bermudez, O. I., Tucker, K.L. (2003). Trends in dietary patterns of Latin American populations. *Cardernos o Saude Publica*, 19:S87–S99.

3. Dracoglu, D., Aydin, R., Baskent, A., Celik, A. (2005). Effects of kinesthesia and balance exercises in knee osteoarthritis. *Journal of Clinical Rheumatology*, 11:303–10.

4. Hefzy, M.S., Kelly, B.P., Cooke, T.D., al-Baddah, A.M., Harrison, L. (1997). Knee kinematics in-vivo of kneeling in deep flexion examined by bi-planar radiographs. *Biomedical Scientific Instrumentation*, 33:453–8.

5. Hemmerich, A., Brown, H., Simth, S., Marthandam, S.S.K., Wyss, U.P. (2006). Hip, knee, and ankle kinematics of high range of motion activities of daily living. *Journal of Orthopaedic Research*, 24:770–781.

6. Mokhtar, N., Elati, J., Chabir, R., Bour, A., Elkari, K., Schlossman, N.P., Caballero, B.,

Aguenaou, H. (2001). Diet culture and obesity in Northern Africa. *Journal of Nutritional Sciences*, 131:887–892.

7. Mulholland, S.J., Wyss, U.P. Activities of daily living in non-Western cultures: range of motion requirements for hip and knee joint implants. *International Journal of Rehabilitation Research*, 24:191–8.

8. Panagiotakos, D.B., Pitsavos, C., Chrysohoou, C., Stefanadis, C., Toutouzas, P. (2002). Primary prevention of acute coronary events through the adoption of a Mediterranean-style diet. *Eastern Mediterranean Health Journal*, 8:1–9.

9. Segal, L., Day, S.E., Chapman, A.B., Osborne, R.H. (2004). Can we reduce disease burden from osteoarthritis? *Medical Journal of Australia*, 180:S11–S17.

10. Shelley, F.M, Clarke, A.E. (1994). *Human and Cultural Geography: A Global Perspective*. Dubuque, Iowa: William C. Brown, pp 214–228.

11. Smith, A.B., Han, Q., Breslin, P.A., Beauchamp, G.K. (2005). Synthesis and assignment of absolute configuration of (-)-oleocanthal: a potent, natural occurring nonsteroidal anti-inflammatory and anti-oxidant agent derived from extra virgin olive oils. *Organic Letters*, 7(22):5075–8.

12. Snodgrass, J.J., Leonard, W.R., Sorensen, M.V., Tarskaia, L.A., Alekseev, V.P., Krivoshapkin, V. (2006). The emergence of obesity of indigenous Siberian. *Journal of Physiological Anthropology*, 25:75–84.

13. Woolf, A.D., Pfleger, B. (2003). Burden of major musculoskeletal conditions. *Bulletin of the World Health Organization*, 81:646–656.

14. World Health Organizaion (2001). The Global Economic and Healthcare Burden of Musculoskeletal Disease.

 Dec 16 2011 **(MORE) ABOUT KATY**

I get a lot of random email questions, not always about alignment. These are either the Greatest Hits or questions that make me question whether or not I should get an unlisted phone number.

> 1. Is Katy your real name?

No. It's Frances, actually. But I thought that Katy made me sound more mature. Just kidding. I am assuming you are wondering if Katy is a nickname? Nope. Katy is my real name. Please note, it is spelled with a Y. Y? Because that's how it's spelled.

> 2. What made you want to go into biomechanics? Can I know more about your education and where you went to school, etc?

I started school at a community college, because I couldn't afford to go straight to a four-year. I did my first two years as a mathematics major. Then I decided that math was too boring and decided to study physics next. I did that for another year. By that time I was already at university

(CSU Fresno), and I was starting to get bored again. I think becoming bored is a bad habit of mine.

I was flipping through the school catalog and I saw two words: Bio-mechanics Option. It was basically a combo degree of math, physics, physiology, and exercise science. I think I screamed. With all of my math and physics already under my belt, I finished my degree quickly, studying anatomy, biology, and physiology to fill in the gaps. I worked in the field for about four years before deciding to go back to graduate school. I got my Masters of Science at CSU Northridge, doing a full biomechanics and kinesiology program there. I loved it. I was home.

3. Why did you choose to study biomechanical science?

I guess you can say that initially, I studied it by default. But when I went back for graduate school, I was really driven to explain disease in a way people could understand and then take action.

4. How did you meet your husband?

My DH is an Ayurvedic practitioner. I made an appointment for a consultation.

He looked at me.

I looked at him.

He recommended an enema.

It kinda went like that.

5. What do you do for exercise? How did you get your figure back so fast?

Over the last few years I have transitioned both my mind and my body to get rid of the paradigm of exercise. So I don't really exercise anymore. Instead I use my body, constantly, in the natural way it was developed to move. I lift. Not repetitively and with the same loads, but my baby and plants and whatever else I can. I walk every day, and not just to cross "take walk" off the list, but to transport my body for a purpose. I don't have furniture in my living room, so I stretch all day, while I'm just living. I'm in the process of building an obstacle course I can go through—climbing over stuff, a little scaling, a little swinging, a little balancing. That kind of stuff.

This was a log not far from my house. First attempt was wobbly with my arms out and head forward, looking down. By my third pass, I could relax my arms and keep my body upright, looking down with my eyes. By the twenty-seventh pass, I was just like Patrick Swayze in *Dirty Dancing*.

And, of course, the big secret is, when your alignment is off, so's your metabolism. If you want a perfectly toned body, just work on your alignment while walking around a lot. Nothing else needed.

6. What do you do for fun?

Well, the nerdy thing is, I do biomechanics for fun. I write my blog for fun. I'm a mother for fun. I'm a wife for fun. I walk for fun. I go on Facebook for fun. I don't separate the duty from the play. This whole thing is a gift that I thoroughly enjoy. And when I am feeling overwhelmed and like things are beginning to get un-fun, I meditate. Which I do, of course, for fun.

7. I'd love a break down of what you "do"/how you move each day, as inspiration. Also would be neat to hear any stories about how your work has influenced your parenting and how your family members receive and utilize your knowledge.

I get up with the sun. I'd like to say I do this by choice but it's really because you-know-who likes to get up with the sun. I work on my laptop, while stretching and balancing, for about an hour while the wee one works at his standing workstation too. I make coffee or tea and make breakfast. I typically eat a Core Defender bar while waiting for breakfast to cook because I am always very hungry.

Then I crawl around and play and stretch on the living-room jungle gym for an hour or so, until it's naptime. Finn's naptime, not mine. And while he sleeps for an hour, I bang away (again standing) as much work on my computer as possible. We usually take a family mile-long walk to the post office to mail stuff. And then a quarter-mile family walk to my sister's house to play with her kids for an hour. Then home, for another nap. Again, not mine (boo). More work for an hour and typically one of us will go to the store for fresh food for dinner.

Three o'clock is really the end of day here. Everything after 3:00 is play-stretch time in the living room. Lots of chatting and dancing and soundtracks to movies like *Grease, The Sound of Music* (Can I get a whoop whoop for "The Lonely Goatherd"?), and *Footloose*. Dinner. Quick writing for fifteen to twenty minutes, and then maybe a movie in bed before heading to sleep!

> 8 What kind of shoes do you wear on a regular basis, the ones you reach for every day?

Most of the time you will find me barefoot, in my pink Vibrams (image top right) or in my (new) warmer Sockwas (image middle right).

I'm on the move a lot, but if I do happen to do something for TV or a formal lecture, I tend to wear my Kalso Earth footwear. I like boots. Oh, and as you might remember, I farm in these (also from Earth) (image bottom right):

> 9. I would like to know what kind of reactions you get when you tell people you don't like strollers. I'm guessing they start thinking about stuff like the scene in *Away We Go*! Haha!

Well, believe it or not, I don't go around telling people my opinion of things. Of course, it's different if someone (like you) comes to my blog or to my class or to my course. If you've come to me for information, I will shove it down. your. throat. Because you've asked me to. I don't, however, walk up to people and say, Hey You. Yeah You. With the tucked pelvis and the stroller. You're doing it all wrong.

One thing to understand is, people can rarely hear information they haven't sought for themselves. This is a tough lesson to learn for many of our new graduates. They think that other people think just like them, and that's not the case. Chances are you've tried to tell your partner, kids, coworker, or person in the elevator some little tidbit that you learned here. And worse, you probably use the phrase "Well, Katy says..." a lot. This origination of phrase (and blog title) stems from our whole-body-course students who, when unable to explain the physics of the situation, reverted to the five-year-old explanation of "Well, Katy says..."

I'd like you to be clear that the reason you stand with your weight in your heels is not because I said so. It is because you understand why you're supposed to do it. Optimal health only works when you understand the whys and not just the hows.

Now, that doesn't mean that when the stroller and tucked pelvis walks by that I don't mutter and curse. I'm not that pious. And, P.S., I love *Away We Go*. It's one of my favorite movies of all time. That and *Top Gun*.

10. What do you love to eat, above all else? In what position do you like to eat it?

I don't know if I could pick a "thing" to eat. Can I pick an entire country of things? I'd have to go with Greece. If there was one food I couldn't live without, it would be olive oil. And maybe avocados.

I typically eat standing because I eat constantly, all day long, and getting down on the floor seems like a lot of work. But, when we eat together, we just sit on a blanket on the living-room floor.

11. Please provide some more shoe suggestions, but for under $100.

Sockwas. Puma Sabadella Ballet Flats. Earth Glides. SoleRebels. And a bunch more. Try Google with the words "minimal footwear less than $100." I don't posses any magical searching skills, FYI.

12. What is your favorite doughnut?

It used to be the apple fritter when I was a kid. Then it went to maple bars when I was a teenager. Now I think I'd pick the rainbow sprinkles. I've never had a Krispy Kreme.

13. How in the world do you get so much done with an infant?!?

Well, before I had any kids people always said I got a more done than what seemed normal. I guess I read and write very fast. I figure stuff out quickly. But those things aside, it probably has to do with the fact that I don't have a television. Television will eat away at all of your hopes and dreams.

14. The furniture (or lack thereof!) in your house. I think you mentioned

once you'd post a pic sometime of your main living area—I would be curious to see how you do it.

Like this. Turns out, to not have furniture in your house, you just have to 1) not put it there or 2) have it removed.

15. What bad alignment habits do you work to correct on yourself? Aches and pains? Favorite philosophy or approach to life? What pets have you had throughout your life? Do you have a favorite animal because that seems to say a lot about a person. Are you a very social person, or prefer your own company?

I work, constantly, on my left hip. My entire left side, really. And my neck and shoulders are always tight! As for philosophies/approaches to life, I would have to say that I try to follow the teachings of the Big J.C. and the Golden Rule. Karma. Buddhism. Go with the flow. Do unto others. Resistance is futile. All you need is love. I think these are all the same thing.

My totem animal has always been a whale. Specifically, a humpback whale. I'd love to know what that says about me if anyone knows. I would say that I am a balance of a very social person and a very introverted person. I love to have parties and host meals often, but I also really need a lot of solo time. Mostly because I am a chronic thinker and am always developing new theories that need to be written and researched. I would totally be a recluse, I think, if I didn't enjoy teaching people so much and eating really good food. If we were all living in Greece instead of on Facebook, we'd be discussing alignment matters while eating baklava and olives in my backyard.

16. I would like to know more about your pregnancy, if you had any ailments or pains and how you dealt with them!

My pregnancy was awesome for the most part. I studied pregnancy pain in graduate school so I already knew that that these aches and pains were not a result of being pregnant, but a result of being weak, misaligned, and loaded with the extra mass.

I was, however, terribly nauseated for about four months. This picture, taken at month two, kinda sums it up:

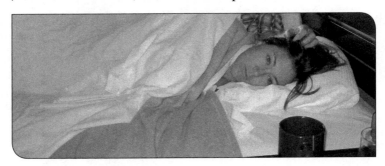

This is how I felt, pretty much all day long. When I was about six weeks preggo, the hubs went to get food. We had already discussed earlier that I needed more protein. He came back with a selection of groceries that have since been heckled. I couldn't remember exactly what he brought home, so I had to check my inbox for an email I sent my SIL Here's that exchange:

Me: Yesterday was a new low. I mastered the art of throwing up food while I was practically eating it. Disgusting. MC tried to help by bringing me more food after not eating the bulk of the day and staying in bed. This is what he brought home (really):

1. Chili

2. Tiny Shrimp in a Can

3. Spicy Chicken Nuggets that you can Heat Up in the Oven

oh, and roses. MC says to mention the roses.

S.I.L.: I'm still laughing about the chili, tiny shrimp in a can, and spicy chicken nuggets that you can heat up in the oven. Was he serious? I'm about to throw up just looking at those words.

Me: I think he had a "protein" thing in his mind. These all made me throw up just looking at them, especially the tiny shrimp in a can. Who has ever bought tiny shrimp in a can, unless they wanted to feed their pet turtle or something? I wouldn't eat those even if I wasn't pregnant.

17. Any phobias? I know someone deathly afraid of chickens and how

about your most embarrassing moment? ☺

Hmmm. Toss up. It could be the time I decided to take a yoga class at a gym while out of town and mistakenly entered the wrong-gender locker room and stripped down for a shower. It wasn't until I was standing there with a hand towel (forgot my full towel, of course) when I head men's voices (a lot of them) coming from around the corner.

It could be that, or it could be the year my hair looked like this:

My sister calls this the Broccoli Haircut. Which would also explain why I have a phobia of broccoli.

18. I want to know your every move and every thought during one whole day.

"Damn, it's early. Feet straight. Damn, I'm hungry. Feet straight. Damn I'm tired. And hungry. Feet straight." Repeat ten more times. "Thank God it's time for bed."

19. How did your parents respond when you told them what you wanted to do when you grew up?

I don't really have a family like that. I went to college on my own and was the first to do so. I'm not really sure they, like many, really "get" what I do. I think my mom tells people I "fix things" in the body. My dad told me to put him down for one DVD when they came out, as you would a candy bar for a school fundraiser. Very endearing. But they have (and maybe even do once in awhile) some videos. And, my dad was just here and read an article on foot biomechanics I wrote for *Idea Fitness Journal*. He then jumped up (or as much as an eighty-four-year-old can jump) and said he never knew that he had so many muscles in his feet and that he needed to "strengthen his intrinsic muscles" right away. We gave him some Sockwas to wear while he was here:

20. I'm interested in your Writing Story. Clearly, you are an excellent writer, how did that come? Do you credit it to being a good reader?

Thanks! I've always been a very good reader[1] but I wasn't always a very good writer. When I look back at my earlier writing, I had a very strange way of organizing sentences and paragraphs that made it very difficult to understand. I don't think this is the case any longer. Writing for a blog over the last few years has definitely made me create easy-to-comprehend pieces, even when the topics aren't that simple. And, surprisingly enough, this has become my career trademark. If I was a writing teacher, I'd give everyone a daily writing assignment of explaining something in one hundred words. There is a power to editing that I didn't fully understand until working with pop magazines. And, my husband is a copyeditor and all-around wordsmith, which has absolutely rubbed off. His alignment's pretty good now too!

21. Do you have an "aha" moment (or moments) and if so when and what was it that led you this direction instead of following the masses?

Hmm. I wouldn't say that my path was created from the perspective of not wanting to do what others had done. I believe my path was created by wanting to find the most correct solution for a problem. I am not a good student or follower of other things because, as Richard Feynman once said, "It doesn't matter how beautiful your theory is or how smart you are or what your name is. If it disagrees with experience, it's wrong. That's all there is to it."

So many people are following paradigms for reasons that don't add up to their experience with it (i.e., I'm doing this for my health—yet their health sucks). If you give me a statement that I can find a false example for, the statement isn't valid, and I move on. It's handy to have a really great evaluation tool, but it means that I'm constantly refining to find that one statement that has no false examples. And, you have to be willing to be wrong without seeing it as a personal failure. Being wrong does not make you bad. It just gives you a chance to be better. So, long story long, I have "a-ha" moments multiple times a day.

I'm a multiple *a-ha*-er.

---

1. One of my favorite stories to tell is how my fourth-grade teacher took my *Clan of the Cave Bear* book away because she was sure my mother didn't know I was reading it. Mom was the one who gave it to me because there was nothing left to read.

Oh, and, teachers and parents, want to make sure a child is a huge nerd and is harassed by her peers? When the first day of fourth grade starts and you ask for students to put a book sticker next to their name for every twenty-five pages they documented reading, make sure you let them put eighty-seven stickers in a row. Even if it takes four loooong rows and goes off the board and most other kids have six stickers. Just go ahead and let her do that, okay?

RESOURCES

Want More?

If you've been bitten by the alignment bug, don't worry! There are additional resources available in the form of articles, books, DVDs, online courses and downloads.

To ready Katy's most recent blog posts, visit www.katysays.com.

For the most up-to-date offerings in video-based courses and DVD products, please visit the Restorative Exercise Institute's website www.restorativeexercise.com.

CREDITS

The bulk of the images in this book were the poor efforts of the author. All of the good photos were most likely taken by Cecilia Ortiz.

The image of a duck on page 55 was taken by Eli Davis and graciously donated by his wife Megan.

The images of the leg extension machine on page 88 were taken by Natalie Wieneroider.

The rest were provided by countless "volunteers" dragged by the author from whatever they were doing at the time. She hopes she remembered to thank them!

A Special Thanks/Shout Out to Michael Curran for his gracious (read: unpaid) editing services for all KatySays.com posts.

INDEX

rotator cuff injury, 189, 196-197, 291, 369, 390
running, *see* jogging
sciatic pain, 37, 65, 76, 337
self-efficacy theory, 337-340
SI joint issues, 37, 72, 337, 396
side bend, 332-334
sinuses, 198-199, 315
skiing, cross-country, 282
sleep position, 64, 245-246, 309, 369, 381, 392, 394, 396
smoking, 30, 35-36, 50, 250, 275, 276, 318
Sockwa, 264, 429, 430, 434
soda, 19-20, 190, 192, 314
Soft Star Shoes, 235, 408
spasm, 77, 138, 246, 343-344
stomach sucking, 96, 104, 144, 146-148, 154, 156, 162, 221, 316-319, 395, 407, 409
strap stretch, 80-81
stretch marks, 108-109
strollers, 85, 217, 228, 237, 240, 429
swallowing, 162, 187-188, 200, 325
sway backed, 75-76, 146,
swimming, 52, 107, 212, 213, 215, 258, 289-292, 329, 348, 358, 371, 400, 417
swinging, *see* hanging (by the arms), *or see* arm swing (during gait)
teeth, 186-187, 193, 200-201, 209
testicles, 95-96, 311
TMJ disorder, 404
toe gripping, 19, 36-37, 402
treadmill, 107, 144, 172-173, 259, 272, 282, 285, 289, 301
twist, 73-74, 77, 138-140, 409
A User's Guide to the Brain, 324
vagina, 94-95, 98, 114-115, 123, 218, 221, 251, 253, 315, 316, 337, 394
Valsalva, 220, 315-319
varicose veins, 153, 297, 390, 407
Vibram Five Fingers, 41, 43, 59, 264, 429
Vipassana, see meditation
voice, 201-204,
weight management, 23-24, 35, 67, 137, 283, 288, 290, 311, 361, 423-424, 426
    during pregnancy, *see* chapter Pregnancy, Childbirth, Babies, and
    Children, 207-255
whiskers, 329-330

Biomechanist Katy Bowman, MS, is the director of the Restorative Exercise Institute, the creator and talent behind the Aligned and Well™ DVD line, and author of *Every Woman's Guide to Foot Pain Relief: The New Science of Healthy Feet*. She is a regular contributor and expert for national health, fitness, and wellness publications and TV segments—and does all these things while her two children nap, infrequently.

She and her family have split their time between their farm on Washington state's Olympic Peninsula and the "ground-zero" of wellness-through-alignment in Ventura, California, and are now planning to spend 2015 abroad, working with alignment students throughout the world. Yes, with the kidlets in tow.